T0187929

Special Integrals of Gradshteyn and Ryzhik

the Proofs - Volume II

MONOGRAPHS AND RESEARCH NOTES IN MATHEMATICS

Series Editors

John A. Burns
Thomas J. Tucker
Miklos Bona
Michael Ruzhansky

Published Titles

Application of Fuzzy Logic to Social Choice Theory, John N. Mordeson, Davender S. Malik and Terry D. Clark

Blow-up Patterns for Higher-Order: Nonlinear Parabolic, Hyperbolic Dispersion and Schrödinger Equations, Victor A. Galaktionov, Enzo L. Mitidieri, and Stanislav Pohozaev

Cremona Groups and Icosahedron, Ivan Cheltsov and Constantin Shramov

Difference Equations: Theory, Applications and Advanced Topics, Third Edition, Ronald E. Mickens

Dictionary of Inequalities, Second Edition, Peter Bullen

Iterative Optimization in Inverse Problems, Charles L. Byrne

Line Integral Methods for Conservative Problems, Luigi Brugnano and Felice Iaverno

Lineability: The Search for Linearity in Mathematics, Richard M. Aron, Luis Bernal González, Daniel M. Pellegrino, and Juan B. Seoane Sepúlveda

Modeling and Inverse Problems in the Presence of Uncertainty, H. T. Banks, Shuhua Hu, and W. Clayton Thompson

Monomial Algebras, Second Edition, Rafael H. Villarreal

Nonlinear Functional Analysis in Banach Spaces and Banach Algebras: Fixed Point Theory Under Weak Topology for Nonlinear Operators and Block Operator Matrices with Applications, Aref Jeribi and Bilel Krichen

Partial Differential Equations with Variable Exponents: Variational Methods and Qualitative Analysis, Vicenţiu D. Rădulescu and Dušan D. Repovš

A Practical Guide to Geometric Regulation for Distributed Parameter Systems, Eugenio Aulisa and David Gilliam

Signal Processing: A Mathematical Approach, Second Edition, Charles L. Byrne

Sinusoids: Theory and Technological Applications, Prem K. Kythe

Special Integrals of Gradshteyn and Ryzhik: the Proofs – Volume I, Victor H. Moll

Special Integrals of Gradshteyn and Ryzhik: the Proofs – Volume II, Victor H. Moll

Forthcoming Titles

Actions and Invariants of Algebraic Groups, Second Edition, Walter Ferrer Santos
  and Alvaro Rittatore

Analytical Methods for Kolmogorov Equations, Second Edition, Luca Lorenzi

Complex Analysis: Conformal Inequalities and the Bierbach Conjecture, Prem K. Kythe

Computational Aspects of Polynomial Identities: Volume I, Kemer's Theorems, 2nd Edition
  Belov Alexey, Yaakov Karasik, Louis Halle Rowen

Geometric Modeling and Mesh Generation from Scanned Images, Yongjie Zhang

Groups, Designs, and Linear Algebra, Donald L. Kreher

Handbook of the Tutte Polynomial, Joanna Anthony Ellis-Monaghan and Iain Moffat

Line Integral Methods and Their Applications, Luigi Brugnano and Felice Iaverno

Microlocal Analysis on R^n and on NonCompact Manifolds, Sandro Coriasco

Practical Guide to Geometric Regulation for Distributed Parameter Systems,
  Eugenio Aulisa and David S. Gilliam

Reconstructions from the Data of Integrals, Victor Palamodov

*Stochastic Cauchy Problems in Infinite Dimensions: Generalized and Regularized
  Solutions*, Irina V. Melnikova and Alexei Filinkov

Symmetry and Quantum Mechanics, Scott Corry

MONOGRAPHS AND RESEARCH NOTES IN MATHEMATICS

Special Integrals of Gradshteyn and Ryzhik

the Proofs - Volume II

Victor H. Moll

Tulane University
New Orleans, Louisiana, USA

CRC Press
Taylor & Francis Group
Boca Raton London New York

CRC Press is an imprint of the
Taylor & Francis Group, an informa business

A CHAPMAN & HALL BOOK

CRC Press
Taylor & Francis Group
6000 Broken Sound Parkway NW, Suite 300
Boca Raton, FL 33487-2742

First issued in paperback 2019

© 2016 by Taylor & Francis Group, LLC
CRC Press is an imprint of Taylor & Francis Group, an Informa business

No claim to original U.S. Government works

ISBN-13: 978-1-4822-5653-6 (hbk)
ISBN-13: 978-0-367-37727-4 (pbk)

This book contains information obtained from authentic and highly regarded sources. Reasonable efforts have been made to publish reliable data and information, but the author and publisher cannot assume responsibility for the validity of all materials or the consequences of their use. The authors and publishers have attempted to trace the copyright holders of all material reproduced in this publication and apologize to copyright holders if permission to publish in this form has not been obtained. If any copyright material has not been acknowledged please write and let us know so we may rectify in any future reprint.

Except as permitted under U.S. Copyright Law, no part of this book may be reprinted, reproduced, transmitted, or utilized in any form by any electronic, mechanical, or other means, now known or hereafter invented, including photocopying, microfilming, and recording, or in any information storage or retrieval system, without written permission from the publishers.

For permission to photocopy or use material electronically from this work, please access www.copyright.com (http://www.copyright.com/) or contact the Copyright Clearance Center, Inc. (CCC), 222 Rosewood Drive, Danvers, MA 01923, 978-750-8400. CCC is a not-for-profit organization that provides licenses and registration for a variety of users. For organizations that have been granted a photocopy license by the CCC, a separate system of payment has been arranged.

Trademark Notice: Product or corporate names may be trademarks or registered trademarks, and are used only for identification and explanation without intent to infringe.

**Visit the Taylor & Francis Web site at
http://www.taylorandfrancis.com**

**and the CRC Press Web site at
http://www.crcpress.com**

Contents

Introduction

This represents the second volume of the project to evaluate all entries in the classical table of integrals by I. Gradshteyn and I. S. Ryzhik. The first volume appeared as [53]. The present collection contains papers that have appeared in Revista Scientia. The author wishes to thank the editors of this journal for their collaboration in this project.

Chapter 1: Complete elliptic integrals, **20**, 2011, 45–59; with S. Boettner.

Chapter 2: The Riemann zeta function, **20**, 2011, 61–71; with T. Amdeberhan and K. Boyadzhiev.

Chapter 3: Some automatic proofs, **20**, 2011, 93–111; with C. Koutschan.

Chapter 4: The error function, **21**, 2011, 25–42; with M. Albano, T. Amdeberhan, E. Beyerdsted.

Chapter 5: Hypergeometric functions, **21**, 2011, 43–54; with K. Kohl.

Chapter 6: Hyperbolic functions, **22**, 2011, 109–127; with K. Bodadzhiev.

Chapter 7: Bessel K-functions, **22**, 2011, 129–151; with L. Glasser, K. Kohl, C. Koutschan, A. Straub.

Chapter 8: Combinations of logarithms and rational functions, **23**, 2012, 1–18; with L. Medina.

Chapter 9: Polylogarithm functions, **23**, 2012, 45–51; with K. McInturff.

Chapter 10: Evaluation by series, **23**, 2012, 53–65.

Chapter 11: The exponential integral, to appear 2015; with K. Bodadzhiev.

Chapter 12: More logarithmic integrals, to appear 2015; with L. Medina.

Chapter 13: Confluent hypergeometric and Whittaker functions, to appear 2015; with A. Dixit.

Chapter 14: Evaluation of entries in Gradshteyn and Ryzhik employing the method of brackets, **25**, 2014, 65–84; with I. Gonzalez and K. Kohl.

The author was supported in part by NSF-DMS 1112656.

CHAPTER 1

Complete elliptic integrals

1.1. Introduction

Elliptic integrals were at the center of analysis at the end of 19th century. The **complete elliptic integral of the first kind** defined by

$$(1.1.1) \qquad \mathbf{K}(k) := \int_0^1 \frac{dx}{\sqrt{(1-x^2)(1-k^2x^2)}}$$

is a function of the so-called **modulus** k^2. The corresponding **complete elliptic integral of the second kind** is defined by

$$(1.1.2) \qquad \mathbf{E}(k) := \int_0^1 \sqrt{\frac{1-k^2x^2}{1-x^2}}\, dx.$$

The total collection of complete elliptic integrals contains one more, the so-called **complete elliptic integral of the third kind,** defined by

$$(1.1.3) \qquad \mathbf{\Pi}(n,k) := \int_0^1 \frac{dx}{(1-n^2x^2)\sqrt{(1-x^2)(1-k^2x^2)}}.$$

The **complementary integrals** are defined by

$$(1.1.4) \qquad \mathbf{K}'(k) := \mathbf{K}(k')$$

where $k' = \sqrt{1-k^2}$ is the so-called **complementary modulus.**

The change of variables $x = \sin t$ yields the trigonometric versions

$$(1.1.5) \quad \mathbf{K}(k) = \int_0^{\pi/2} \frac{dt}{\sqrt{1-k^2\sin^2 t}} \text{ and } \mathbf{E}(k) = \int_0^{\pi/2} \sqrt{1-k^2\sin^2 t}\, dt,$$

with a similar expression for $\mathbf{\Pi}(n,k)$.

In general, an elliptic integral is one of the form

$$(1.1.6) \qquad I := \int_a^b \frac{P(x)\, dx}{y},$$

where y^2 is a cubic or quartic polynomial in x. The integral is called **complete** if a and b are roots of $y = 0$. It is clear that $\mathbf{K}(k)$ is elliptic. The same is true for $\mathbf{E}(k)$, written in the form

$$(1.1.7) \qquad \mathbf{E}(k) := \int_0^1 \frac{(1-k^2x^2)\, dx}{\sqrt{(1-x^2)(1-k^2x^2)}}.$$

1

1.2. Some examples

In this section we offer some evaluations from [**35**] that follow directly from the definitions. Some special values are offered first. The evaluation of these integrals is facilitated by Legendre's relation

$$(1.2.1) \qquad \mathbf{K}(k)\mathbf{E}'(k) + \mathbf{K}'(k)\mathbf{E}(k) - \mathbf{K}(k)\mathbf{K}'(k) = \frac{\pi}{2}.$$

The reader will find this identity as Exercise 4 in Section 2.4 of [**45**].

EXAMPLE 1.2.1.

$$(1.2.2) \qquad \mathbf{K}\left(\sqrt{-1}\right) = \frac{1}{4\sqrt{2\pi}}\Gamma^2\left(\frac{1}{4}\right).$$

The proof is direct. The integral is
$$(1.2.3)$$
$$\mathbf{K}(\sqrt{-1}) = \int_0^1 \frac{dx}{\sqrt{1-x^4}} = \frac{1}{4}\int_0^1 y^{-3/4}(1-y)^{-1/2}\,dy = \frac{\Gamma(1/4)\,\Gamma(1/2)}{4\Gamma(3/4)}.$$

The result now follows from the symmetry rule

$$(1.2.4) \qquad \Gamma(a)\Gamma(1-a) = \frac{\pi}{\sin \pi a}$$

for the gamma function and the special value $\Gamma(1/2) = \sqrt{\pi}$. This example appears as entry **3.166.16** in [**35**]. Entry **3.166.18** states that

$$(1.2.5) \qquad \int_0^1 \frac{x^2\,dx}{\sqrt{1-x^4}} = \frac{1}{\sqrt{2\pi}}\Gamma^2\left(\frac{3}{4}\right).$$

The proof consists of a reduction to a special value of the beta function. The change of variables $t = x^4$ gives

$$(1.2.6) \qquad \int_0^1 \frac{x^2\,dx}{\sqrt{1-x^4}} = \frac{1}{4}\int_0^1 t^{-1/4}(1-t)^{-1/2}\,dt.$$

This integral is $\frac{1}{4}B\left(\frac{3}{4}, \frac{1}{2}\right)$. The simplified result is obtained as above.

Formula (1.1.7) with $k = \sqrt{-1}$ shows that

$$(1.2.7) \qquad \mathbf{E}(\sqrt{-1}) = \int_0^1 \frac{1+x^2}{\sqrt{1-x^4}}\,dx.$$

The values given above show that

$$(1.2.8) \qquad \mathbf{E}(\sqrt{-1}) = \frac{1}{4\sqrt{2\pi}}\left[\Gamma^2\left(\frac{1}{4}\right) + 4\Gamma^2\left(\frac{3}{4}\right)\right].$$

EXAMPLE 1.2.2.

$$(1.2.9) \qquad \mathbf{K}\left(\frac{1}{\sqrt{2}}\right) = \frac{1}{4\sqrt{\pi}}\Gamma^2\left(\frac{1}{4}\right).$$

This appears as entry **8.129.1** in [**35**]. This value comes from the previous example and the identity

$$(1.2.10) \qquad \mathbf{K}(\sqrt{-1}k) = \frac{1}{\sqrt{1+k^2}}K\left(\frac{k}{\sqrt{1+k^2}}\right),$$

with $k = 1$. The identity (1.2.10) follows by the change of variables

$$x \mapsto x/\sqrt{1 + k^2(1 - x^2)}$$

in the left-hand side integral.

The values of the modulus k for which $\mathbf{K'}/\mathbf{K}$ is the square root of an integer are of considerable interest. These are called the **singular values**. The previous example shows that $1/\sqrt{2}$ is the simplest of them: in this case

$$(1.2.11) \qquad \frac{\mathbf{K'}}{\mathbf{K}}\left(\frac{1}{\sqrt{2}}\right) = 1.$$

A list of the first few values k_r for which

$$(1.2.12) \qquad \frac{\mathbf{K'}}{\mathbf{K}}(k_r) = \sqrt{r}$$

is given in [**18**] and it starts with

$$k_2 = \sqrt{2}-1, \; k_3 = \frac{\sqrt{2}(\sqrt{3}-1)}{4}, \; k_4 = 3-2\sqrt{2}, \; k_5 = \frac{1}{2}\left(\sqrt{\sqrt{5}-1} - \sqrt{3-\sqrt{5}}\right).$$

1.3. An elementary transformation

Elementary manipulations can be employed to evaluate certain entries in [**35**]. For instance, direct integration by parts on the integrals defining the functions $\mathbf{K}$ and $\mathbf{E}$ produces

$$(1.3.1) \qquad \int_0^1 \frac{x \; \arcsin x}{\sqrt{(1 - k^2 x^2)^3}} \, dx = \frac{1}{k^2}\left(\frac{\pi}{2k'} - \mathbf{K}(k)\right)$$

and

$$(1.3.2) \qquad \int_0^1 \frac{x \; \arcsin x}{\sqrt{1 - k^2 x^2}} \, dx = \frac{1}{k^2}\left(\mathbf{E}(k) - \frac{\pi}{2}k'\right).$$

This last evaluation appears as entry **4.522.4** in [**35**].

On the other hand, several entries in [**35**] may be evaluated also by integration by parts, choosing the inverse trigonometric term to be differentiated. Such a procedure gives

$$(1.3.3) \qquad \int_0^1 \frac{x \arccos x \, dx}{\sqrt{1 - k^2 x^2}} = \frac{1}{k^2}\left(\frac{\pi}{2} - \mathbf{E}(k)\right),$$

which appears as entry **4.522.5**,

$$(1.3.4) \qquad \int_0^1 \frac{x \arcsin x \, dx}{\sqrt{k'^2 + k^2 x^2}} = \frac{1}{k^2}\left(\frac{\pi}{2} - \mathbf{E}(k)\right),$$

which appears as entry **4.522.6**, and finally **4.522.7**,

$$(1.3.5) \qquad \int_0^1 \frac{x \arccos x \, dx}{\sqrt{k'^2 + k^2 x^2}} = \frac{1}{k^2}\left(-\frac{\pi}{2}k' + \mathbf{E}(k)\right).$$

In this section we derive a different type of elementary transformation for integrals and use it to obtain the value of some elliptic integrals appearing in [**35**].

LEMMA 1.3.1. *Let f be an odd periodic function of period a. Then*

$$(1.3.6) \qquad \int_0^\infty \frac{f(x)}{x}\, dx = \frac{\pi}{a} \int_0^{a/2} \frac{f(x)}{\tan \frac{\pi x}{a}}\, dx.$$

PROOF. The result follows by splitting the integral as

$$\int_0^\infty \frac{f(x)}{x}\, dx = \sum_{k=0}^\infty \int_0^a \frac{f(x)}{x + ka}\, dx$$

$$= \sum_{k=0}^\infty \int_0^{a/2} f(x) \left[\frac{1}{x + ka} - \frac{1}{(k+1)a - x} \right] dx$$

and using the partial fraction decomposition

$$(1.3.7) \qquad \tan \frac{\pi b}{2} = \frac{4b}{\pi} \sum_{j=1}^\infty \frac{1}{(2j-1)^2 - b^2},$$

given as entry **1.421.1** in [**35**]. □

COROLLARY 1.3.2. *Let f be an even function with period a. Then*

$$(1.3.8) \qquad \int_0^\infty \frac{f(x)}{x} \sin \frac{\pi x}{a}\, dx = \frac{\pi}{a} \int_0^{a/2} f(x)\, dx.$$

In particular, for $a = \pi$,

$$(1.3.9) \qquad \int_0^\infty \frac{f(x)}{x} \sin x\, dx = \int_0^{\pi/2} f(x)\, dx.$$

PROOF. Apply the lemma to the function $f(x) \sin \frac{\pi x}{a}$, which is odd and has period $2a$. The result follows from the half-angle formula

$$(1.3.10) \qquad \tan \frac{x}{2} = \frac{\sin x}{1 + \cos x}$$

and the value

$$(1.3.11) \qquad \int_0^a f(x) \cos \frac{\pi x}{a}\, dx = 0.$$

□

A similar results holds for odd functions. These appear as entry **3.033** in [**35**].

COROLLARY 1.3.3. *Let f be an odd function with period a. Then*

$$(1.3.12) \qquad \int_0^\infty \frac{f(x)}{x} \sin \frac{\pi x}{a}\, dx = \frac{\pi}{a} \int_0^{a/2} f(x) \cos \frac{\pi x}{a}\, dx.$$

In particular, for $a = \pi$,

$$(1.3.13) \qquad \int_0^\infty \frac{f(x)}{x} \sin x\, dx = \int_0^{\pi/2} f(x) \cos x\, dx.$$

EXAMPLE 1.3.4. The function $f(x) \equiv 1$ and $a = \pi$ in Corollary 1.3.2 gives the classical integral

$$(1.3.14) \qquad \int_0^\infty \frac{\sin x}{x} \, dx = \frac{\pi}{2}.$$

This is entry **3.721.1** in [**35**]. The reader will find in [**37, 38**] a couple of articles by G. H. Hardy with an evaluation of the many proofs of this identity. These papers are available in volume 5 of his *Complete Works*.

EXAMPLE 1.3.5. Entry **3.842.3** of [**35**] consists of four evaluations, the first of which is

$$(1.3.15) \qquad \int_0^\infty \frac{\sin x}{\sqrt{1 - k^2 \sin^2 x}} \frac{dx}{x} = \mathbf{K}(k).$$

This follows from Corollary 1.3.2 by choosing $a = \pi$ and $f(x) = 1/\sqrt{1 - k^2 \sin^2 x}$. A different proof of this evaluation is offered in Section 1.6 below.

EXAMPLE 1.3.6. A second integral appearing in **3.842.3** is

$$(1.3.16) \qquad \int_0^\infty \frac{\sin x}{\sqrt{1 - k^2 \cos^2 x}} \frac{dx}{x} = \mathbf{K}(k).$$

It also follows from Corollary 1.3.2. This is also true for entry **3.841.1**

$$(1.3.17) \qquad \int_0^\infty \sin x \sqrt{1 - k^2 \sin^2 x} \, \frac{dx}{x} = \mathbf{E}(k)$$

and its companion entry **3.841.2**

$$(1.3.18) \qquad \int_0^\infty \sin x \sqrt{1 - k^2 \cos^2 x} \, \frac{dx}{x} = \mathbf{E}(k).$$

EXAMPLE 1.3.7. The elementary method introduced here may be used to evaluate all integrals of the type

$$(1.3.19) \qquad I_{m,n}(k) := \int_0^\infty \frac{\sin^n x \cos^m x}{\sqrt{1 - k^2 \sin^2 x}} \frac{dx}{x}$$

and the companion family

$$(1.3.20) \qquad J_{m,n}(k) := \int_0^\infty \frac{\sin^n x \cos^m x}{\sqrt{1 - k^2 \cos^2 x}} \frac{dx}{x}.$$

All entries in Sections **3.844** and **3.846** match one of these forms.

EXAMPLE 1.3.8. Many other evaluations can be produced by this method. For instance,

$$(1.3.21) \qquad \int_0^\infty \frac{\sin x \log(1 - k^2 \sin^2 x)}{\sqrt{1 - k^2 \sin^2 x}} \frac{dx}{x} = \int_0^{\pi/2} \frac{\log(1 - k^2 \sin^2 x)}{\sqrt{1 - k^2 \sin^2 x}} \, dx.$$

The integral on the left appears as entry **4.432.1** and the one on the right is entry **4.414.1** in [**35**]. A proof of the identity

$$(1.3.22) \qquad \int_0^{\pi/2} \frac{\log(1 - k^2 \sin^2 x)}{\sqrt{1 - k^2 \sin^2 x}} \, dx = \mathbf{K}(k) \ln k'$$

is given in Example 1.7.5.

1.4. Some principal value integrals

The method described above can be employed to evaluate some entries of [35] provided the integrals are interpreted as Cauchy principal values.

EXAMPLE 1.4.1. The first example is

$$(1.4.1) \qquad \int_0^\infty \frac{\tan x}{\sqrt{1 - k^2 \sin^2 x}} \frac{dx}{x} = \mathbf{K}(k),$$

which appears as one of the four entries in **3.842.3** of [35].

Let $I_1(k)$ denote the integral and introduce the notation

$$(1.4.2) \qquad f(x) = \frac{\tan x}{\sqrt{1 - k^2 \sin^2 x}}.$$

Then $f(x)$ is odd and it has period π. The principal value of the integral is given by

$$(1.4.3) \qquad I_1(k) = \lim_{\epsilon \to 0} \sum_{j=0}^\infty \left(\int_0^{\pi/2 - \epsilon} \frac{f(x)}{x} \, dx + \int_{\pi/2 + \epsilon}^\pi \frac{f(x)}{x + j\pi} \, dx \right).$$

The substitution $y = \pi - x$ in the second integral above produces

$$I_1(k) = \lim_{\epsilon \to 0} \sum_{j=0}^\infty \int_0^{\pi/2 - \epsilon} \left(\frac{1}{x} + \frac{1}{x - (j+1)\pi} \right) f(x) \, dx$$

$$= \lim_{\epsilon \to 0} \int_0^{\pi/2 - \epsilon} \left(\frac{1}{x} + \sum_{j=1}^\infty \frac{2x}{x^2 - j^2 \pi^2} \right) f(x) \, dx.$$

The series corresponds to the partial fraction expansion of the cotangent function. This completes the evaluation of (1.4.1). The reader will note that this proof is very similar to that of Lemma 1.3.1.

The value

$$(1.4.4) \qquad \int_0^\infty \frac{\tan x}{\sqrt{1 - k^2 \cos^2 x}} \frac{dx}{x} = \mathbf{K}(k),$$

which also appears in **3.842.3**, is established using the same type of argument. This completes the evaluation of the integrals in that entry of [35].

EXAMPLE 1.4.2. Entry **3.841.3** of [35]

$$(1.4.5) \qquad \int_0^\infty \tan x \sqrt{1 - k^2 \sin^2 x} \frac{dx}{x} = \mathbf{E}(k)$$

and its companion **3.841.4**

$$(1.4.6) \qquad \int_0^\infty \tan x \sqrt{1 - k^2 \cos^2 x} \frac{dx}{x} = \mathbf{E}(k)$$

can be established by the method described in the previous example.

1.5. The hypergeometric connection

The identites among elliptic integrals often make use of the series representations

(1.5.1) $$\mathbf{K}(k) = \frac{\pi}{2}\,_2F_1\left(\begin{matrix}\frac{1}{2},\frac{1}{2}\\1\end{matrix}\bigg|k^2\right) = \frac{\pi}{2}\sum_{j=0}^{\infty}\frac{\left(\frac{1}{2}\right)_j\left(\frac{1}{2}\right)_j}{j!}\frac{k^{2j}}{j!},$$

and

(1.5.2) $$\mathbf{E}(k) = \frac{\pi}{2}\,_2F_1\left(\begin{matrix}-\frac{1}{2},\frac{1}{2}\\1\end{matrix}\bigg|k^2\right) = \frac{\pi}{2}\sum_{j=0}^{\infty}\frac{\left(-\frac{1}{2}\right)_j\left(\frac{1}{2}\right)_j}{j!}\frac{k^{2j}}{j!},$$

where $_2F_1$ is the classical hypergeometric function

(1.5.3) $$_2F_1\left(\begin{matrix}a,b\\c\end{matrix}\bigg|x\right) = \sum_{j=0}^{\infty}\frac{(a)_j\,(b)_j}{(c)_j\,j!}x^j$$

and

(1.5.4) $$(a)_j = a(a+1)(a+2)\cdots(a+j-1)$$

is the Pochhammer symbol. The value $(a)_0 = 1$ is adopted.

1.6. Evaluation by series expansions

In this section we describe a method to evaluate many of the elliptic integrals appearing in [**35**].

EXAMPLE 1.6.1. The first example is entry **3.842.3**

(1.6.1) $$\int_0^{\infty}\frac{\sin x}{\sqrt{1-k^2\sin^2 x}}\frac{dx}{x} = \mathbf{K}(k),$$

which has been evaluated in Section 1.3.

Define

(1.6.2) $$I_1(k^2) := \int_0^{\infty}\frac{\sin x}{\sqrt{1-k^2\sin^2 x}}\frac{dx}{x}.$$

To evaluate the integral, let $m = k^2$ and expand the integrand in power series using

(1.6.3) $$\left(\frac{d}{dm}\right)^j\frac{\sin x}{x\sqrt{1-m\sin^2 x}} = \left(\frac{1}{2}\right)_j\frac{\sin^{2j+1}x}{x}(1-m\sin^2 x)^{-1/2-j}.$$

Therefore,

(1.6.4) $$I_1(m) = \sum_{j=0}^{\infty}\left(\frac{1}{2}\right)_j\frac{m^j}{j!}\int_0^{\infty}\frac{\sin^{2j+1}x}{x}\,dx.$$

The remaining integral is entry **3.821.7** in [**35**]:

(1.6.5) $$\int_0^{\infty}\frac{\sin^{2j+1}x}{x}\,dx = \frac{(2j-1)!!}{(2j)!!}\frac{\pi}{2}.$$

The value of the integral (1.6.1) now follows from the series representation of $\mathbf{K}(k)$ given in (1.5.1).

Proof of (1.6.5). Start with

$$(1.6.6) \qquad \sin^{2j+1} x = 2^{-2j} \sum_{\nu=0}^{j} (-1)^{j-\nu} \binom{2j+1}{\nu} \sin(2j - 2\nu + 1)x$$

and the integral in Example 1.3.4 in the form

$$(1.6.7) \qquad \int_0^\infty \frac{\sin \alpha x}{x}\, dx = \frac{\pi}{2}$$

for $\alpha > 0$, to obtain

$$(1.6.8) \qquad \int_0^\infty \frac{\sin^{2j+1} x}{x}\, dx = \frac{\pi}{2^{2j+1}} \sum_{\nu=0}^{j} (-1)^{j-\nu} \binom{2j+1}{\nu}.$$

It follows that

$$(1.6.9) \qquad I_1(m) = \frac{\pi}{2} \sum_{j=0}^{\infty} \left(\frac{1}{2}\right)_j \frac{(-1)^j\, m^j}{2^{2j}\, j!} \times \sum_{\nu=0}^{j} (-1)^\nu \binom{2j+1}{\nu}.$$

The result now follows from the next lemma.

LEMMA 1.6.2. *Let $j, k \in \mathbb{N}$. Then*

$$(1.6.10) \qquad \sum_{\nu=0}^{k} (-1)^j \binom{2j+1}{\nu} = (-1)^k \binom{2j}{k}.$$

PROOF. The proof is by induction on k. The case $k = 0$ is clear. The induction hypothesis is used to produce

$$(1.6.11) \qquad \sum_{\nu=0}^{k} (-1)^\nu \binom{2j+1}{\nu} = (-1)^{k-1} \binom{2j}{k-1} + (-1)^k \binom{2j+1}{k},$$

and an elementary calculation reduces this to $(-1)^k \binom{2j}{k}$. This completes the proof of (1.6.5). $\qquad\square$

Second proof of (1.6.5): apply the identity (1.3.13) to the function $f(x) = \sin^{2j} x$ to obtain

$$(1.6.12) \qquad \int_0^\infty \frac{\sin^{2j+1} x}{x}\, dx = \int_0^{\pi/2} \sin^{2j} x\, dx.$$

This last integral is the classical Wallis formula given by

$$(1.6.13) \qquad \int_0^{\pi/2} \sin^{2j} x\, dx = \frac{\pi}{2} \frac{\left(\frac{1}{2}\right)_j}{j!}.$$

The reader will find information about this formula in [6].

EXAMPLE 1.6.3. Entry **3.841.1** in [**35**]

$$(1.6.14) \qquad \int_0^\infty \sin x \sqrt{1 - k^2 \sin^2 x}\, \frac{dx}{x} = \mathbf{E}(k)$$

is established by the same method employed above. The proof starts with the expansion of the integrand using

$$(1.6.15) \qquad \left(\frac{d}{dm}\right)^j \frac{\sin x}{x} \sqrt{1 - m \sin^2 x} = \left(-\tfrac{1}{2}\right)_j \frac{\sin^{2j+1} x}{x} (1 - m \sin^2 x)^{1/2 - j}$$

and then identifying the result with (1.5.2).

EXAMPLE 1.6.4. Entry **3.842.4** in [**35**] states that

$$(1.6.16) \qquad I_2(k) := \int_0^{\pi/2} \frac{x \sin x \cos x}{\sqrt{1 - k^2 \sin^2 x}}\, dx = -\frac{\pi k'}{2k^2} + \frac{E(k)}{k^2}.$$

The parameter k' is the complementary modulus $k' = \sqrt{1 - k^2}$.

Write $m = k^2$ and expand the integrand in series using

$$(1.6.17) \qquad \left(\frac{d}{dm}\right)^j \frac{x \sin x \cos x}{\sqrt{1 - m \sin^2 x}} = \left(\frac{1}{2}\right)_j \frac{x \sin^{2j+1} x \cos x}{\sqrt{1 - m \sin^2 x}}.$$

Therefore

$$(1.6.18) \qquad I_2(m) = \sum_{j=0}^\infty \left(\frac{1}{2}\right)_j \frac{m^j}{j!} \int_0^{\pi/2} x \sin^{2j+1} x \cos x\, dx.$$

Integration by parts gives

$$(1.6.19) \qquad \int_0^{\pi/2} x \sin^{2j+1} x \cos x\, dx = \frac{\pi}{4(j+1)} - \frac{1}{4(j+1)} B\left(j + \tfrac{3}{2}, \tfrac{1}{2}\right),$$

where

$$(1.6.20) \quad B(u, v) = \int_0^1 t^{u-1}(1 - t)^{v-1}\, dt = 2 \int_0^{\pi/2} \sin^{2u-1} \varphi \cos^{2v-1} \varphi\, d\varphi$$

is the classical beta function. It follows that

$$(1.6.21) \quad I_2(m) = \frac{\pi}{4} \sum_{j=0}^\infty \left(\frac{1}{2}\right)_j \frac{m^j}{(j+1)!} - \frac{1}{4} \sum_{j=0}^\infty \left(\frac{1}{2}\right)_j B\left(j + \frac{3}{2}, \frac{1}{2}\right) \frac{m^j}{(j+1)!}.$$

The two series are now treated separately.

The first sum is computed by the binomial theorem

$$(1.6.22) \qquad (1 - x)^{-a} = \sum_{j=0}^\infty \frac{(a)_j}{j!} x^j$$

as

$$(1.6.23) \qquad \frac{\pi}{4} \sum_{j=0}^\infty \left(\frac{1}{2}\right)_j \frac{m^j}{(j+1)!} = \frac{\pi}{2(1 + \sqrt{1-m})} = \frac{\pi}{2m}(1 - \sqrt{1-m}).$$

The second sum is

$$-\frac{1}{4}\sum_{j=0}^{\infty}\left(\frac{1}{2}\right)_j B\left(j+\frac{3}{2},\frac{1}{2}\right)\frac{m^j}{(j+1)!} = -\frac{\sqrt{\pi}}{4}\sum_{j=0}^{\infty}\left(\frac{1}{2}\right)_j \frac{\Gamma(j+\frac{3}{2})}{(j+1)!\,\Gamma(j+2)}m^j$$

$$= -\frac{\pi}{8}\sum_{j=0}^{\infty}\left(\frac{1}{2}\right)_j\left(\frac{1}{2}\right)_j\frac{m^j}{(j+1)!}$$

$$= \frac{\pi}{2m}\sum_{j=0}^{\infty}\left(-\frac{1}{2}\right)_{j+1}\left(\frac{1}{2}\right)_{j+1}\frac{m^{j+1}}{(j+1)!}$$

$$= \frac{\pi}{2m}\left[{}_2F_1\left(\begin{matrix}-\frac{1}{2} & \frac{1}{2}\\ 1\end{matrix};m\right)-1\right].$$

The hypergeometric representation (1.5.2) and (1.6.21) give

$$(1.6.24)\qquad\qquad I_2(m) = -\frac{\pi\sqrt{1-m}}{2m} + \frac{\mathbf{E}(k)}{m}$$

as claimed.

1.7. A small correction to a formula in Gradshteyn and Ryzhik

In this section we present the evaluation of some elliptic integrals in [**35**]. In particular, a small error in formula **4.395.1** is corrected.

PROPOSITION 1.7.1. *Let* $k' = \sqrt{1-k^2}$ *be the complementary modulus. Then*

$$(1.7.1)\qquad\qquad \int_0^{\infty}\frac{\ln x\,dx}{\sqrt{(1+x^2)(k'^2+x^2)}} = \frac{1}{2}\mathbf{K}(k)\,\ln k'.$$

PROOF. Let $m = k'^2$ and use

$(1.7.2)$

$$\left(\frac{d}{dm}\right)^j\frac{\ln x}{\sqrt{(1+x^2)(m+x^2)}} = (-1)^j\left(\frac{1}{2}\right)_j\frac{\ln x}{\sqrt{(1+x^2)(m+x^2)^{j+1/2}}}$$

to expand the integrand around $m = -1$. It follows that

$(1.7.3)$

$$\int_0^{\infty}\frac{\ln x\,dx}{\sqrt{(1+x^2)(k'^2+x^2)}} = \sum_{j=0}^{\infty}\frac{(-1)^j}{j!}\left(\frac{1}{2}\right)_j\int_0^{\infty}\frac{\ln x\,dx}{(1+x^2)^{j+1}}(m-1)^j.$$

This last integral is given by

$$\int_0^{\infty}\frac{\ln x\,dx}{(1+x^2)^{j+1}} = \frac{1}{4}\int_0^{\infty}\frac{\ln x\,dx}{\sqrt{x}\,(1+x)^{j+1}}$$

$$= \frac{1}{4}\frac{d}{d\alpha}B(\alpha,j-\alpha+1)\Big|_{\alpha=1/2}$$

$$= \frac{1}{4}B\left(\tfrac{1}{2},j+\tfrac{1}{2}\right)\left[\psi\left(\tfrac{1}{2}\right)-\psi\left(j+\tfrac{1}{2}\right)\right]$$

$$= \frac{\pi}{2j!}\left(\frac{1}{2}\right)_j\sum_{i=0}^{j-1}\frac{1}{2i+1}.$$

Therefore, the left-hand side of (11.4.4) satisfies

$$(1.7.4) \qquad \int_0^\infty \frac{\ln x \, dx}{\sqrt{(1+x^2)(k'^2+x^2)}} = \frac{\pi}{2} \sum_{j=0}^\infty \frac{\left(\frac{1}{2}\right)_j^2}{j!^2} \sum_{i=0}^{j-1} \frac{1}{2i+1}(1-m)^j.$$

The series expansion for the complete elliptic integral now shows that the right-hand side of (11.4.4) is given by

$$\frac{1}{4} \ln m \, \mathbf{K}(\sqrt{1-m}) = \frac{\pi}{8} \left[\sum_{j=1}^\infty \frac{(1-m)^j}{j} \right] \times \left[\sum_{j=0}^\infty \frac{\left(\frac{1}{2}\right)_j^2}{j!^2}(1-m)^j \right]$$

$$= \frac{\pi}{8} \sum_{j=0}^\infty \left[\sum_{i=0}^{j-1} \frac{1}{j-i} \frac{\left(\frac{1}{2}\right)_i^2}{i!^2} \right] (1-m)^j.$$

The result follows from the identity established in the next lemma. $\qquad\square$

LEMMA 1.7.2. *Let $j \in \mathbb{N}$. Define*

$$(1.7.5) \qquad a_r = \frac{\left(\frac{1}{2}\right)_r^2}{r!^2}.$$

Then

$$(1.7.6) \qquad \sum_{i=0}^{j-1} \frac{a_i}{j-i} = 4a_j \sum_{i=0}^{j-1} \frac{1}{2i+1}.$$

PROOF. The relations

$$(1.7.7) \qquad (-x)_k = (-1)^k(x-k+1)_k \text{ and } \left(\tfrac{1}{2}\right)_{n-k} \left(\tfrac{1}{2}-n\right)_k = (-1)^k \left(\tfrac{1}{2}\right)_n$$

can be used to rewrite the left-hand side as

$$\sum_{i=0}^{j-1} \frac{\left(\frac{1}{2}\right)_i^2}{i!^2} \frac{1}{j-i} = \sum_{k=0}^{j-1} \frac{\left(\frac{1}{2}\right)_{j-k-1}^2}{(j-k-1)!^2} \frac{1}{k+1}$$

$$= \frac{\left(\frac{1}{2}\right)_j^2}{j!^2} \sum_{k=0}^{j-1} \frac{(-j)_{k+1}^2}{\left(\frac{1}{2}-j\right)_{k+1}^2} \frac{1}{k+1}.$$

Thus the assertion of the lemma is equivalent to

$$(1.7.8) \qquad \sum_{k=0}^{j-1} \frac{(-j)_{k+1}^2}{\left(\frac{1}{2}-j\right)_{k+1}^2} \frac{1}{k+1} = \sum_{k=0}^{j-1} \frac{4}{2k+1}.$$

Next apply the fact that $(x)_{k+1} = x(x+1)_k$ to obtain

$$\sum_{k=0}^{j-1} \frac{(-j)_{k+1}^2}{\left(\frac{1}{2}-j\right)_{k+1}^2} \frac{1}{k+1} = \frac{j^2}{\left(\frac{1}{2}-j\right)^2} \sum_{k=0}^{j-1} \frac{(1-j)_k^2}{\left(\frac{3}{2}-j\right)_k^2} \frac{1}{k+1}$$

$$= \frac{j^2}{\left(\frac{1}{2}-j\right)^2} \sum_{k=0}^{j-1} \frac{(1-j)_k^2(1)_k^2}{\left(\frac{3}{2}-j\right)_k^2(2)_k \, k!}.$$

The right-hand side is a balanced $_4F_3$ series and it can be transformed using

$$_4F_3\left(\begin{matrix} x, y, z, -m \\ u, v, w \end{matrix}\middle| 1\right) = \frac{(v-z)_m\,(w-z)_m}{(v)_m\,(w)_m}$$

$$\times\, _4F_3\left(\begin{matrix} u-x, u-y, z, -m \\ 1-v+z-m, 1-w+z-m, u \end{matrix}\middle| 1\right).$$

See [**15**], page 56. Now let $y = z = 1$, $x = 1 - j$, $m = j - 1$, $u = v = \frac{3}{2} - j$ and $w = 2$. It follows that

$$_4F_3\left(\begin{matrix} 1, 1, 1-j, 1-j \\ \frac{3}{2}-j, \frac{3}{2}-j, 2 \end{matrix}\middle| 1\right) = \frac{(\frac{1}{2}-j)_{j-1}\,(1)_{j-1}}{(\frac{3}{2}-j)_{j-1}\,(2)_{j-1}}\, _4F_3\left(\begin{matrix} 1, \frac{1}{2}, -j+\frac{1}{2}, 1-j \\ -j+\frac{3}{2}, \frac{3}{2}, 1-j \end{matrix}\middle| 1\right).$$

The last hypergeometric term is now simplified

$$\frac{2j-1}{j}\sum_{k=0}^{j-1}\frac{\left(\frac{1}{2}\right)_k\,\left(-j+\frac{1}{2}\right)_k}{\left(\frac{3}{2}\right)_k\,\left(-j+\frac{3}{2}\right)_k} = \frac{(2j-1)^2}{j}\sum_{k=0}^{j-1}\frac{1}{(2k+1)(2j-1-2k)}$$

$$= \frac{(2j-1)^2}{2j^2}\sum_{k=0}^{j-1}\left(\frac{1}{2k+1}+\frac{1}{2j-1-2k}\right)$$

$$= \frac{(2j-1)^2}{j^2}\sum_{k=0}^{j-1}\frac{1}{2k+1},$$

as claimed. $\square$

An automatic proof. The result of Lemma 1.7.2 also admits an automatic proof as described in [**55**]. Define the functions $F(i,j)$ and $G(i,j)$, respectively, as

(1.7.9) $$F(i,j) = \frac{\left(\frac{1}{2}\right)_i^2\, j!^2}{\left(\frac{1}{2}\right)_j^2\, i!^2}\frac{1}{j-i} \quad \text{and} \quad G(i,j) = -\frac{\left(\frac{1}{2}\right)_i^2\, j!^2}{\left(\frac{1}{2}\right)_{j+1}^2\, i!^2}\frac{i^2}{j-i+1}.$$

The stated result is equivalent to the identity $a(j) = b(j)$, where

(1.7.10) $$a(j) = \sum_{i=0}^{j-1} F(i,j) \quad \text{and} \quad b(j) = \sum_{i=0}^{j-1}\frac{1}{2j+1}.$$

The Zeilberger algorithm finds the non-homogeneous recurrence

(1.7.11) $$F(i+1,j) - F(i,j) = G(i+1,j) - G(i,j).$$

Summing this for i from 0 to $j-1$ and using the telescoping of the right-hand side produces

$$\sum_{i=0}^{j-1} F(i,j+1) - \sum_{i=0}^{j-1} F(i,j) = \sum_{i=0}^{j-1} G(i+1,j) - \sum_{i=0}^{j-1} G(i,j)$$

$$= G(j,j) - G(0,j)$$

$$= -\frac{4j^2}{(2j+1)^2}.$$

Now observe that

$$
\begin{aligned}
a(j+1) - a(j) &= \frac{4(j+1)^2}{(2j+1)^2} + \sum_{i=0}^{j-1} F(i, j+1) - \sum_{i=0}^{j-1} F(i, j) \\
&= \frac{4(j+1)^2}{(2j+1)^2} - \frac{4j^2}{(2j+1)^2} \\
&= \frac{4}{2j+1}.
\end{aligned}
$$

The sequence $b(j)$ satisfies the same recurrence. Therefore $a(j) - b(j)$ is a constant. Since $a(1) = b(1) = 4$, this constant vanishes. This establishes the result.

The next result corrects entry **4.395.1** in [**35**].

COROLLARY 1.7.3. *The value*

$$
(1.7.12) \qquad \int_0^\infty \frac{\ln \tan \theta \, d\theta}{\sqrt{1 - k^2 \sin^2 \theta}} = -\frac{1}{2} \ln k' \mathbf{K}(k)
$$

holds.

PROOF. Let $x \mapsto \tan \theta$ in (1.7.1). $\qquad \square$

EXAMPLE 1.7.4. Entry **4.242.1** states

$$
(1.7.13) \qquad \int_0^\infty \frac{\ln x \, dx}{\sqrt{(a^2 + x^2)(b^2 + x^2)}} = \frac{1}{2a} \mathbf{K}\left(\frac{\sqrt{a^2 - b^2}}{a}\right) \ln ab.
$$

Formula (11.4.4) corresponds to the special case $a = 1$. The change of variables $x = at$ produces

$$
\begin{aligned}
\int_0^\infty \frac{\ln x \, dx}{\sqrt{(a^2 + x^2)(b^2 + x^2)}} &= \frac{1}{b} \int_0^\infty \frac{\ln t \, dt}{\sqrt{(1 + t^2)(c^2 + t^2)}} \\
&\quad + \frac{\ln a}{b} \int_0^\infty \frac{dt}{\sqrt{(1 + t^2)(1 + c^2 t^2)}}
\end{aligned}
$$

with $c = b/a$. The first integral is evaluated using (11.4.4) and letting $t = \tan \varphi$ to see that the second integral is $\mathbf{K}(\sqrt{1 - c^2})$. This establishes the result.

EXAMPLE 1.7.5. The techniques illustrated here are now employed to prove entry **4.414.1** in [**35**]:

$$
(1.7.14) \qquad \int_0^{\pi/2} \frac{\ln(1 - k^2 \sin^2 x)}{\sqrt{1 - k^2 \sin^2 x}} \, dx = \mathbf{K}(k) \ln k'.
$$

Let $m = k^2$ and observe that

(1.7.15)

$$
\frac{d}{dm} \frac{\alpha_j + \beta_j \ln(1 - m \sin^2 x)}{(1 - m \sin^2 x)^{j+1/2}} \sin^{2j} x = \frac{\alpha_{j+1} + \beta_{j+1} \ln(1 - m \sin^2 x)}{(1 - m \sin^2 x)^{j+3/2}} \sin^{2j+2} x
$$

where the parameters α_j, β_j satisfy

(1.7.16) $$\alpha_{j+1} = (j + \tfrac{1}{2})\alpha - j - \beta_j \text{ and } \beta_{j+1} = (j + \tfrac{1}{2})\beta_j.$$

Now choose $\alpha_0 = 0$ and $\beta_0 = 1$ to obtain

(1.7.17) $$\left(\frac{d}{dm}\right)^j \frac{\ln(1 - m \sin^2 x)}{\sqrt{1 - m \sin^2 x}} = \frac{\alpha_j + \beta_j \ln(1 - m \sin^2 x)}{(1 - m \sin^2 x)^{j+1/2}} \sin^{2j} x.$$

Expand the integrand of (1.7.14) around $m = 0$ and use

(1.7.18) $$\int_0^\infty \sin^{2j} x \, dx = \frac{\pi}{2} \frac{\left(\frac{1}{2}\right)_j}{j!^2}$$

and the expressions

(1.7.19) $$\alpha_j = \left(\tfrac{1}{2}\right)_j \sum_{i=0}^{j-1} \frac{2}{2i+1} \text{ and } \beta_j = \left(\tfrac{1}{2}\right)_j$$

to see that

(1.7.20) $$\int_0^{\pi/2} \frac{\ln(1 - k^2 \sin^2 x)}{\sqrt{1 - k^2 \sin^2 x}} \, dx = \pi \sum_{j=0}^\infty \frac{\left(\frac{1}{2}\right)_j^2}{j!^2} \left(\sum_{i=0}^{j-1} \frac{1}{2i+1}\right) m^j.$$

The result now follows from the evaluation given in the proof of Proposition 1.7.1.

CHAPTER 2

The Riemann zeta function

2.1. Introduction

The table of integrals [35] contains a large variety of definite integrals that involve the *Riemann zeta* function

$$(2.1.1) \qquad \zeta(s) = \sum_{n=1}^{\infty} \frac{1}{n^s}.$$

The series converges for $\operatorname{Re} s > 1$.

This is a classical function that plays an important role in the distribution of prime numbers. The reader will find in [27] a historical description of the fundamental properties of $\zeta(s)$. The textbook [19] presents interesting information about the major open question related to $\zeta(s)$: all its non-trivial zeros are on the vertical line $\operatorname{Re} s = \frac{1}{2}$. This is the famous *Riemann hypothesis*.

In this section we summarize elementary properties of ζ that will be employed in the evaluation of definite integrals.

The zeta function at the even integers. The values of $\zeta(s)$ at the *even* integers are given in terms of the *Bernoulli numbers* defined by the generating function

$$(2.1.2) \qquad \frac{u}{e^u - 1} = \sum_{k=0}^{\infty} \frac{B_k}{k!} u^k.$$

It turns out that $B_{2n+1} = 0$ for $n > 1$. The relation

$$(2.1.3) \qquad \zeta(2n) = (-1)^{n-1} \frac{(2\pi)^{2n}}{2(2n)!} B_{2n}$$

can be found in [17]. The sign of B_{2n} is $(-1)^{n-1}$, so we can write (2.1.3) as

$$(2.1.4) \qquad \zeta(2n) = \frac{(2\pi)^{2n}}{2(2n)!} |B_{2n}|,$$

which looks more compact. The case of $\zeta(2n + 1)$ is more complicated. No simple expression, such as (2.1.4), is known.

There are other series that can be expressed in terms of $\zeta(s)$. We present here the case of the alternating zeta series.

PROPOSITION 2.1.1. *Assume $s > 1$. Then*

$$(2.1.5) \qquad \sum_{n=1}^{\infty} \frac{(-1)^n}{n^s} = (2^{1-s} - 1)\zeta(s).$$

PROOF. Split the sum (2.1.1) according to the parity of n. Then

$$\sum_{n=1}^{\infty} \frac{(-1)^n}{n^s} = \sum_{k=1}^{\infty} \frac{1}{(2k)^s} - \sum_{k=1}^{\infty} \frac{1}{(2k-1)^s}$$

$$= 2^{-s} \sum_{k=1}^{\infty} \frac{1}{k^s} - \left(\sum_{k=1}^{\infty} \frac{1}{k^s} - \sum_{k=1}^{\infty} \frac{1}{(2k)^s} \right).$$

The identity (2.1.5) has been established. $\qquad\qquad\square$

NOTE 2.1.2. The expression (2.1.5), written as

$$(2.1.6) \qquad \zeta(s) = \frac{1}{2^{1-s} - 1} \sum_{n=1}^{\infty} \frac{(-1)^k}{k^s},$$

provides a *continuation* of $\zeta(s)$ to $0 < \operatorname{Re} s$, with the natural exception at $s = 1$.

PROPOSITION 2.1.3. *Let $a > 1$. Then*

$$(2.1.7) \qquad \sum_{k=0}^{\infty} \frac{1}{(2k+1)^a} = \frac{2^a - 1}{2^a} \zeta(a).$$

PROOF. This simply comes from

$$\sum_{k=0}^{\infty} \frac{1}{(2k+1)^a} = \sum_{k=1}^{\infty} \frac{1}{k^a} - \sum_{k=1}^{\infty} \frac{1}{(2k)^a}.$$

$\qquad\qquad\square$

2.2. A first integral representation

The first integral in [**35**] that is evaluated in terms of the Riemann zeta function is **3.411.1**:

$$(2.2.1) \qquad \int_0^{\infty} \frac{x^{s-1}\, dx}{e^{px} - 1} = \frac{\Gamma(s)\zeta(s)}{p^s}.$$

Here Γ is the *gamma function* defined by

$$(2.2.2) \qquad \Gamma(s) = \int_0^{\infty} t^{s-1} e^{-t}\, dt.$$

To verify (2.2.1), observe that the parameter p can be scaled out of the integral. Indeed, the change of variables $t = px$ shows that (2.2.1) is equivalent to the case $p = 1$:

$$(2.2.3) \qquad \int_0^{\infty} \frac{t^{s-1}\, dt}{e^t - 1} = \Gamma(s)\zeta(s).$$

To prove this, expand the integrand as

$$(2.2.4) \qquad \frac{1}{e^t - 1} = \frac{e^{-t}}{1 - e^{-t}} = \sum_{k=0}^{\infty} e^{-(k+1)t}.$$

Therefore,

$$(2.2.5) \qquad \int_0^{\infty} \frac{x^{s-1} \, dx}{e^x - 1} = \sum_{k=0}^{\infty} \int_0^{\infty} t^{s-1} e^{-(k+1)t} \, dt.$$

The change of variables $v = (1 + k)t$ yields the result.

EXAMPLE 2.2.1. The evaluation of **3.411.2**,

$$(2.2.6) \qquad \int_0^{\infty} \frac{x^{2n-1} \, dx}{e^{px} - 1} = (-1)^{n-1} \left(\frac{2\pi}{p} \right)^{2n} \frac{B_{2n}}{4n},$$

can be reduced to the case $p = 1$ by the scaling $t = px$ and it follows from (2.1.3). Using (2.1.4), we write it as

$$(2.2.7) \qquad \int_0^{\infty} \frac{x^{2n-1} \, dx}{e^x - 1} = \frac{(2\pi)^{2n}}{4n} |B_{2n}|.$$

EXAMPLE 2.2.2. The evaluation of **3.411.3**

$$(2.2.8) \qquad \int_0^{\infty} \frac{x^{s-1} \, dx}{e^{px} + 1} = \frac{(1 - 2^{1-s})\Gamma(s)}{p^s} \zeta(s),$$

is first reduced, via $t = px$, to the case $p = 1$

$$(2.2.9) \qquad \int_0^{\infty} \frac{t^{s-1} \, dx}{e^t + 1} = (1 - 2^{1-s})\Gamma(s)\zeta(s),$$

and this is evaluated expanding the integrand and integrating term by term to obtain

$$(2.2.10) \qquad \int_0^{\infty} \frac{t^{s-1}}{e^t + 1} \, dt = \frac{1}{\Gamma(s)} \sum_{k=0}^{\infty} \frac{(-1)^k}{(k + 1)^s}.$$

The result now follows from (2.1.5).

EXAMPLE 2.2.3. The special case $s = 2n$ in (2.2.8) yields

$$(2.2.11) \qquad \int_0^{\infty} \frac{t^{2n-1} \, dt}{e^t + 1} = (1 - 2^{1-2n}) \frac{(2\pi)^{2n}}{4n} |B_{2n}|.$$

The integral **3.411.4**

$$(2.2.12) \qquad \int_0^{\infty} \frac{x^{2n-1} \, dx}{e^{px} + 1} = (1 - 2^{1-2n}) \left(\frac{2\pi}{p} \right)^{2n} \frac{|B_{2n}|}{4n}$$

is reduced to (2.2.11) by the usual scaling.

2.3. Integrals involving partial sums of $\zeta(s)$

In this section we consider in a unified form a series of definite integrals in [**35**] whose values involve partial sums of the Riemann zeta function. We begin with the evaluation of **3.411.6**: expanding the integrand we obtain

$$(2.3.1) \qquad \int_0^\infty \frac{x^{a-1} e^{-\beta x}}{1 - \delta e^{-\gamma x}} \, dx = \sum_{k=0}^\infty \delta^k \int_0^\infty x^{a-1} e^{-x(\beta + \gamma k)} \, dx$$

$$= \frac{\Gamma(a)}{\gamma^a} \sum_{k=0}^\infty \delta^k \left(k + \frac{\beta}{\gamma} \right)^{-a}.$$

The sum is identified as the *Lerch function* defined by

$$(2.3.2) \qquad \Phi(z, s, v) = \sum_{n=0}^\infty (v + n)^{-s} z^n.$$

Therefore

$$(2.3.3) \qquad \int_0^\infty \frac{x^{a-1} e^{-\beta x} \, dx}{1 - \delta e^{-\gamma x}} = \frac{\Gamma(a)}{\gamma^a} \Phi(\delta, a, \beta/\gamma).$$

Integrals involving the Lerch Φ-function will be discussed in a future publication. Here we simply observe that **3.411.22**

$$(2.3.4) \qquad \int_0^\infty \frac{x^{p-1} \, dx}{e^{rx} - q} = \frac{\Gamma(p)}{r^p} \Phi(q, p, 1)$$

follows directly from (2.3.1) after writing

$$(2.3.5) \qquad \int_0^\infty \frac{x^{p-1} \, dx}{e^{rx} - q} = \int_0^\infty \frac{x^{p-1} e^{-rx} \, dx}{1 - q e^{-rx}}.$$

We now discuss several special cases of (2.3.1).

EXAMPLE 2.3.1. The case $\delta = 1$ in (2.3.1) is related to the *Hurwitz zeta function* defined by

$$(2.3.6) \qquad \zeta(z, q) = \sum_{n=0}^\infty \frac{1}{(n + q)^z}.$$

Replacing $\delta = 1$ in (2.3.1) gives

$$(2.3.7) \qquad \int_0^\infty \frac{x^{a-1} e^{-\beta x}}{1 - e^{-\gamma x}} \, dx = \frac{\Gamma(a)}{\gamma^a} \zeta(a, \beta/\gamma).$$

This appears as **3.411.7**.

EXAMPLE 2.3.2. We now consider the special case of (2.3.7) in which β/γ is a positive integer, say, $\beta = m\gamma$. Then we obtain

$$(2.3.8) \qquad \int_0^\infty \frac{x^{a-1} e^{-m\gamma x} \, dx}{1 - e^{-\gamma x}} = \frac{\Gamma(a)}{\gamma^a} \sum_{k=0}^\infty \frac{1}{(m + k)^a}.$$

Now observe that

$$(2.3.9) \qquad \sum_{k=0}^{\infty} \frac{1}{(m+k)^a} = \sum_{k=1}^{\infty} \frac{1}{k^a} - \sum_{k=1}^{m-1} \frac{1}{k^a},$$

so that

$$(2.3.10) \qquad \int_0^{\infty} \frac{x^{a-1} e^{-m\gamma x} \, dx}{1 - e^{-\gamma x}} = \frac{\Gamma(a)}{\gamma^a} \left(\zeta(a) - \sum_{k=1}^{m-1} \frac{1}{k^a} \right).$$

We restate the previous result.

PROPOSITION 2.3.3. *Let* $a, \gamma \in \mathbb{R}^+$ *and* $m \in \mathbb{N}$. *Then*

$$(2.3.11) \qquad \int_0^{\infty} \frac{x^{a-1} e^{-m\gamma x} \, dx}{1 - e^{-\gamma x}} = \frac{\Gamma(a)}{\gamma^a} \left(\zeta(a) - \sum_{k=1}^{m-1} \frac{1}{k^a} \right).$$

EXAMPLE 2.3.4. The values $a = 2$, $\gamma = 1$ and $m = 1$ in (2.3.10) give

$$(2.3.12) \qquad \int_0^{\infty} \frac{x e^{-x} \, dx}{e^x - 1} = \frac{\pi^2}{6} - 1,$$

using $\Gamma(2) = 1$ and $\zeta(2) = \pi^2/6$. This appears as **3.411.9** in [**35**].

EXAMPLE 2.3.5. The case $a = 3$, $\gamma = 1$ and $m \in \mathbb{N}$ gives **3.411.14**

$$(2.3.13) \qquad \int_0^{\infty} \frac{x^2 e^{-mx}}{1 - e^{-x}} \, dx = 2 \left(\zeta(3) - \sum_{k=1}^{m-1} \frac{1}{k^3} \right).$$

EXAMPLE 2.3.6. The case $a = 4$, $\gamma = 1$ and $m \in \mathbf{N}$ gives **3.411.17**

$$(2.3.14) \qquad \int_0^{\infty} \frac{x^3 e^{-mx}}{1 - e^{-x}} \, dx = \frac{\pi^4}{15} - 6 \sum_{k=1}^{m-1} \frac{1}{k^4}.$$

Here we have used $\Gamma(4) = 6$ and $\zeta(4) = \pi^4/90$.

EXAMPLE 2.3.7. Formula **3.411.25** is

$$(2.3.15) \qquad \int_0^{\infty} x \frac{1 + e^{-x}}{e^x - 1} \, dx = \int_0^{\infty} \frac{x e^{-x} \, dx}{1 - e^{-x}} + \int_0^{\infty} \frac{x e^{-2x} \, dx}{1 - e^{-x}}.$$

The first integral corresponds to $a = 2$, $\gamma = 1$, $m = 1$ and the second one to $a = 2$, $\gamma = 1$, $m = 2$. Therefore

$$(2.3.16) \qquad \int_0^{\infty} x \frac{1 + e^{-x}}{e^x - 1} \, dx = \Gamma(2) \left(\zeta(2) + \zeta(2) - 1 \right) = \frac{\pi^2}{3} - 1.$$

EXAMPLE 2.3.8. The final example in this section is **3.411.21**

$$(2.3.17) \qquad \int_0^{\infty} x^{n-1} \frac{1 - e^{-mx}}{1 - e^x} \, dx = (n-1)! \sum_{k=1}^{m} \frac{1}{k^n}.$$

We now show that the *correct formula* is

$$(2.3.18) \qquad \int_0^{\infty} x^{n-1} \frac{1 - e^{-mx}}{1 - e^x} \, dx = -(n-1)! \sum_{k=1}^{m} \frac{1}{k^n}.$$

To establish this, we write

$$(2.3.19) \qquad \int_0^\infty x^{n-1}\frac{1-e^{-mx}}{1-e^x}\,dx = \int_0^\infty \frac{x^{n-1}e^{-(m+1)x}}{1-e^{-x}}\,dx - \int_0^\infty \frac{x^{n-1}e^{-x}}{1-e^{-x}}\,dx.$$

The first integral corresponds to $a = n$, $\gamma = 1$ and $m+1$ instead of m, so that

$$(2.3.20) \qquad \int_0^\infty \frac{x^{n-1}e^{-(m+1)x}}{1-e^{-x}}\,dx = \Gamma(n)\left(\zeta(n) - \sum_{k=1}^m \frac{1}{k^n}\right).$$

The second integral corresponds to $a = n$, $\gamma = 1$ and $m = 1$. Therefore

$$(2.3.21) \qquad \int_0^\infty \frac{x^{n-1}e^{-x}}{1-e^{-x}}\,dx = \Gamma(n)\zeta(n).$$

Formula (2.3.18) has been established.

2.4. The alternate version

The alternating version of (2.3.1) gives

$$(2.4.1) \qquad \int_0^\infty \frac{x^{a-1}e^{-\beta x}}{1+\delta e^{-\gamma x}}\,dx = \frac{\Gamma(a)}{\gamma^a}\sum_{k=0}^\infty (-1)^k \delta^k \left(k + \frac{\beta}{\gamma}\right)^{-a},$$

which in the case $\delta = 1$ provides

$$(2.4.2) \qquad \int_0^\infty \frac{x^{a-1}e^{-\beta x}\,dx}{1+e^{-\gamma x}} = \frac{\Gamma(a)}{\gamma^a}\sum_{k=0}^\infty (-1)^k (k + \beta/\gamma)^{-a}.$$

In particular, if $\beta = m\gamma$, with $m \in \mathbb{N}$, we have

$$(2.4.3) \qquad \int_0^\infty \frac{x^{a-1}e^{-m\gamma x}\,dx}{1+e^{-\gamma x}} = \frac{\Gamma(a)}{\gamma^a}\sum_{k=0}^\infty \frac{(-1)^k}{(k+m)^a}.$$

Using (2.1.5) we obtain the next proposition.

PROPOSITION 2.4.1. *Let $a, \gamma \in \mathbb{R}^+$ and $m \in \mathbb{N}$. Then*

$$(2.4.4) \qquad \int_0^\infty \frac{x^{a-1}e^{-m\gamma x}\,dx}{1+e^{-\gamma x}} = \frac{(-1)^m\Gamma(a)}{\gamma^a}\left((2^{1-a}-1)\zeta(a) - \sum_{k=1}^{m-1}\frac{(-1)^k}{k^a}\right).$$

The next examples come from (2.4.3).

EXAMPLE 2.4.2. The case $a = n$, $\gamma = 1$ and $m = p+1$ gives **3.411.8**

$$(2.4.5) \qquad \int_0^\infty \frac{x^{n-1}e^{-px}\,dx}{1+e^x} = (-1)^p\Gamma(n)\left[(1-2^{1-n})\zeta(n) + \sum_{k=1}^p \frac{(-1)^k}{k^n}\right].$$

The reader will check that the answer can be written as

$$(2.4.6) \qquad \int_0^\infty \frac{x^{n-1}e^{-px}\,dx}{1+e^{-x}} = (n-1)!\sum_{k=1}^\infty \frac{(-1)^{k-1}}{(p+k)^n}.$$

EXAMPLE 2.4.3. The case $a = 2$, $c = 1$, and $m = 2$ gives **3.411.10**

$$(2.4.7) \qquad \int_0^\infty \frac{xe^{-2x}}{1 + e^{-x}}\, dx = 1 - \frac{\pi^2}{12}.$$

EXAMPLE 2.4.4. The case $a = 2$, $c = 1$, and $m = 3$ gives **3.411.11**

$$(2.4.8) \qquad \int_0^\infty \frac{xe^{-3x}}{1 + e^{-x}}\, dx = \frac{\pi^2}{12} - \frac{3}{4}.$$

EXAMPLE 2.4.5. The case $a = 2$, $c = 1$, and $m = 2n$ gives **3.411.12**

$$(2.4.9) \qquad \int_0^\infty \frac{xe^{-(2n-1)x}}{1 + e^{-x}}\, dx = -\frac{\pi^2}{12} + \sum_{k=1}^{2n-1} \frac{(-1)^{k-1}}{k^2}.$$

EXAMPLE 2.4.6. The case $a = 2$, $c = 1$, and $m = 2n + 1$ gives **3.411.13**

$$(2.4.10) \qquad \int_0^\infty \frac{xe^{-2nx}}{1 + e^{-x}}\, dx = \frac{\pi^2}{12} + \sum_{k=1}^{2n} \frac{(-1)^k}{k^2}.$$

EXAMPLE 2.4.7. The case $a = 3$, $c = 1$, and $m \in \mathbb{N}$ gives **3.411.15**

$$(2.4.11) \qquad \int_0^\infty \frac{x^2 e^{-nx}}{1 + e^{-x}}\, dx = (-1)^{n+1}\left(\frac{3}{2}\zeta(3) + 2\sum_{k=1}^{n-1} \frac{(-1)^k}{k^3}\right).$$

EXAMPLE 2.4.8. The case $a = 4$, $c = 1$, and $m \in \mathbb{N}$ gives **3.411.18**

$$(2.4.12) \qquad \int_0^\infty \frac{x^3 e^{-nx}}{1 + e^{-x}}\, dx = (-1)^{n+1}\left(\frac{7\pi^4}{120} + 6\sum_{k=1}^{n-1} \frac{(-1)^k}{k^4}\right).$$

EXAMPLE 2.4.9. Similar manipulations produce **3.411.26**

$$(2.4.13) \qquad \int_0^\infty xe^{-x}\frac{1 - e^{-x}}{1 + e^{-3x}}\, dx = \frac{2\pi^2}{27}.$$

2.5. The logarithmic scale

The integrals described in Section 2.4 can be transformed into logarithmic integrals via the change of variables $t = e^{-cx}$. For example, (2.3.1) becomes

$$(2.5.1) \qquad \int_0^1 \frac{t^{\beta-1} \ln^{a-1} t\, dt}{1 - \delta t} = (-1)^{a-1}\Gamma(a) \sum_{k=0}^\infty \frac{\delta^k}{(k + \beta)^a}$$

and the special case $\delta = 1$ replaces (2.3.7) with

$$(2.5.2) \qquad \int_0^1 \frac{t^{\beta-1} \ln^{a-1} t\, dt}{1 - t} = (-1)^{a-1}\Gamma(a) \sum_{k=0}^\infty \frac{1}{(k + \beta)^a}.$$

In the special case that $m \in \mathbb{N}$, the formula (2.3.11) becomes

$$(2.5.3) \qquad \int_0^1 \frac{t^{m-1} \ln^{a-1} t\, dt}{1 - t} = (-1)^{a-1}\Gamma(a)\left(\zeta(a) - \sum_{k=1}^{m-1} \frac{1}{k^a}\right),$$

in particular, for $m = 1$, we have

$$(2.5.4) \qquad \int_0^1 \frac{\ln^{a-1} t \, dt}{1-t} = (-1)^{a-1} \Gamma(a) \zeta(a).$$

Finally, the change of variables $t = s^\gamma$ in (2.5.2) produces

$$(2.5.5) \qquad \int_0^1 \frac{s^{\beta-1} \ln^{a-1} s \, dt}{1 - s^\gamma} = (-1)^{a-1} \Gamma(\gamma) \sum_{k=0}^\infty \frac{1}{(\gamma k + \beta)^a}.$$

We now present examples of these formulas that appear in [**35**].

EXAMPLE 2.5.1. Formula (2.5.4) appears in [**35**] only for a even. This is the case where the value of $\zeta(a)$ reduces via (2.1.3). We find **4.231**.2 for $a = 2$

$$(2.5.6) \qquad \int_0^1 \frac{\ln x \, dx}{1-x} = -\frac{\pi^2}{6},$$

and **4.262.2**

$$(2.5.7) \qquad \int_0^1 \frac{\ln^3 x \, dx}{1-x} = -\frac{\pi^4}{15},$$

which uses $\Gamma(4) = 6$ and $\zeta(4) = \pi^4/90$. The next example is **4.264.2**

$$(2.5.8) \qquad \int_0^1 \frac{\ln^5 x \, dx}{1-x} = -\frac{8\pi^6}{63},$$

which uses $\Gamma(6) = 120$ and $\zeta(6) = \pi^6/945$. The final example is **4.266.2**

$$(2.5.9) \qquad \int_0^1 \frac{\ln^7 x \, dx}{1-x} = -\frac{8\pi^8}{15},$$

which uses $\Gamma(8) = 5040$ and $\zeta(8) = \pi^8/9450$.

EXAMPLE 2.5.2. The choice $a = 4$ and $m = n + 1$ in (2.5.3) produces **4.262.5**:

$$(2.5.10) \qquad \int_0^1 \frac{x^n \ln^3 x}{1-x} \, dx = -\frac{\pi^4}{15} + 6 \sum_{k=1}^n \frac{1}{k^4}.$$

EXAMPLE 2.5.3. The choice $a = 4$, $\beta = 2n+1$, and $\gamma = 2$ in (2.5.5) gives **4.262.6**

$$(2.5.11) \qquad \int_0^1 \frac{x^{2n} \ln^3 x}{1-x^2} \, dx = -\frac{\pi^4}{16} + 6 \sum_{k=1}^n \frac{1}{(2k+1)^4}.$$

In this calculation we have used (2.1.7) to produce the value

$$(2.5.12) \qquad \sum_{k=0}^\infty \frac{1}{(2k+1)^4} = \frac{\pi^4}{96}.$$

EXAMPLE 2.5.4. The choice $a = 3$ and $m = n+1$ in (2.5.3) gives **4.261.12**

$$(2.5.13) \qquad \int_0^1 \frac{x^n \ln^2 x}{1-x} \, dx = 2 \left(\zeta(3) - \sum_{k=1}^n \frac{1}{k^3} \right).$$

EXAMPLE 2.5.5. The choice $a = 3$, $\beta = 2n + 1$, and $\gamma = 2$ gives **4.261.13**

$$(2.5.14) \qquad \int_0^1 \frac{x^{2n} \ln^2 x}{1 - x^2}\, dx = \frac{7\zeta(3)}{4} - 2 \sum_{k=0}^{n-1} \frac{1}{(2k+1)^3}.$$

2.6. The alternating logarithmic scale

There is a corresponding list of formulas for logarithmic integrals that produce alternating series. For example (2.5.1) becomes

$$(2.6.1) \qquad \int_0^1 \frac{t^{\beta-1} \ln^{a-1} t\, dt}{1 + \delta t} = (-1)^{a-1}\Gamma(a) \sum_{k=0}^{\infty} \frac{(-1)^k \delta^k}{(k+\beta)^a}$$

and the case $\delta = 1$ gives

$$(2.6.2) \qquad \int_0^1 \frac{t^{\beta-1} \ln^{a-1} t\, dt}{1 + t} = (-1)^{a-1}\Gamma(a) \sum_{k=0}^{\infty} \frac{(-1)^k}{(k+\beta)^a}.$$

In the special case that $m \in \mathbb{N}$, we have

$$(2.6.3) \qquad \int_0^1 \frac{t^{m-1} \ln^{a-1} t\, dt}{1 + t} = (-1)^{a+m}\Gamma(a) \left(\frac{2^{a-1} - 1}{2^{a-1}} \zeta(a) + \sum_{k=1}^{m-1} \frac{(-1)^k}{k^a} \right),$$

in particular, for $m = 1$, we have

$$(2.6.4) \qquad \int_0^1 \frac{\ln^{a-1} t\, dt}{1 + t} = (-1)^{a+1} \frac{2^{a-1} - 1}{2^{a-1}} \Gamma(a)\zeta(a).$$

Finally, (2.5.5) produces

$$(2.6.5) \qquad \int_0^1 \frac{s^{\beta-1} \ln^{a-1} s\, ds}{1 + s^\gamma} = (-1)^{a-1}\Gamma(a) \sum_{k=0}^{\infty} \frac{(-1)^k}{(\gamma k + \beta)^a}.$$

We now present examples of these formulas that appear in [**35**].

EXAMPLE 2.6.1. The choice $a = 2$ in (2.6.4) produces **4.231.1**:

$$(2.6.6) \qquad \int_0^1 \frac{\ln x}{1 + x}\, dx = -\frac{\pi^2}{12}.$$

The table contains formulas that use (2.6.4) only for a even; in that form, the integrals are expressible as powers of π. For example, **4.262.1**

$$(2.6.7) \qquad \int_0^1 \frac{\ln^3 x}{1 + x}\, dx = -\frac{7\pi^4}{120},$$

uses $\Gamma(4) = 6$ and $\zeta(4) = \pi^4/90$ Similarly, **4.264.1**

$$(2.6.8) \qquad \int_0^1 \frac{\ln^5 x}{1 + x}\, dx = -\frac{31\pi^6}{252}$$

uses $\Gamma(6) = 120$ and $\zeta(6) = \pi^6/945$. The final example of this form is **4.266.1**

$$(2.6.9) \qquad \int_0^1 \frac{\ln^7 x}{1 + x}\, dx = -\frac{127\pi^8}{240},$$

which employs $\Gamma(8) = 5040$ and $\zeta(8) = \pi^8/9450$. The next cases in this list would be

$$(2.6.10) \qquad \int_0^1 \frac{\ln^9 x}{1+x}\,dx = -\frac{511\pi^{10}}{132}$$

and

$$(2.6.11) \qquad \int_0^1 \frac{\ln^{11} x}{1+x}\,dx = -\frac{1414477\pi^{12}}{32760},$$

which do not appear in [**35**].

EXAMPLE 2.6.2. The choice $a = 2n+1$ in (2.6.4) gives **4.271.1**

$$(2.6.12) \qquad \int_0^1 \frac{\ln^{2n} x}{1+x}\,dx = \frac{2^{2n}-1}{2^{2n}}(2n)!\,\zeta(2n+1).$$

EXAMPLE 2.6.3. The choice $a = 2n$ in (2.6.4) gives **4.271.2**

$$(2.6.13) \qquad \int_0^1 \frac{\ln^{2n-1} x}{1+x}\,dx = -\frac{2^{2n-1}-1}{2^{2n-1}}(2n-1)!\,\zeta(2n),$$

and using (2.1.3) gives

$$(2.6.14) \qquad \int_0^1 \frac{\ln^{2n-1} x}{1+x}\,dx = -\frac{2^{2n-1}-1}{2n}|B_{2n}|\pi^{2n}.$$

2.7. Integrals over the whole line

The change of variables $x = \frac{1}{p}e^{-t}$ in (2.2.1) gives entry **3.333.1**

$$(2.7.1) \qquad \int_{-\infty}^{\infty} \frac{e^{-sx}\,dx}{\exp(e^{-x})-1} = \Gamma(s)\zeta(s).$$

The same change of variable in (2.2.8) gives entry **3.333.2**

$$(2.7.2) \qquad \int_{-\infty}^{\infty} \frac{e^{-sx}\,dx}{\exp(e^{-x})+1} = (1-2^{1-s})\Gamma(s)\zeta(s).$$

The exceptional case,

$$(2.7.3) \qquad \int_{-\infty}^{\infty} \frac{e^{-x}\,dx}{\exp(e^{-x})+1} = \ln 2,$$

mentioned in entry **3.333.2**, is elementary.

CHAPTER 3

Some automatic proofs

3.1. Introduction

The volume [**35**] is one of the most widely used table of integrals. This work, now in its 7th edition, has been edited and amplified several times. The initial work of the authors I. Gradshteyn and I. M. Ryzhik is now supplemented by entries proposed by a large number of researchers.

This paper is part of a project, initiated in [**49**], with the goal of establishing the validity of these formulas and to place them in context. The previous papers in this project contain evaluation of entries in [**35**] by traditional analytical methods. Symbolic languages, mostly *Mathematica*, have so far only been used to check the entries in search of possible errors (e.g., by numerical evaluation). The methodology employed here is different: the computer package **HolonomicFunctions** is employed to deliver *computer-generated proofs* of some entries in [**35**]. The examples are chosen to illustrate the different capabilities of the package. Note that Mathematica (version 7) fails on most of these examples.

3.2. The class of holonomic functions

The computer algebra methods employed here originate in Zeilberger's holonomic systems approach [**24, 42, 69**]. They can be seen as a generalization of the Almkvist–Zeilberger algorithm [**2**] to integrands that are not necessarily hyperexponential. The basic idea is that of the representation of a function (or a sequence) as solutions of differential (or difference) equations together with some initial conditions. These equations are required to be linear, homogeneous and with polynomial coefficients. It is convenient to present them in operator notation: D_x is used for the (partial) derivative with respect to x and S_n for the shift in n. The main advantage of this notation is that the differential equations and recurrences under consideration turn into polynomials, which are the basic objects in computer algebra.

Consider first the case of functions of two variables: a continuous one x and a discrete one n. Define $\mathbb{O}$ to be the algebra generated by the operators D_x and S_n with coefficients that are rational functions in x and n. This is a non-commutative algebra and the rules $D_x x = x D_x + 1$ and $S_n n = n S_n + S_n$ must be incorporated into $\mathbb{O}$. Such structures are usually called *Ore algebras*. A similar definition can be made in the case of many continuous and discrete variables.

The operator $P \in \mathbb{O}$ is said to *annihilate* the function f if $P(f) = 0$. For example, $\sin x$ is annihilated by the operator $D_x^2 + 1$ and the Fibonacci numbers F_n by $S_n^2 - S_n - 1$. The latter is nothing but the recurrence $F_n = F_{n-1} + F_{n-2}$ used to define these numbers. Given a function f, the set

$$(3.2.1) \qquad \mathrm{Ann}_{\mathbb{O}}(f) := \{P \in \mathbb{O} : P(f) = 0\}$$

represents the set of all the equations satisfied by f. Naturally, a given operator P may annihilate many different functions. For instance, $D_x^2 + 1$ also annihilates $\cos x$. Thus $\mathrm{Ann}_{\mathbb{O}}(f)$ does not determine f uniquely: initial conditions must be included. The term *equation* in the present context refers to an equation of the form $P(f) = 0$ with $P \in \mathbb{O}$. Many classical functions, such as rational or algebraic functions, exponentials, logarithms, and some of the trigonometric functions, as well as a multitude of special functions satisfy equations of the type described above.

Observe that $\mathrm{Ann}_{\mathbb{O}}(f)$ is a left ideal in the algebra $\mathbb{O}$, called the *annihilating ideal of* f. Indeed, given $P \in \mathbb{O}$, that represents the equation $P(f) = 0$ satisfied by f, its differentiation yields a new equation for f, represented by $D_x P$. Similarly, if f depends on the discrete index n and satisfies a linear recurrence $Q(f) = 0$ for $Q \in \mathbb{O}$, then f also satisfies the shifted recurrence, represented by $S_n Q$.

The annihilating left ideal can be described by a suitable set of generators. The concept of Gröbner bases is then employed to decide the *ideal membership problem*: in our context, to decide whether a function satisfies a given equation. These bases are also used to obtain a unique representation of the residue classes modulo an ideal.

In the holonomic systems approach all manipulations are carried out with these implicit function descriptions, e.g., an algorithm for computing a definite integral requires as input an implicit description of the integrand (viz. the Gröbner basis of an annihilating ideal of the integrand) and will return an implicit description for the integral. The initial values are usually considered afterwards.

In this paper we deal with the so-called *holonomic functions*. Apart from some technical aspects, the most important necessary condition for a function to be holonomic is that there exists an annihilating ideal of dimension zero (a concept that depends on the choice of the underlying algebra $\mathbb{O}$). Equivalently, for each continuous variable x for which D_x belongs to $\mathbb{O}$, there must exist an ordinary differential equation, and similarly, a pure recurrence equation for each discrete variable under consideration. For example, the function $\tan x$ is not holonomic since it does not satisfy any linear differential equation with polynomial coefficients. Similarly, if f is an arbitrary smooth function, then

$$(3.2.2) \qquad g(z, m) := \left(\frac{d}{dz} \right)^m f(z)$$

is annihilated by the operator $D_z - S_m$. But this single operator is not enough for $g(z, m)$ to be holonomic. In the general situation, there exists neither a pure differential equation (without shifts) nor a pure recurrence (without D_x). However, in special instances, e.g., $f(z) = \sqrt{z}$, the function $g(z, m)$ turns out to be holonomic, being annihilated in this case by the pure operators $2xD_x - m$ and $S_m^2 - x$.

The symbolic framework employed here includes algorithms for *basic arithmetic* that are referred to as *closure properties*, i.e., given two annihilating left ideals for holonomic functions f and g, respectively, it is possible to compute such an ideal for $f + g$, fg, and $P(f)$, where P is an operator in the underlying Ore algebra. Furthermore, certain substitutions are allowed: an algebraic expression in some continuous variables may be substituted for a continuous variable, and a $\mathbb{Q}$-linear combination of discrete variables may be substituted for a discrete variable. Finally, the definite integral of a holonomic function is again holonomic. Note that quotients and compositions of holonomic functions are not holonomic in general.

The question of deciding if a given function is holonomic is non-trivial. Most of the results are indirect. For instance, Flajolet et al. [29] use the fact that a univariate holonomic function has finitely many singularities to illustrate the fact that the generating function of partitions $P(z) = \prod_n (1 - z^n)^{-1}$ is non-holonomic. In this same paper, the authors establish conjectures of S. Gerhold [30] on the non-holonomicity of the sequences $\log n$ and p_n, the nth prime number. The closure properties described above indicate that the function $\sin x \cos x$ is holonomic whereas $\sin x / \cos x = \tan x$ is not. Likewise, $\sin(\sqrt{1 - x^2})$ and F_{2m+k} (with F_n denoting the nth Fibonacci number) are holonomic, whereas $\cos(\sin x)$ and F_{n^2} are not.

The main tool for computing definite integrals with the holonomic systems approach is a technique called *creative telescoping*. It consists in finding annihilating operators of a special form. For example, in the computation of the definite integral $F = \int_a^b f\, dx$, assume that the integrand f contains additional variables other than x. Then an operator T in the annihilating ideal of f of the form $T = P + D_x Q$ is desired, with the condition that P does not contain x or D_x. It is then straightforward to produce an equation for the integral F:

$$
\begin{aligned}
0 &= \int_a^b T(f)\, dx \\
&= \int_a^b P(f)\, dx + \int_a^b \frac{d}{dx} Q(f)\, dx \\
&= P(F) + \left[Q(f) \right]_{x=a}^{x=b}.
\end{aligned}
$$

The operator P is called the *principal part* or the *telescoper*, and Q is called the *delta part*. If the summand coming from the delta part does not simplify to zero, the resulting equation can be homogenized. The operator T is usually found by making an ansatz with undetermined coefficients. Using the fact that reduction with the Gröbner basis yields a unique representation of the remainder, a linear system for the unknowns is obtained by coefficient comparison when equating the remainder to zero. This algorithm has been proposed by Chyzak [24, 25]. An algorithm due to Takayama [64] uses elimination techniques for computing only the principal part P. It therefore can only be applied if it can be ensured a priori that the delta part will vanish; this situation is called *natural boundaries*.

The integrals presented in the rest of the paper illustrate these concepts. These evaluations were obtained using the Mathematica package **Holonomic-Functions**, developed by the author in [42]. It can be downloaded from the webpage

http://www.risc.uni-linz.ac.at/research/combinat/software/

for free, and also a Mathematica notebook containing all examples of this paper is available there. The commands required for the use of this package are described as they are needed, but more information is provided in [43]. The whole paper is organized as a single Mathematica session which we start by loading the package:

In[1]:= << **HolonomicFunctions.m**

HolonomicFunctions package by Christoph Koutschan, RISC-Linz,

Version 1.3 (25.01.2010)

⟶ Type ?HolonomicFunctions for help

3.3. A first example: The indefinite form of Wallis' integral

The first example considered here deals with the indefinite integral

$$(3.3.1) \qquad I_n(x) = \int \frac{dx}{(1 + x^2)^n}.$$

Entry **2.148.3** in [35] states the recurrence

$$(3.3.2) \qquad I_n(x) = \frac{1}{2n - 2} \frac{x}{(1 + x^2)^{n-1}} + \frac{2n - 3}{2n - 2} I_{n-1}(x).$$

Note that this entry does not provide a closed-form evaluation, but a recursive description. Hence this example is very much in the spirit of the methods described in Section 3.2. The recursive nature of 2.148.3 becomes even more striking after shifting $n \mapsto n + 1$ and rearranging to produce

$$(3.3.3) \qquad 2n\, I_{n+1}(x) - (2n - 1)I_n(x) = \frac{x}{(1 + x^2)^n}.$$

The package **HolonomicFunctions** is now used to compute this integral directly. The command **Annihilator[expr, ops]** produces annihilating operators for **expr** with respect to the Ore operators **ops**:

In[2]:= **Annihilator[Integrate[1/(1 + x^2)^n, x], S[n]]**

Out[2]= $\{(2nx^2 + 2n + 2x^2 + 2)S_n^2 + (-2nx^2 - 4n - x^2 - 1)S_n + (2n - 1)\}$

In order to verify that this result agrees with (3.3.2), it is required to produce a homogeneous version of the latter. This is achieved by left multiplying by an annihilating operator of the inhomogeneous part of (3.3.3). The expression $x(1+x^2)^{-n}$ is clearly annihilated by $(x^2+1)S_n - 1$ and hence we obtain

$$((x^2+1)S_n - 1) \cdot (2nS_n - (2n-1)) =$$
$$2(n+1)(x^2+1)S_n^2 + (-2nx^2 - 4n - x^2 - 1)S_n + (2n-1),$$

matching the operator computed by the program. An alternative procedure is to employ an option that produces inhomogeneous relations:

In[3]:= **Annihilator**$\big[$**Integrate**$[1/(1+x2)n, x]$, **S**$[n]$, **Inhomogeneous** $\rightarrow$ **True**$\big]$

Out[3]= $\left\{ \{2nS_n + (1-2n)\}, \left\{ -x\left(x^2+1\right)^{-n} \right\} \right\}$

The first part of the result is the operator to be applied to the integral. Adding the second part and equating to zero yields (3.3.3).

3.4. A differential equation for hypergeometric functions in two variables

The second example appears as entry **9.181.1** in [**35**]. It concerns differential equations for the hypergeometric function

$$(3.4.1) \qquad F_1(\alpha, \beta, \beta', \gamma; x, y) = \sum_{m=0}^{\infty} \sum_{n=0}^{\infty} \frac{x^m y^n (\beta)_m (\beta')_n (\alpha)_{m+n}}{m!n!(\gamma)_{m+n}}$$

for $|x| < 1$ and $|y| < 1$. This is defined in **9.180.1**.

Again, the nature of this example is that no closed form is desired. The result is a system of partial differential equations. These equations are now derived completely automatically from (3.4.1). To achieve this, Takayama's algorithm is employed. Details about this integration algorithm are provided in [**42**].

First observe that both sums have natural boundaries; therefore, Takayama's algorithm may be applied. The input is an annihilating ideal for the summand. This is obtained with the command **Annihilator** described in the previous section. The computation is direct since the summand is hypergeometric and hyperexponential in all variables. The first step

In[4]:= **ann = Annihilator**$\big[$**Pochhammer**$[\alpha, m+n]$ **Pochhammer**$[\beta, m]$
      Pochhammer$[b, n]/($**Pochhammer**$[\gamma, m+n] \, m! \, n!) \, x$m y$n$,
      $\{$**S**$[m]$, **S**$[n]$, **Der**$[x]$, **Der**$[y]\}\big]$

Out[4]= $\{ yD_y - n, xD_x - m,$
      $(mn + m + n^2 + n\gamma + n + \gamma)S_n + (-bmy - bny - by\alpha - mny - n^2y - ny\alpha),$
      $(m^2 + mn + m\gamma + m + n + \gamma)S_m + (-m^2x - mnx - mx\alpha - nx\beta - x\alpha\beta)\}$

finds the annihilating ideal for the summand in the hypergeometric function (where the parameter β' has been replaced by b). The second step

In[5]:= **pde = Takayama[ann, $\{m, n\}$]**

Out[5]= $\{ (xy^2 - xy - y^3 + y^2)D_y^2 + (bx^2 - bx)D_x$
      $+ (bxy - by^2 + xy\alpha - xy\beta + xy + x\beta - x\gamma - y^2\alpha - y^2 + y\gamma)D_y + (bx\alpha - by\alpha),$

$$(x-y)D_x D_y - bD_x + \beta D_y,$$
$$(x^3 - x^2 y - x^2 + xy)D_x^2 + (bxy - by + x^2\alpha + x^2\beta + x^2 - xy\alpha - xy\beta - xy - x\gamma + y\gamma)D_x$$
$$+ (y\beta - y^2\beta)D_y + (x\alpha\beta - y\alpha\beta)\}$$

performs the double summation and computes a Gröbner basis for the ideal containing all the differential equations satisfied by the series F_1. Observe that the two equations given in **9.181.1** are not among these generators. Thus it is required to verify whether the differential equations given in [**35**] are members of the ideal. This is achieved by reducing them with the Gröbner basis and checking whether the remainder is zero; the necessary command is **OreReduce**.

The program allows for a better result. The desired equations can be produced automatically by observing that the first is free of β' and the second does not involve β. The elimination of a parameter can be done either by another Gröbner basis computation (i.e., elimination by rewriting) or by using the command **FindRelation**, which performs elimination by ansatz.

The command **FindRelation[ann, opts]** computes relations in the annihilating ideal **ann** specified by the options **opts**. In this case, the option **Eliminate** forces the coefficients to be free of the given variables.

In[6]:= **FindRelation[pde, Eliminate → β]**

Out[6]= $\{(xy - x)D_x D_y + (y^2 - y)D_y^2 + bxD_x + (by + y\alpha + y - \gamma)D_y + b\alpha\}$

Alternatively, the command **OreGroebnerBasis[$\{P_1, \ldots, P_k\}$, alg]** translates the operators $P_1, \ldots, P_k$ into the Ore algebra **alg** and then computes their left Gröbner basis.

In[7]:= **OreGroebnerBasis[pde, OreAlgebra[b, Der[y], Der[x]],**
$\qquad$ **MonomialOrder → EliminationOrder[1]]** // **First**

Out[7]= $(xy - y)D_y D_x + (x^2 - x)D_x^2 + y\beta D_y + (x\alpha + x\beta + x - \gamma)D_x + \alpha\beta$

This is precisely the form in which these differential equations are given in [**35**].

3.5. An integral involving Chebyshev polynomials

The symbolic algorithms implemented in **HolonomicFunctions** do not provide closed-form expressions for definite integrals. Their main use in the evaluation of integrals is based on the fact that, in many examples, it is possible to produce a computer-generated proof of the stated identities. In this sense, *both sides* of an identity are required. The program yields an automatic proof of its validity. The example presented here appears as entry **7.349** in [**35**]:

$$(3.5.1) \qquad \int_{-1}^{1} (1-x^2)^{-1/2} T_n(1-x^2 y) \, dx = \frac{\pi}{2} \big(P_{n-1}(1-y) + P_n(1-y) \big).$$

Here $T_n(x)$ is the Chebyshev polynomial of the first kind defined by

$$T_n(x) = \cos(n \arccos x)$$

and the answer contains the Legendre polynomial $P_n(x)$ defined by

$$P_n(x) = \frac{1}{2^n n!} \frac{d^n}{dx^n} \left(x^2 - 1\right)^n.$$

This simple example is chosen to describe in more detail what the software does in the background.

The starting point is the computation of an annihilating ideal for the integrand in (3.5.1). The integrand is referred to as $f(n, x, y)$. For this purpose, recall the recurrence

(3.5.2) $$T_{n+2}(z) - 2zT_{n+1}(z) + T_n(z) = 0$$

and the differential equation

(3.5.3) $$(z^2 - 1)T_n''(z) + zT_n'(z) - n^2 T_n(z) = 0$$

for the Chebyshev polynomials. These basic relations are stored in a database that the software can access. An easy argument shows that f satisfies the recurrence (3.5.2) if z is replaced by $1 - x^2 y$. Observe that the factor $(1 - x^2)^{-1/2}$ (which is constant with respect to n) does not change the recurrence. The same substitution is performed in (3.5.3) and considering $T_n(1 - x^2 y)$ as a function in y yields

$$\frac{(1 - x^2 y)^2 - 1}{x^4} \frac{\partial^2}{\partial y^2} T_n(1 - x^2 y) + \frac{1 - x^2 y}{-x^2} \frac{\partial}{\partial y} T_n(1 - x^2 y) - n^2 T_n(1 - x^2 y) = 0.$$

Multiplication by x^2 produces the annihilating operator

$$(x^2 y^2 - 2y)D_y^2 + (x^2 y - 1)D_y - n^2 x^2$$

for the integrand f. Once again, the square root term does not play a role since it is free of the variable y. Finally, observe that

$$\frac{df}{dx} = \frac{-2xy}{\sqrt{1 - x^2}} T_n'(1 - x^2 y) + \frac{x}{(1 - x^2)^{3/2}} T_n(1 - x^2 y)$$

$$\frac{df}{dy} = \frac{-x^2}{\sqrt{1 - x^2}} T_n'(1 - x^2 y),$$

giving rise to the operator

$$xD_x - 2yD_y - \frac{x^2}{1 - x^2},$$

which also annihilates f.

Similar operations, with fewer ad hoc tricks but more algorithmic steps, are performed by typing the command

In[8]:= **Annihilator[ChebyshevT[n, 1−x²y]/Sqrt[1−x²], {S[n], Der[x], Der[y]}]**

Out[8]= $\{(x^3 - x)D_x + (2y - 2x^2 y)D_y + x^2,$
$\quad nS_n + (x^2 y^2 - 2y)D_y + (nx^2 y - n),$
$\quad (x^2 y^2 - 2y)D_y^2 + (x^2 y - 1)D_y - n^2 x^2\}$

Since the command **Annihilator** always returns a Gröbner basis, the above operators differ slightly from the ones that were derived by hand. But these can be obtained by simple combining and rewriting.

The next step in the evaluation of the integral (3.5.1) employs a new command: **CreativeTelescoping**$[f, \mathrm{Der}[x], \mathbf{ops}]$ computes a set of operators $P_i + D_x Q_i$ of the form described in Section 3.2. These operators annihilate the function f and they are chosen in a way that the principal parts P_i form a Gröbner basis in the Ore algebra generated by **ops**.

In[9]:= **CreativeTelescoping**$\big[$**ChebyshevT**$[n, 1 - x\text{^}2\, y]/$**Sqrt**$[1 - x\text{^}2],$ **Der**$[x],$
$\quad \{\mathbf{S}[n], \mathbf{Der}[y]\}\big]$

Out[9]= $\Big\{ \big\{ (2n^2 + 2n)S_n + (2ny^2 - 4ny + y^2 - 2y)D_y + (2n^2y - 2n^2 + ny - 2n),$

$\quad (y^2 - 2y)D_y^2 + (y - 2)D_y - n^2 \big\},$

$\quad \Big\{ \dfrac{y\,(x^4 y - x^2 y - 2x^2 + 2)}{x} D_y + y\,(nx^3 - nx),\ \dfrac{x^2 - 1}{x} D_y \Big\} \Big\}$

The first list contains two principal parts P_1 and P_2, and the second list contains the corresponding delta parts Q_1 and Q_2. It is easily verified that the latter do not contribute:

In[10]:= **ApplyOreOperator**$\big[$**Last**$[\%],$ **ChebyshevT**$[n, 1 - x^2 y]/$**Sqrt**$[1 - x^2]\big]$ //
$\quad$ **Simplify**

Out[10]= $\Big\{ -nx\sqrt{1 - x^2}\,y\big(\text{ChebyshevT}\,[n, 1 - x^2 y] + (2 - x^2 y)\,\text{ChebyshevU}\,[n -$

$\quad 1, 1 - x^2 y]\big),\ nx\sqrt{1 - x^2}\,\text{ChebyshevU}\,[n - 1, 1 - x^2 y] \Big\}$

In[11]:= $(\mathbf{Limit}[\#, x \to 1] - \mathbf{Limit}[\#, x \to -1])\,\&\ /@\ \%$
Out[11]= $\{0, 0\}$

It follows that P_1 and P_2 annihilate the integral. All the previous steps are performed in the background by typing

In[12]:= **Annihilator**$\big[$**Integrate**$[$**ChebyshevT**$[n, 1 - x^2 y]/$**Sqrt**$[1 - x^2], \{x, -1, 1\}],$
$\quad \{\mathbf{S}[n], \mathbf{Der}[y]\}\big]$

Out[12]= $\big\{ (2n^2 + 2n)S_n + (2ny^2 - 4ny + y^2 - 2y)D_y + (2n^2 y - 2n^2 + ny - 2n),$
$\quad (y^2 - 2y)D_y^2 + (y - 2)D_y - n^2 \big\}$

The next step is the computation of an annihilating ideal for the right-hand side of (3.5.1). In this process the fact that the sum of the two Legendre polynomials can be written as $Q(P_{n-1}(1 - y))$ with $Q = S_n + 1$ is employed. This observation produces simpler results (see also equation (3.10.3) for a more detailed discussion of this issue).

In[13]:= **rhs** = **Annihilator**$\big[$**ApplyOreOperator**$[\mathbf{S}[n] + 1, \mathbf{LegendreP}[n - 1, 1 -$
$\quad y]],$
$\quad \{\mathbf{S}[n], \mathbf{Der}[y]\}\big]$

Out[13]= $\big\{ (2n^2 + 2n)S_n + (2ny^2 - 4ny + y^2 - 2y)D_y + (2n^2 y - 2n^2 + ny - 2n),$
$\quad (y^2 - 2y)D_y^2 + (y - 2)D_y - n^2 \big\}$

This produces the same annihilating ideal for the right-hand side as the one produced for the left-hand side. The desired identity is now obtained by comparing some initial values. The necessary cases can be read off from the shape of the Gröbner basis. They correspond to the monomials that lie under the stairs which are formed by the leading monomials of this basis. The required command is

In[14]:= **UnderTheStaircase[rhs]**

Out[14]= $\{1, D_y\}$

In other words, we identify those instances (shifts and derivatives) of the function that cannot be reduced by using the annihilating operators in **rhs**. Hence it is required to check whether $L_0(0) = R_0(0)$ and $L'_0(0) = R'_0(0)$. Here $L_n(y)$ and $R_n(y)$ denote the left and right side of (3.5.1). This is left as an exercise for the reader.

3.6. An integral involving a hypergeometric function

The example of this section appears as entry **7.512.5** in [**35**]: for $\mathrm{Re}\, r > 0$, $\mathrm{Re}\, s > 0$, and $\mathrm{Re}\,(c + s - a - b) > 0$,

$$(3.6.1) \quad \int_0^1 x^{r-1}(1-x)^{s-1}{}_2F_1(a,b;c;x)\,dx = \frac{\Gamma(r)\Gamma(s){}_3F_2(a,b,r;c,r+s;1)}{\Gamma(r+s)}$$

where

$$(3.6.2) \quad {}_pF_q(a_1,\ldots,a_p;b_1,\ldots,b_q;x) := \sum_{k=0}^{\infty} \frac{(a_1)_k \cdots (a_p)_k}{(b_1)_k \cdots (b_q)_k k!} x^k$$

is the classical hypergeometric series.

The strategy of the proof is to compute a system of recurrences for each side of the identity. These recurrences then reduce the problem to checking finitely many initial values. For this purpose, the parameters a, b, c, r, s are assumed to be integers. The first step is to compute an annihilating ideal of the left-hand side:

In[15]:= **lhs = Annihilator[Integrate[x^(r − 1) (1 − x)^(s − 1)**
Hypergeometric2F1[a, b, c, x], {x, 0, 1}], {S[a], S[b], S[c], S[r], S[s]},
Assumptions → s > 1]

Out[15]= $\{ S_r + S_s - 1,$
$(abc - abr - ac^2 + acr - bc^2 + bcr + c^3 - c^2r)S_c$
$\quad + (-abc + acr + acs + bcr + bcs - cr^2 - 2crs - cs^2)S_s$
$\quad + (ac^2 - acr - acs + bc^2 - bcr - bcs - c^3 + c^2r + crs + cs^2),$
$(-ab - b^2 + bc + bs - b)S_b + (ab - ar - as - br - bs + r^2 + 2rs + s^2)S_s + (as +$
$b^2 - bc + b - rs - s^2),$
$(-a^2 - ab + ac + as - a)S_a + (ab - ar - as - br - bs + r^2 + 2rs + s^2)S_s +$
$(a^2 - ac + a + bs - rs - s^2),$
$(ab - ar - as - a - br - bs - b + r^2 + 2rs + 2r + s^2 + 2s + 1)S_s^2$
$\quad + (-ab + ar + 2as + a + br + 2bs + b - cr - cs - 2rs - r - 2s^2 - 2s - 1)S_s$
$\quad + (-as - bs + cs + s^2)\}$

The restriction $s > 1$ is a technicality: the automatic simplification does not succeed if $s \geq 1$ is imposed. The special case $s = 1$ can be done separately in the very same manner. Annihilating operators for the right side are obtained in a similar fashion:

In[16]:= **rhs = Annihilator[Gamma[r] Gamma[s]/Gamma[r + s]**
   HypergeometricPFQ[{a, b, r}, {c, r + s}, 1], {S[a], S[b],
   S[c], S[r], S[s]}]

DFiniteSubstitute::divzero :   Division by zero happened during algebraic substitution (caused by a singularity in the original annihilator). The result is not guaranteed to be correct. Please check whether this substitution and the output given below make sense.

Out[16]= $\{ S_r + S_s - 1,$
$(-abc + abr + ac^2 - acr + bc^2 - bcr - c^3 + c^2 r)S_c$
$\quad + (abc - acr - acs - bcr - bcs + cr^2 + 2crs + cs^2)S_s$
$\quad + (-ac^2 + acr + acs - bc^2 + bcr + bcs + c^3 - c^2 r - crs - cs^2),$
$(ab + b^2 - bc - bs + b)S_b + (-ab + ar + as + br + bs - r^2 - 2rs - s^2)S_s + (-as - b^2 + bc - b + rs + s^2),$
$(a^2 + ab - ac - as + a)S_a + (-ab + ar + as + br + bs - r^2 - 2rs - s^2)S_s$
$\quad + (-a^2 + ac - a - bs + rs + s^2),$
$(ab - ar - as - a - br - bs - b + r^2 + 2rs + 2r + s^2 + 2s + 1)S_s^2$
$\quad + (-ab + ar + 2as + a + br + 2bs + b - cr - cs - 2rs - r - 2s^2 - 2s - 1)S_s$
$\quad + (-as - bs + cs + s^2)\}$

This unfavorable warning comes from the factor $(1 - x)$ that appears in the leading coefficient of the hypergeometric differential equation, making $x = 1$ a singular value. To avoid this issue, the necessary relations are derived by a different approach. Applying Takayama's algorithm to (3.6.2) yields

In[17]:= **smnd = Pochhammer[a, k] Pochhammer[b, k] Pochhammer[r, k]**
   /Pochhammer[c, k]/Pochhammer[r + s, k]/k!;

In[18]:= **tak = Takayama[Annihilator[smnd, {S[k], S[a], S[b], S[c], S[r], S[s]}],**
   {k}];

In[19]:= **rhs2 = DFiniteTimes[Annihilator[Gamma[r] Gamma[s]/Gamma[r + s]],**
   {S[a], S[b], S[c], S[r], S[s]}], tak];

In[20]:= **GBEqual[rhs, rhs2]**
Out[20]= True

The lengthy output has been suppressed, but the last line shows that the annihilating ideal is identical to the one obtained before. Therefore, both sides of identity (3.6.1) are annihilated by the same operator ideal.

The next step is to determine the initial values required to complete the proof. A range of parameters is fixed first, say $a \geq 0, b \geq 0, c \geq 1, r \geq 1, s \geq 1$. Taking the recurrences as the defining equations, it is now required to find the values needed to determine a multivariate sequence uniquely. In the univariate case, this corresponds to the first d values when d is the order of the recurrence. The multivariate analogs are all monomials that lie under the stairs of the Gröbner basis.

Morever, as in the univariate case, the vanishing of some leading coefficient in the recurrences has to be investigated. In the univariate case, this question reduces to finding the non-negative integer roots of the leading coefficient. In the multivariate case, this analysis is more intriguing. It can even happen that there are infinitely many such singular points. Therefore a command that automatically determines all these critical points has been implemented:

In[21]:= **sing = AnnihilatorSingularities[lhs, {0, 0, 1, 1, 1},**
        Assumptions → $c + s - a - b > 0$]

Out[21]= $\{\{\{a \to 0, b \to 0, c \to 1, r \to 1, s \to 1\}, \text{True}\}$,
        $\{\{a \to 0, b \to 0, c \to 1, r \to 1, s \to 2\}, \text{True}\}$,
        $\{\{a \to 0, b \to 0, c \to 2, r \to 1, s \to 1\}, \text{True}\}$,
        $\{\{a \to 0, b \to 0, c \to 2, r \to 1, s \to 2\}, \text{True}\}$,
        $\{\{a \to 0, b \to 1, c \to 1, r \to 1, s \to 1\}, \text{True}\}$,
        $\{\{a \to 0, b \to 1, c \to 1, r \to 1, s \to 2\}, \text{True}\}$,
        $\{\{a \to 0, b \to 1, c \to 2, r \to 1, s \to 1\}, \text{True}\}$,
        $\{\{a \to 0, b \to 1, c \to 2, r \to 1, s \to 2\}, \text{True}\}$,
        $\{\{a \to 1, b \to 0, c \to 1, r \to 1, s \to 1\}, \text{True}\}$,
        $\{\{a \to 1, b \to 0, c \to 1, r \to 1, s \to 2\}, \text{True}\}$,
        $\{\{a \to 1, b \to 0, c \to 2, r \to 1, s \to 1\}, \text{True}\}$,
        $\{\{a \to 1, b \to 0, c \to 2, r \to 1, s \to 2\}, \text{True}\}$,
        $\{\{a \to 1, b \to 1, c \to 1, r \to 1, s \to 2\}, \text{True}\}$,
        $\{\{a \to 1, b \to 1, c \to 1, r \to 1, s \to 3\}, \text{True}\}$,
        $\{\{a \to 1, b \to 1, c \to 2, r \to 1, s \to 1\}, \text{True}\}$,
        $\{\{a \to 1, b \to 1, c \to 2, r \to 1, s \to 2\}, \text{True}\}\}$

The last entry in each case states the condition under which the corresponding points are singular. It is seen that there are many more such points than the staircase of the Gröbner basis would indicate:

In[22]:= **UnderTheStaircase[lhs]**
Out[22]= $\{1, S_s\}$

It is now routine to check the identity (3.6.1) for the above 16 cases, since in most of them the hypergeometric function in the integral reduces to a simple polynomial:

In[23]:= $\big($**MyInt$[x(r-1)(1-x)(s-1)$ Hypergeometric2F1$[a, b, c, x], \{x, 0, 1\}]$ ==**

        Gamma$[r]$ Gamma$[s]$/Gamma$[r + s]$
        HypergeometricPFQ$[\{a, b, r\}, \{c, r + s\}, 1]\big)$
        /. (First /@ sing) /. MyInt → Integrate

Out[23]= {True, True, True, True, True, True, True, True, True, True, True, True, True, True, True, True}

Hence entry **7.512.5** in [**35**] has been verified.

3.7. An integral involving Gegenbauer polynomials

The next identity appears as entry **7.322** in [**35**]: it states that

$$\int_0^{2a} e^{-bx}(x(2a-x))^{\nu-\frac{1}{2}} C_n^{(\nu)}\left(\frac{x}{a}-1\right)\, dx$$

$$= \frac{\pi(-1)^n e^{-ab}\left(\frac{a}{2b}\right)^\nu \Gamma(n+2\nu) I_{n+\nu}(ab)}{n!\,\Gamma(\nu)}$$

where $C_n^{(\nu)}(x)$ denote the Gegenbauer polynomials and $I_\nu(x)$ the modified Bessel function of the first kind. The former special function is defined via the generating function

$$(3.7.1) \qquad\qquad (1-2x\alpha+\alpha^2)^{-\nu} = \sum_{n=0}^\infty C_n^{(\nu)}(x)\alpha^n$$

and the latter is

$$(3.7.2) \qquad\qquad I_\nu(x) = \sum_{k=0}^\infty \frac{1}{k!\,\Gamma(\nu+k+1)}\left(\frac{x}{2}\right)^{\nu+2k}.$$

The computation of the annihilating ideal for the integral requires some human intervention: the problem is that in this instance the inhomogeneous part cannot be evaluated automatically. Hence the option **Inhomogeneous→ True** once again is used and the simplifications are done "by hand." It turns out that all inhomogeneous parts evaluate to 0.

In[24]:= **{lhs, inh} = Annihilator[Integrate[**
  $((x(2a-x))$)ʌ$(\nu-1/2)$ GegenbauerC[n, ν, x/a−1])/E(bx), {x, 0, 2a}],
  {Der[a], Der[b], S[n], S[ν]},
  Assumptions → $\nu \geq 1$,
  Inhomogeneous → True];

In[25]:= **Simplify[ReleaseHold[inh /. Limit → myLimit]] /. myLimit → Limit,**
  Assumptions → $\nu \geq 1$] // Simplify

Out[25]= $\{0,0,0,0\}$

The right-hand side can be handled in completely automatic fashion and we obtain exactly the same differential-difference operators as for the other side:

In[26]:= **rhs = Annihilator[**
  $(-1)^n$Pi Gamma[$2\nu+n$]/n!/Gamma[ν]$(a/(2b))^\nu$BesselI[$\nu+n$, ab]/E^
  (ab),
  {Der[a], Der[b], S[n], S[ν]}]

Out[26]= $\{(an^2 + 2an\nu + 2an + 2a\nu + a)S_n + 2b\nu S_\nu,$
  $(bn^2 + 4bn\nu + bn + 4b\nu^2 + 2b\nu)D_b - 2b^2\nu S_\nu$
  $\quad + (abn^2 + 4abn\nu + abn + 4ab\nu^2 + 2ab\nu - n^3 - 4n^2\nu - n^2 - 4n\nu^2 - 2n\nu),$
  $(an^2 + 4an\nu + an + 4a\nu^2 + 2a\nu)D_a - 2b^2\nu S_\nu$
  $\quad + (abn^2 + 4abn\nu + abn + 4ab\nu^2 + 2ab\nu - n^3 - 6n^2\nu - n^2 - 12n\nu^2 - 4n\nu -$
  $8\nu^3 - 4\nu^2),$
  $(4b^2\nu^2 + 4b^2\nu)S_\nu^2 + (4n^3\nu + 20n^2\nu^2 + 24n^2\nu + 32n\nu^3 + 76n\nu^2 + 44n\nu + 16\nu^4 + 56\nu^3$

$$+64\nu^2+24\nu)S_\nu+(-a^2n^4-8a^2n^3\nu-6a^2n^3-24a^2n^2\nu^2-36a^2n^2\nu-11a^2n^2$$
$$-\,32a^2n\nu^3-72a^2n\nu^2-44a^2n\nu-6a^2n-16a^2\nu^4-48a^2\nu^3-44a^2\nu^2-$$
$$12a^2\nu)\}$$

In[27]:= **GBEqual[lhs, rhs]**

Out[27]= True

This is already a strong indication that the identity is correct, but the initial values have to be compared. There are two monomials under the stairs of the Gröbner basis **lhs** (or **rhs** which is the same):

In[28]:= **UnderTheStaircase[lhs]**

Out[28]= $\{1, S_\nu\}$

Hence the initial values for $\nu = 0$ and $\nu = 1$ have to be compared. The values for a, b, and n can be prescribed, since the corresponding operators D_a, D_b, and S_n had been included in the algebra:

In[29]:= $(x(2a - x))\hat{\ }(\nu - 1/2)$ **GegenbauerC**$[n, \nu, x/a - 1]/$**E**$\hat{\ }(bx)$
    $/.\ \{a \to 1, b \to 1, n \to 1\}\ /.\ \{\{\nu \to 0\}, \{\nu \to 1\}\}$

Out[29]= $\left\{0, 2(x - 1)\sqrt{(2 - x)x}\,e^{-x}\right\}$

In[30]:= **Integrate**$[\%, \{x, 0, 2\}]$

Out[30]= $\left\{0, -\dfrac{2\pi \text{BesselI}[2, 1]}{e}\right\}$

In[31]:= $(-1)^n$**Pi Gamma**$[2\nu+n]/n!/$**Gamma**$[\nu]\ (a/(2b))^\nu$**BesselI**$[\nu+n, ab]/$**E**$\hat{\ }$
    (ab)
    $/.\ \{a \to 1, b \to 1, n \to 1\}\ /.\ \{\{\nu \to 0\}, \{\nu \to 1\}\}$

Out[31]= $\left\{0, -\dfrac{\pi \text{BesselI}[2, 1]}{e}\right\}$

Obviously, the right-hand side has a factor 2 missing. *Hence a misprint in the book has been found!*

3.8. The product of two Bessel functions

For $a > b > 0$ and $\text{Re}\,(m + n) > -1$, formula **6.512.1** states that

$$\int_0^\infty J_m(ax)J_n(bx)\,dx = \frac{b^n}{a^{n+1}}\frac{\Gamma\left(\frac{m+n+1}{2}\right)}{\Gamma(n+1)\Gamma\left(\frac{m-n+1}{2}\right)}{}_2F_1$$
$$\times\left(\frac{m+n+1}{2}, \frac{n-m+1}{2}; n+1; \frac{b^2}{a^2}\right)$$

where $J_n(x)$ denotes the Bessel function. This classical special function is defined by the series

$$(3.8.1) \qquad J_\nu(z) = \left(\frac{z}{2}\right)^\nu \sum_{k=0}^\infty \frac{(-1)^k}{k!\,\Gamma(\nu + k + 1)}\left(\frac{z}{2}\right)^{2k}.$$

Some problems appeared in the computation of an annihilating ideal for the left-hand side, namely, the software complains that it cannot evaluate some delta part. To obtain further details about these problems, the creative telescoping relations were explicitly computed (the delta parts are shown in the output):

In[32]:= **{ann, delta}** = **CreativeTelescoping**[BesselJ[m, ax] BesselJ[n, bx],
        Der[x], {**Der**[a], **Der**[b], **S**[m], **S**[n]}];

In[33]:= **delta**

Out[33]= $\{ -x, 2bx(an+a)S_m S_n - 2b(n+1)(m+n+1)S_n + 2b^2(nx+x),$
    $abxS_m S_n + b^2 x, -2ax(bm+b)S_m S_n + 2a(m+1)(m+n+1)S_m - 2(b^2 mx + b^2 x),$
    $-abxS_m + b^2 xS_n, b^2 xS_m - abxS_n, ab^2 x^2 S_m - b^3 x^2 S_n - b^2(mx - nx - x)\}$

The first delta part already reveals the difficulties involved: according to the derivation in Section 3.2, the boundary condition to evaluate is

$$\left[xJ_m(ax)J_n(bx) \right]_{x=0}^{x=\infty}.$$

The Bessel function $J_n(x)$ is asymptotically equivalent to $\sqrt{2/(\pi x)}$ as $x \to \infty$ (see [**1**, 9.2.1]). Therefore the limit in the first delta part does not exist. Moreover, some other delta parts involve x^2, which makes the situation even worse.

One way to overcome these difficulties consists in going back to the roots of the holonomic systems approach. In his original paper [**69**], Zeilberger suggested finding an operator whose coefficients are completely free of the integration variable x. This is more than necessary, since usually it does no harm if x occurs in the delta part. Once such an operator is found, it is immediate to rewrite it into the form $P + D_x Q$. This method was called the "slow algorithm" by Zeilberger himself, and this points to the reason why it is rarely used in practice. In the example described here, this technique could be useful, since the occurrence of x in the delta part is exactly the problem encountered. It is also desired to have no derivatives with respect to a or b in the delta parts, since they cause the same difficulties. Such operators can be found by means of the command **FindRelation**, where the first condition can be encoded by the option **Eliminate**, and the second by the option **Support**: all monomials up to total degree 2 are given, but $D_a D_x$ and $D_b D_x$ are omitted:

In[34]:= **ops** = {**Der**[x], **Der**[a], **Der**[b], **S**[m], **S**[n]};

In[35]:= **supp** = **Complement**[**Join**[{**1**}, ops, **Flatten**[**Outer**[**Times**, ops, ops]]],
        {**Der**[x] **Der**[a], **Der**[x]**Der**[b]}];

In[36]:= **rels** = **FindRelation**[**Annihilator**[BesselJ[m, ax] BesselJ[n, bx], ops],
        Eliminate $\to x$, **Support** $\to$ **supp**]

Out[36]= $\{ -ab^2 mD_a S_m + (a^2 bn + a^2 b)D_a S_n + (a^2(-b) - a^2 bn)D_b S_m + ab^2 mD_b S_n$
    $+ (a^2 n^2 + a^2 n - b^2 m^2 - b^2 m)S_m,$
    $(2m+2)D_x S_m + (am - an + a)S_m^2 + (2bm + 2b)S_m S_n + (a(-m) - an - a),$
    $(2n+2)D_x S_n + (2an + 2a)S_m S_n + (b(-m) + bn + b)S_n^2 + (b(-m) - bn - b),$
    $-abD_a S_n + abD_b S_m - anS_m + bmS_n,$
    $-a^2 b^2 D_a^2 + a^2 b^2 D_b^2 - ab^2 D_a + a^2 bD_b + (b^2 m^2 - a^2 n^2)\}$

The principal and delta parts of these five relations have to be separated manually. Observe that D_x is not invertible in the underlying Ore algebra. Therefore this operation has to be performed on the level of Mathematica expressions:

In[37]:= **pps = OrePolynomialSubstitute[rels, {Der[x] → 0}];**

In[38]:= **deltas = Together[(Normal /@ (rels − pps))/Der[x]]**

Out[38]= $\{0, 2(m + 1)S[m], 2(n + 1)S[n], 0, 0\}$

Now all the inhomogeneous parts vanish: the limits for $x \to \infty$ as well as the evaluations at $x = 0$. The latter is true because at least one of the orders of the two Bessel functions becomes ≥ 1 (recall that we impose $m \geq 0$ and $n \geq 0$). Hence the principal parts annihilate the integral. In order to compare them against the right-hand side, the Gröbner basis of the ideal generated by them is computed:

In[39]:= **lhs = OreGroebnerBasis[pps, OreAlgebra[Der[a], Der[b], S[m], S[n]],**
MonomialOrder → DegreeLexicographic]

Out[39]= $\{aD_a + bD_b + 1,$
$(b^2m^2 - b^2n^2 - 2b^2n - b^2)S_n^2 + (2a^2bn + 2a^2b - 2b^3n - 2b^3)D_b$
$\quad + (-2a^2n^2 - 2a^2n + b^2m^2 + b^2n^2 - b^2),$
$(-abm - abn - ab)S_mS_n + (b^3 - a^2b)D_b + (a^2n + b^2m + b^2),$
$(a^2(-m^2) - 2a^2m + a^2n^2 - a^2)S_m^2 + (2a^2bm + 2a^2b - 2b^3m - 2b^3)D_b$
$\quad + (a^2m^2 + 2a^2m + a^2n^2 + a^2 - 2b^2m^2 - 4b^2m - 2b^2),$
$(a^2b - b^3)D_bS_n + (abn - abm)S_m + (a^2n + a^2 - b^2m - b^2)S_n,$
$(a^2b - b^3)D_bS_m + (b^2m - a^2n)S_m + (abm - abn)S_n,$
$(b^4 - a^2b^2)D_b^2 + (3b^3 - a^2b)D_b + (a^2n^2 + b^2(-m^2) + b^2)\}$

In[40]:= **rhs = Annihilator[**
$b^n a^{-n-1}$**Gamma[$(m + n + 1)/2$]/Gamma[$n + 1$]/Gamma[$(m -$**
$n + 1)/2$**]**
Hypergeometric2F1[$(m + n + 1)/2, (n - m + 1)/2, n + 1, b^2/a^2$],
{Der[a], Der[b], S[m], S[n]}];

In[41]:= **GBEqual[lhs, rhs]**

Out[41]= True

The proof is completed by checking four initial values. This is left to the reader.

3.9. An example involving parabolic cylinder functions

The first entry in Section **3.953** states that

$$\int_0^\infty x^{\mu-1}e^{-\gamma x - \beta x^2}\sin(ax)\,dx = -\frac{i}{2(2\beta)^{\mu/2}}\exp\left(\frac{\gamma^2 - a^2}{8\beta}\right)\Gamma(\mu)$$

$$\times \left\{\exp\left(-\frac{ia\gamma}{4\beta}\right)D_{-\mu}\left(\frac{\gamma - ia}{\sqrt{2\beta}}\right) - \exp\left(\frac{ia\gamma}{4\beta}\right)D_{-\mu}\left(\frac{\gamma + ia}{\sqrt{2\beta}}\right)\right\}$$

for $\operatorname{Re}\mu > -1$, $\operatorname{Re}\beta > 0$, and $a > 0$. The symbol $D_s(z)$ denotes the parabolic cylinder function defined by

$$D_s(z): \quad = \quad 2^{s/2}\sqrt{\pi}e^{-\frac{z^2}{4}}$$

$$\times \left(\frac{1}{\Gamma\left(\frac{1-s}{2}\right)}\,_1F_1\left(-\frac{s}{2},\frac{1}{2},\frac{z^2}{2}\right) - \frac{\sqrt{2}z}{\Gamma\left(-\frac{s}{2}\right)}\,_1F_1\left(\frac{1-s}{2},\frac{3}{2},\frac{z^2}{2}\right)\right).$$

Alternatively, this function is defined as a certain solution of the differential equation

$$4y''(x) + \left(4s - x^2 + 2\right)y(x) = 0.$$

For convenience of typing, the Greek letters are replaced by Roman ones ($\beta = b$, $\gamma = c$, $\mu = m$), and the complicated right-hand side is stored as an extra variable.

In[42]:= **rexpr = −I/(2(2b)^(m/2)) ∗ Exp[(c^2 − a^2)/(8b)] ∗ Gamma[m]**
(Exp[−Iac/(4b)] ∗ ParabolicCylinderD[−m, (c − Ia)/Sqrt[2b]]
− Exp[Iac/(4b)] ∗ ParabolicCylinderD[−m, (c + Ia)/Sqrt[2b]]);

A short look at the expressions involved in this identity suggests acting on a, b, c with a partial derivative, and on m with the shift operator. Observe that the identity holds for $a = 0$ as well. Annihilating ideals for both sides are readily computed, but unfortunately, they do not agree.

In[43]:= **lhs = Annihilator[**
Integrate[x^(m − 1) Exp[−cx − bx^2] Sin[ax], {x, 0, Infinity}],
{S[m], Der[a], Der[b], Der[c]},
Assumptions → Re[m] > −1 && Re[b] > 0 && a ≥ 0]

Out[43]= $\{aD_a + 2bD_b + cD_c + m,\ S_m + D_c,\ D_c^2 + D_b,$
$4b^2D_b^2 + 4bcD_bD_c + (-a^2 + 4bm + 6b - c^2)D_b + (2cm + 2c)D_c + (m^2 + m)\}$

In[44]:= **rhs = Annihilator[rexpr, {S[m], Der[a], Der[b], Der[c]}]**

Out[44]= $\{aD_a + 2bD_b + cD_c + m,\ D_c^2 + D_b,$
$4b^2D_b^2 + 4bcD_bD_c + (-a^2 + 4bm + 6b - c^2)D_b + (2cm + 2c)D_c + (m^2 + m),\ S_m^2 + D_b\}$

It turns out that the latter is a subideal of the previous one:

In[45]:= **Length[UnderTheStaircase[#]]& /@ {lhs, rhs}**
Out[45]= $\{4, 8\}$

In[46]:= **OreReduce[rhs, lhs]**
Out[46]= $\{0, 0, 0, 0\}$

Why are fewer relations found for the right-hand side? The reason is the factor $\sqrt{b}$ that appears in the argument of the parabolic cylinder function. **Annihilator** computes operators with coefficients that are polynomials in the variables corresponding to the Ore operators in the algebra. This can increase the order of the resulting operators, as the following two examples demonstrate.

In[47]:= **Annihilator[ParabolicCylinderD[−m, Sqrt[b]], S[m]]**
Out[47]= $\{(m + 1)S_m^2 + \sqrt{b}S_m - 1\}$

In[48]:= **Annihilator**[**ParabolicCylinderD**$[-m, \mathbf{Sqrt}[b]]$, $\{\mathbf{Der}[b], \mathbf{S}[m]\}$,
         MonomialOrder $\rightarrow$ **Lexicographic**]

Out[48]= $\{(m^2 + 5m + 6)S_m^4 + (-b - 2m - 3)S_m^2 + 1, \ 4bD_b + (-2m^2 - 2m)S_m^2 + (b + 2m)\}$

However, the structure of the identity in question is special in the sense that, if computed with $\sqrt{b}$ in the coefficients, then all occurrences of it would disappear in the final result. This can be reproduced by introducing a new variable $s = \sqrt{b}$ in the original expression, and in the end the corresponding closure property "substitution" is performed. This produces the same annihilating ideal as for the left side:

In[49]:= **rexpr1** = **Simplify**[**rexpr** /. $b \rightarrow s\verb|^|2, s > 0$];

In[50]:= **rhs1** = **Annihilator**[**rexpr1**, $\{\mathbf{S}[m], \mathbf{Der}[a], \mathbf{Der}[s], \mathbf{Der}[c]\}$]

Out[50]= $\{aD_a + sD_s + cD_c + m, \ S_m + D_c, \ 2sD_c^2 + D_s,$
      $2s^3 D_s^2 + 4cs^2 D_s D_c + (-a^2 - c^2 + 4ms^2 + 4s^2)D_s + (4cms + 4cs)D_c + (2m^2s +$
      $2ms)\}$

In[51]:= **rhs1** = **DFiniteSubstitute**$[\mathbf{rhs1}, \{s \rightarrow \mathbf{Sqrt}[b]\}$,
         Algebra $\rightarrow$ **OreAlgebra**$[\mathbf{S}[m], \mathbf{Der}[a], \mathbf{Der}[b], \mathbf{Der}[c]]]$

Out[51]= $\{aD_a + 2bD_b + cD_c + m, \ S_m + D_c, \ D_c^2 + D_b,$
      $4b^2 D_b^2 + 4bcD_b D_c + (-a^2 + 4bm + 6b - c^2)D_b + (2cm + 2c)D_c + (m^2 + m)\}$

In[52]:= **GBEqual**[**lhs**, **rhs1**]

Out[52]= True

Alternatively, the operator D_b can be set aside. Since in this problem, the initial values to be checked are simple enough with symbolic m and b, the shift in m will not be included. Considering m as a parameter has the additional advantage that the proof is valid for any m and not only for integer values.

In[53]:= **lhs** = **Annihilator**[
        Integrate$[x\verb|^|(m-1) \ \mathbf{Exp}[-cx - bx\verb|^|2] \ \mathbf{Sin}[ax], \{x, 0, \mathbf{Infinity}\}]$,
        $\{\mathbf{Der}[a], \mathbf{Der}[c]\}$, **Assumptions** $\rightarrow$ $\mathbf{Re}[m] > -1 \ \&\& \ \mathbf{Re}[b] > 0$
        $\&\&a \geq 0$]

Out[53]= $\{2bD_c^2 - aD_a - cD_c - m, \ 2bD_a^2 + aD_a + cD_c + m\}$

In[54]:= **rhs** = **Annihilator**[**rexpr**, $\{\mathbf{Der}[a], \mathbf{Der}[c]\}$]

Out[54]= $\{2bD_c^2 - aD_a - cD_c - m, \ 2bD_a^2 + aD_a + cD_c + m\}$

As already mentioned, the leading monomials in the Gröbner basis indicate the initial values to be compared. The integrals that remain to be computed now are simpler than the original one.

In[55]:= **uts** = **UnderTheStaircase**[**lhs**]

Out[55]= $\{1, D_c, D_a, D_a D_c\}$

In[56]:= **ApplyOreOperator**$[\mathbf{uts}, x\verb|^|(m-1) \ \mathbf{Exp}[-cx - bx\verb|^|2] \ \mathbf{Sin}[ax]]$
      /. $\{a \rightarrow 0, c \rightarrow 0\}$

Out[56]= $\left\{0, 0, x^m e^{-bx^2}, x^{m+1} \left(-e^{-bx^2}\right)\right\}$

In[57]:= **Integrate**$\big[\%, \{x, 0, \textbf{Infinity}\}, \textbf{Assumptions} \rightarrow \textbf{Re}[m] > -1$
&& **Re**$[b] > 0\big]$

Out[57]= $\left\{ 0, 0, \dfrac{1}{2} b^{-\frac{m}{2}-\frac{1}{2}} \text{Gamma} \left[\dfrac{m+1}{2} \right], -\dfrac{1}{2} b^{-\frac{m}{2}-1} \text{Gamma} \left[\dfrac{m}{2} + 1 \right] \right\}$

In[58]:= **FullSimplify**$\big[\textbf{ApplyOreOperator}[\text{uts}, \text{rexpr}] \;/.\; \{a \rightarrow 0, c \rightarrow 0\}\big]$

Out[58]= $\left\{ 0, 0, \dfrac{1}{2} b^{-\frac{m}{2}-\frac{1}{2}} \text{Gamma} \left[\dfrac{m+1}{2} \right], -\dfrac{1}{2} b^{-\frac{m}{2}-1} \text{Gamma} \left[\dfrac{m}{2} + 1 \right] \right\}.$

3.10. An elementary trigonometric integral

The final example is entry **4.535.1** in [**35**]:

$$(3.10.1) \qquad \int_0^1 \frac{\arctan px}{1 + p^2 x} \, dx = \frac{1}{2p^2} \arctan p \, \ln\left(1 + p^2\right).$$

In the computation of a differential equation for each side of the identity, an overshoot concerning the order is observed for the left-hand side (for brevity, only the leading monomials of both operators are displayed).

In[59]:= **lhs = Annihilator**$\big[\textbf{Integrate}[\textbf{ArcTan}[px]/(1+xp2), \{x, 0, 1\}], \textbf{Der}[p]\big];$

In[60]:= **rhs = Annihilator**$\big[(1/(2p\hat{~}2))\, \textbf{ArcTan}[p]\, \textbf{Log}[1 + p\hat{~}2], \textbf{Der}[p]\big];$

In[61]:= **LeadingPowerProduct** $/@$ **Flatten**$[\{\text{lhs}, \text{rhs}\}]$

Out[61]= $\{D_p^5, D_p^4\}$

It turns out that the fifth-order differential equation is a left multiple of the other one. An explanation for this non-agreement is desirable. This is obtained by considering the inhomogeneous part that remains after creative telescoping:

In[62]:= $\{\{\textbf{op}\}, \{\textbf{inh}\}\}$ = **Annihilator**$\big[\textbf{Integrate}[\textbf{ArcTan}[px]/(1 + xp\hat{~}2),$
$\{x, 0, 1\}], \textbf{Der}[p], \textbf{Inhomogeneous} \rightarrow \textbf{True}\big]$

Out[62]= $\left\{ \{(p^4 + p^2)D_p^2 + (6p^3 + 4p)D_p + (6p^2 + 2)\}, \right.$
$\left. \left\{ \dfrac{(-p^7 - 3p^5 - 3p^3 - p)\text{ArcTan}[p] - p^6 - p^4 + p^2 + 1}{p\,(p^2 + 1)^3} - \dfrac{1}{p} \right\} \right\}$

In[63]:= **FullSimplify**[inh]

Out[63]= $-\text{ArcTan}[p] - \dfrac{2p}{p^2 + 1}$

Hence the inhomogeneous differential equation that remains after telescoping is

(3.10.2)

$$\left(p^4 + p^2\right) f''(p) + \left(6p^3 + 4p\right) f'(p) + \left(6p^2 + 2\right) f(p) = \frac{2p}{p^2 + 1} + \arctan p,$$

which has to be homogenized. For this purpose, an annihilating operator for the inhomogeneous part (i.e., the right side of (3.10.2)) is computed and then left-multiplied to the operator **op**. By default, the closure property "addition" is used for such expressions. But a careful inspection shows that it can also be written as an operator application:

$$(3.10.3) \qquad (2pD_p + 1)(\arctan p) = \frac{2p}{p^2 + 1} + \arctan p.$$

In such situations the latter closure property is preferable, as the following computations demonstrate:

In[64]:= **Annihilator$\left[2(p/(1 + p\text{^}2)) + \text{ArcTan}[p], \text{Der}[p]\right]$**

Out[64]= $\left\{ (p^5 + 2p^3 + p)D_p^3 + (7p^4 + 6p^2 - 1)D_p^2 + 8p^3 D_p \right\}$

In[65]:= **Annihilator$\left[\text{ApplyOreOperator}[2p\,\text{Der}[p] + 1, \text{ArcTan}[p]], \text{Der}[p]\right]$**

Out[65]= $\left\{ (p^4 - 2p^2 - 3)D_p^2 + (2p^3 - 14p)D_p \right\}$

This leads to the same fourth-order differential equation previously obtained for the right-hand side:

In[66]:= **First$[\%] **\text{op}$**

Out[66]= $\left(p^8 - p^6 - 5p^4 - 3p^2\right) D_p^4 + \left(16p^7 - 32p^5 - 72p^3 - 24p\right) D_p^3$
$+ \left(74p^6 - 224p^4 - 270p^2 - 36\right) D_p^2 + \left(108p^5 - 444p^3 - 264p\right) D_p$
$+ \left(36p^4 - 192p^2 - 36\right)$

In[67]:= **GBEqual$[\%, \text{rhs}]$**

Out[67]= True

At this point, the known evaluation of the integral is ignored, and the differential equation is employed to find it. The Mathematica command **DSolve** delivers the following four independent solutions:

In[68]:= **DSolve$\left[\text{ApplyOreOperator}[\text{First}[\text{rhs}], f[p]] == 0, f[p], p\right][[1, 1, 2]]$**

Out[68]= $\dfrac{C[1]}{p^2} + \dfrac{C[2]\,\text{ArcTan}[p]}{p^2} + \dfrac{C[3]\,\text{Log}\left[p^2 + 1\right]}{2p^2} - \dfrac{C[4]\,\text{ArcTan}[p]\,\text{Log}\left[p^2 + 1\right]}{6p^2}$

Four constants need to be determined. From the fact that the integral is 0 for $p = 0$, the first two solutions are excluded, since they tend to infinity as $p \to 0$. The fourth solution tends to 0 for $p \to 0$, but the third one does not; hence C[3] must also vanish. The last constant can be determined from the first derivative with respect to p, evaluated at $p = 0$:

$$\int_0^1 \frac{d}{dp}\left(\frac{\arctan px}{1 + p^2 x} \right)\Bigg|_{p=0} dx = \int_0^1 x\,dx = \frac{1}{2}.$$

The remaining constant C[4] is seen to be 3 and the evaluation (3.10.1) has been rediscovered.

CHAPTER 4

The error function

4.1. Introduction

The *error* function defined by

$$(4.1.1) \qquad \mathrm{erf}(u) := \frac{2}{\sqrt{\pi}} \int_0^u e^{-x^2} \, dx$$

is one of the basic non-elementary special functions. The coefficient $2/\sqrt{\pi}$ is a normalization factor that has the effect of giving $\mathrm{erf}(\infty) = 1$ in view of the *normal integral*

$$(4.1.2) \qquad \int_0^\infty e^{-x^2} \, dx = \frac{\sqrt{\pi}}{2}.$$

The reader will find in [**17**] different proofs of this evaluation.

In this paper we produce evaluations of entries in [**35**] that contain this function. The methods are elementary. The reader will find in [**23**] a more advanced approach to the question of symbolic integration around this function.

4.2. Elementary integrals

The table [**35**] contains many integrals involving the error function. This section contains some elementary examples.

LEMMA 4.2.1. *Define*

$$(4.2.1) \qquad F_n(v) = \int_0^v t^n e^{-t^2} \, dt.$$

Then the function F_n satisfies the recurrence

$$(4.2.2) \qquad F_n(v) = -\tfrac{1}{2} v^{n-1} e^{-v^2} + \tfrac{n-1}{2} F_{n-2}(v),$$

with initial conditions given by

$$(4.2.3) \qquad F_0(v) = \frac{\sqrt{\pi}}{2} \, \mathit{erf}(v) \text{ and } F_1(v) = \frac{1}{2} \left(1 - e^{-v^2} \right).$$

PROOF. This follows simply by integration by parts. □

The recurrence (4.2.2) shows that $F_{2n}(v)$ is determined by $F_0(v)$ and $F_{2n+1}(v)$ by $F_1(v)$. An explicit formula for the latter is easy to establish.

LEMMA 4.2.2. *The function* $F_{2n+1}(v)$ *is given by*

$$(4.2.4) \qquad F_{2n+1}(v) = \frac{n!}{2}\left(1 - e^{-v^2}\sum_{j=0}^{n}\frac{v^{2j}}{j!}\right).$$

PROOF. The recurrence and the induction hypothesis give

$$
\begin{aligned}
F_{2n+1}(v) &= -\frac{1}{2}v^{2n}e^{-v^2} + nF_{2n-1}(v) \\
&= -\frac{1}{2}v^{2n}e^{-v^2} + n\times\frac{(n-1)!}{2}\left(1 - e^{-v^2}\sum_{j=0}^{n-1}\frac{v^{2j}}{j!}\right) \\
&= -\frac{1}{2}v^{2n}e^{-v^2} + \frac{n!}{2}\left(1 - e^{-v^2}\sum_{j=0}^{n-1}\frac{v^{2j}}{j!}\right).
\end{aligned}
$$

Now observe that the first term matches the one in the sum for $j = n$. □

The case of an even index follows a similar pattern.

LEMMA 4.2.3. *The function* $F_{2n}(v)$ *has the form*

$$(4.2.5) \qquad F_{2n}(v) = \frac{(2n-1)!!}{2^n}F_0(v) - \frac{1}{2^n}ve^{-v^2}P_n(v),$$

where $P_n(v)$ *satisfies the recurrence*

$$(4.2.6) \qquad P_n(v) = 2^{n-1}v^{2n-2} + (2n-1)P_{n-1}(v)$$

with initial condition $P_0(v) = 0$. *Therefore* $P_n(v)$ *is a polynomial given explicitly by*

$$(4.2.7) \qquad P_n(v) = \sum_{k=0}^{n-1}\frac{(2n-1)!!\,2^k}{(2k+1)!!}v^{2k}.$$

PROOF. The details are left to the reader. □

4.3. Elementary scaling

Several entries in [**35**] are obtained from the expressions in the last section via simple changes of variables. For example, a linear transformation gives

$$(4.3.1) \qquad \int_0^v x^n e^{-q^2 x^2}\,dx = \frac{1}{q^{n+1}}F_n(qv).$$

This section presents some examples of this type.

EXAMPLE 4.3.1. The case $n = 0$ yields entry **3.321.2**

$$(4.3.2) \qquad \int_0^u e^{-q^2 x^2}\,dx = \frac{\sqrt{\pi}}{2q}\operatorname{erf}(qu).$$

EXAMPLE 4.3.2. Entry **3.321.3**

$$(4.3.3) \qquad \int_0^\infty e^{-q^2 x^2}\, dx = \frac{\sqrt{\pi}}{2q}$$

is obtained by simply letting $u \to \infty$ in Example 4.3.1 and using $\mathrm{erf}(\infty) = 1$.

EXAMPLE 4.3.3. Entries **3.321.4, 5, 6, 7** are obtained from the recurrence for F_n:

$$\int_0^u x e^{-q^2 x^2}\, dx = \frac{1}{2q^2}\left(1 - e^{-q^2 u^2}\right),$$

$$\int_0^u x^2 e^{-q^2 x^2}\, dx = \frac{1}{2q^3}\left(\frac{\sqrt{\pi}}{2}\mathrm{erf}(qu) - que^{-q^2 u^2}\right),$$

$$\int_0^u x^3 e^{-q^2 x^2}\, dx = \frac{1}{2q^4}\left(1 - (1 + q^2 u^2)e^{-q^2 u^2}\right),$$

$$\int_0^u x^4 e^{-q^2 x^2}\, dx = \frac{1}{2q^5}\left(\frac{3\sqrt{\pi}}{4}\mathrm{erf}(qu) - \left(\frac{3}{2} + q^2 u^2\right)que^{-q^2 u^2}\right).$$

EXAMPLE 4.3.4. Simple scaling produces other integrals in [**35**]. For example, starting with

$$(4.3.4) \qquad \int_a^\infty e^{-x^2}\, dx = \frac{\sqrt{\pi}}{2}\left(1 - \mathrm{erf}(a)\right),$$

the change of variables $t = qx$ yields

$$(4.3.5) \qquad \int_a^\infty e^{-q^2 t^2}\, dt = \frac{\sqrt{\pi}}{2q}\left(1 - \mathrm{erf}(qa)\right).$$

The integral

$$(4.3.6) \qquad \int_a^\infty e^{-q^2 x^2 - px}\, dx = \frac{\sqrt{\pi}}{2q}e^{p^2/4q^2}\left(1 - \mathrm{erf}\left(\frac{p + 2aq^2}{2q}\right)\right),$$

is now computed by completing the square. The choice $q = 1/2\sqrt{\beta}$ and $p = \gamma$ appears as entry **3.322.1** in [**35**]

$$(4.3.7) \qquad \int_u^\infty \exp\left(-\frac{x^2}{4\beta} - \gamma x\right)\, dx = \sqrt{\pi\beta}e^{\beta\gamma^2}\left(1 - \mathrm{erf}\left(\frac{u}{2\sqrt{\beta}} + \sqrt{\beta}\gamma\right)\right).$$

In order to minimize the choice of Greek letters (clearly a personal choice of the authors), it is suggested writing this entry using the notation in (4.3.6).

EXAMPLE 4.3.5. Entry **3.322.2**

$$(4.3.8) \qquad \int_0^\infty e^{-q^2 x^2 - px}\, dx = \frac{\sqrt{\pi}}{2q}e^{p^2/4q^2}\left(1 - \mathrm{erf}\left(\frac{p}{2q}\right)\right)$$

comes from letting $a \to 0$ in (4.3.6).

EXAMPLE 4.3.6. The choice of parameters $q = 1$ and $a = 1$ in (4.3.6) produces

$$(4.3.9) \qquad \int_1^\infty e^{-x^2 - px} \, dx = \frac{\sqrt{\pi}}{2} e^{p^2/4} \left(1 - \operatorname{erf}\left(\frac{p+2}{2}\right) \right).$$

This appears as entry **3.323.1** (unfortunately, with p instead of q. This is inconsistent with the notation in the rest of the section).

EXAMPLE 4.3.7. The evaluation **3.323.2**

$$(4.3.10) \qquad \int_{-\infty}^\infty \exp\left(-p^2 x^2 \pm qx\right) \, dx = \frac{\sqrt{\pi}}{p} \exp\left(\frac{q^2}{4p^2}\right)$$

follows directly from completing the square

$$(4.3.11) \qquad -p^2 x^2 \pm qx = -p^2 \left(x \mp q/2p^2\right)^2 + \frac{q^2}{4p^2}.$$

4.4. A series representation for the error function

The table [35] contains some series representation for the error function. Entry **3.321.1** contains two of them. The first one is equivalent to

$$(4.4.1) \qquad \operatorname{erf}(u) = \frac{2}{\sqrt{\pi}} \sum_{k=0}^\infty \frac{(-1)^k u^{2k+1}}{k! \, (2k+1)},$$

which comes from term by term integration of the power series of the integrand. The second one is more interesting and is given in the next example.

EXAMPLE 4.4.1. Entry **3.321.1** states that

$$(4.4.2) \qquad \operatorname{erf}(u) = \frac{2}{\sqrt{\pi}} e^{-u^2} \sum_{k=0}^\infty \frac{2^k}{(2k+1)!!} u^{2k+1}.$$

To check this identity, we need to prove

$$\left(\sum_{k=0}^\infty \frac{(-1)^k}{k!} u^{2k} \right) \times \left(\sum_{j=0}^\infty \frac{2^j}{(2j+1)!!} u^{2j+1} \right) = \sum_{r=0}^\infty \frac{(-1)^r}{r!(2r+1)} u^{2r+1}.$$

Multiplying the two series on the left, we conclude that the result follows from the finite sums identity

$$(4.4.3) \qquad \sum_{k=0}^r \frac{(-1)^k 2^{r-k}}{k! \, (2r - 2k + 1)!!} = \frac{(-1)^r}{r! \, (2r+1)}.$$

This can be written as

$$(4.4.4) \qquad \sum_{k=0}^r (-4)^k \binom{r}{k} \binom{2k}{k}^{-1} \frac{2r+1}{2k+1} = 1,$$

which is now established using the WZ-technology [55]. Define

$$(4.4.5) \qquad A(r, k) = (-4)^k \binom{r}{k} \binom{2k}{k}^{-1} \frac{2r+1}{2k+1}$$

and use the WZ-method to produce the companion function

$$(4.4.6) \qquad B(r,k) = (-1)^{k+1} \binom{r}{k-1} \binom{2k}{k}^{-1}.$$

The reader can now verify the relation

$$(4.4.7) \qquad A(r+1,k) - A(r,k) = B(r,k+1) - B(r,k).$$

Both terms A and B have natural boundaries, that is, they vanish outside the summation range. Summing from $k = -\infty$ to $k = +\infty$ and using the telescoping property of the right-hand side shows that

$$(4.4.8) \qquad a_r := \sum_{k=0}^{r} A(r,k)$$

is independent of r. The value $a_0 = 1$ completes the proof.

4.5. An integral of Laplace

The first example in this section reproduces a classical integral due to P. Laplace.

EXAMPLE 4.5.1. Entry **3.325** states that

$$(4.5.1) \qquad \int_0^\infty \exp\left(-ax^2 - bx^{-2}\right) dx = \frac{1}{2}\sqrt{\frac{\pi}{a}} e^{-2\sqrt{ab}}.$$

To evaluate this, we complete the square in the exponent and write

$$(4.5.2) \qquad \int_0^\infty \exp\left(-ax^2 - bx^{-2}\right) dx = e^{-2\sqrt{ab}} \int_0^\infty e^{-(\sqrt{a}x - \sqrt{b}/x)^2} dx.$$

Denote this last integral by J, that is,

$$(4.5.3) \qquad J := \int_0^\infty e^{-(\sqrt{a}x - \sqrt{b}/x)^2} dx.$$

The change of variables $t = \sqrt{b}/\sqrt{a}x$ produces

$$(4.5.4) \qquad J := \frac{\sqrt{b}}{\sqrt{a}} \int_0^\infty e^{-(\sqrt{a}t - \sqrt{b}/t)^2} \frac{dt}{t^2}.$$

The average of these two forms for J produces

$$(4.5.5) \qquad J = \frac{1}{2\sqrt{a}} \int_0^\infty e^{-(\sqrt{a}x - \sqrt{b}/x)^2} \left(\sqrt{a} + \sqrt{b}/x^2\right) dx.$$

The change of variables $u = \sqrt{a}x - \sqrt{b}/x$ now yields

$$(4.5.6) \qquad J = \frac{1}{2\sqrt{a}} \int_{-\infty}^\infty e^{-u^2} du = \frac{\sqrt{\pi}}{2\sqrt{a}},$$

and the evaluation is complete.

The method employed in this evaluation was expanded by O. Schlömilch, who considered the identity

(4.5.7) $$\int_0^\infty f\left((ax - bx^{-1})^2\right) dx = \frac{1}{a} \int_0^\infty f(y^2) \, dy.$$

The reader will find in [3] details about this transformation and the evaluation of many related integrals.

EXAMPLE 4.5.2. Entry **3.472.1** of [35]

(4.5.8) $$\int_0^\infty \left(\exp\left(-a/x^2\right) - 1\right) e^{-\mu x^2} dx = \frac{1}{2}\sqrt{\frac{\pi}{\mu}}\left(e^{-2\sqrt{a\mu}} - 1\right)$$

can be evaluated directly from **3.325**. Indeed,

$$
\begin{aligned}
\int_0^\infty \left(\exp\left(-a/x^2\right) - 1\right) e^{-\mu x^2} dx &= \int_0^\infty \exp\left(-a/x^2 - \mu x^2\right) dx \\
&- \int_0^\infty e^{-\mu x^2} dx = \frac{1}{2}\sqrt{\frac{\pi}{\mu}}\left(e^{-2\sqrt{a\mu}} - 1\right),
\end{aligned}
$$

as required.

EXAMPLE 4.5.3. Entry **3.471.15**

(4.5.9) $$\int_0^\infty x^{-1/2} e^{-ax - b/x} \, dx = \sqrt{\frac{\pi}{a}} e^{-2\sqrt{ab}}$$

can be reduced, via $t = x^{1/2}$, to

(4.5.10) $$I = 2 \int_0^\infty e^{-at^2 - b/t^2} \, dt.$$

The value of this integral is given in (4.5.1).

EXAMPLE 4.5.4. Differentiating with respect to the parameter p shows that the integral

(4.5.11) $$I_n(p) = \int_0^\infty x^{n-1/2} e^{-px - q/x} \, dx$$

satisfies

(4.5.12) $$\frac{\partial I_n}{\partial p} = -I_{n+1}(p).$$

Using this, it is an easy induction exercise, with (4.5.9) as the base case, to verify the evaluation

(4.5.13) $$\int_0^\infty x^{n-1/2} e^{-px - q/x} \, dx = (-1)^n \sqrt{\pi} \left(\frac{\partial}{\partial p}\right)^n \left[p^{-1/2} e^{-2\sqrt{pq}}\right].$$

This is entry **3.471.16** of [35].

4.6. Some elementary changes of variables

Many of the entries in [**35**] can be obtained from the definition

$$
(4.6.1) \qquad \operatorname{erf}(u) := \frac{2}{\sqrt{\pi}} \int_0^u e^{-x^2} \, dx
$$

by elementary changes of variables. Some of these are recorded in this section.

EXAMPLE 4.6.1. The change of variables $x = \sqrt{tq}$ in (4.6.1) produces

$$
(4.6.2) \qquad \int_0^{u^2/q} \frac{e^{-qt}}{\sqrt{t}} \, dt = \sqrt{\frac{\pi}{q}} \operatorname{erf}(u).
$$

Now let $v = u^2/q$ to write the previous integral as

$$
(4.6.3) \qquad \int_0^v \frac{e^{-qt}}{\sqrt{t}} \, dt = \sqrt{\frac{\pi}{q}} \operatorname{erf}(\sqrt{qv}).
$$

This appears as **3.361.1** in [**35**].

EXAMPLE 4.6.2. Let $v \to \infty$ in (4.6.3) and use $\operatorname{erf}(+\infty) = 1$ to obtain **3.361.2**

$$
(4.6.4) \qquad \int_0^\infty \frac{e^{-qt}}{\sqrt{t}} \, dt = \sqrt{\frac{\pi}{q}}.
$$

The change of variables $x = t + a$ produces

$$
(4.6.5) \qquad \int_a^\infty \frac{e^{-qx}}{\sqrt{x-a}} \, dx = e^{-aq} \sqrt{\frac{\pi}{q}}.
$$

The special case $a = -1$ appears as **3.361.3**

$$
(4.6.6) \qquad \int_{-1}^\infty \frac{e^{-qx}}{\sqrt{x+1}} \, dx = e^q \sqrt{\frac{\pi}{q}},
$$

and $a = 1$ appears as **3.362.1**

$$
(4.6.7) \qquad \int_1^\infty \frac{e^{-qx}}{\sqrt{x-1}} \, dx = e^{-q} \sqrt{\frac{\pi}{q}}.
$$

EXAMPLE 4.6.3. The evaluation of **3.461.5**

$$
(4.6.8) \qquad \int_u^\infty e^{-qx^2} \frac{dx}{x^2} = \frac{1}{u} e^{-qu^2} - \sqrt{\pi q}\,(1 - \operatorname{erf}(u\sqrt{q}))
$$

is obtained by integration by parts. Indeed,

$$
(4.6.9) \qquad \int_u^\infty e^{-qx^2} \frac{dx}{x^2} = \frac{1}{u} e^{-qu^2} - 2q \int_u^\infty e^{-qx^2} \, dx,
$$

and this last integral can be reduced to the error function using

$$
(4.6.10) \qquad \int_u^\infty e^{-qx^2} \, dx = \int_0^\infty e^{-qx^2} \, dx - \int_0^u e^{-qx^2} \, dx.
$$

EXAMPLE 4.6.4. The evaluation of **3.466.2**

$$(4.6.11) \qquad \int_0^\infty \frac{x^2 e^{-a^2 x^2}}{x^2 + b^2}\, dx = \frac{\sqrt{\pi}}{2a} - \frac{\pi b}{2} e^{a^2 b^2} \left(1 - \operatorname{erf}(ab)\right)$$

is obtained by writing

$$\frac{\partial}{\partial a} \left(I e^{-a^2 b^2}\right) = -2a e^{-a^2 b^2} \int_0^\infty x^2 e^{-a^2 x^2}\, dx$$

and the integral I can be evaluated via the change of variables $t = ax$, to get

$$\frac{\partial}{\partial a} \left(I e^{-a^2 b^2}\right) = -\frac{\sqrt{\pi}}{2a^2} e^{-a^2 b^2}.$$

Integrate from a to ∞ and use **3.461.5** to obtain

$$(4.6.12) \qquad -I e^{-a^2 b^2} = -\frac{\sqrt{\pi}}{2} \left(\frac{e^{-a^2 b^2}}{a} - b\sqrt{\pi}\left(1 - \operatorname{erf}(ab)\right)\right).$$

Now simplify to produce the result.

EXAMPLE 4.6.5. The evaluation of entry **3.462.5**

$$(4.6.13) \qquad \int_0^\infty x e^{-\mu x^2 - 2\nu x}\, dx = \frac{1}{2\mu} - \frac{\nu}{2\mu} e^{\nu^2/\mu} \sqrt{\frac{\pi}{\mu}} (1 - \operatorname{erf}(\nu/\sqrt{\mu}))$$

can also be obtained in elementary terms. The change of variables $t = \sqrt{\mu} x$ followed by $y = t + c$ with $c = \nu/\sqrt{\mu}$ yields

$$(4.6.14) \qquad I = \frac{e^{c^2}}{\mu} J$$

where

$$(4.6.15) \qquad J = \int_c^\infty (y - c) e^{-y^2}\, dy.$$

The first integrand is a perfect derivative and the second one can be reduced to twice the normal integral to complete the evaluation.

EXAMPLE 4.6.6. The integral in entry **3.462.6**

$$(4.6.16) \qquad \int_{-\infty}^\infty x e^{-px^2 - 2qx}\, dx = \frac{q}{p} \sqrt{\frac{\pi}{p}} \exp(q^2/p)$$

is evaluated by completing the square in the exponent. It produces

$$(4.6.17) \qquad I = e^{q^2/p} \int_{-\infty}^\infty x e^{-p(x - q/p)^2}\, dx$$

and shifting the integrand by $t = x - p/q$ yields

$$(4.6.18) \qquad I = e^{q^2/p} \int_{-\infty}^\infty (t + p/q) e^{-pt^2}\, dt.$$

The first integral is elementary and the second one can be reduced to twice the normal integral to produce the result.

EXAMPLE 4.6.7. Similar arguments as those presented above yield entry **3.462.7**

$$(4.6.19) \quad \int_0^\infty x^2 e^{-\mu x^2 - 2\nu x}\, dx = -\frac{\nu}{2\mu^2} + \sqrt{\frac{\pi}{\mu^5}}\frac{2\nu^2 + \mu}{4} e^{\nu^2/\mu}(1 - \operatorname{erf}(\nu/\sqrt{\mu})),$$

and **3.462.8**

$$(4.6.20) \quad \int_{-\infty}^\infty x^2 e^{-\mu x^2 + 2\nu x}\, dx = \frac{1}{2\mu}\sqrt{\frac{\pi}{\mu}}(1 + 2\nu^2/\mu)e^{\nu^2/\mu}.$$

4.7. Some more challenging elementary integrals

In this section we discuss the evaluation of some entries in [**35**] that are completed by elementary terms. Even though the arguments are elementary, some of them require techniques that should be helpful in more complicated entries.

EXAMPLE 4.7.1. Entry **3.363.1** states that

$$(4.7.1) \quad \int_u^\infty \frac{\sqrt{x - u}}{x} e^{-qx}\, dx = \sqrt{\frac{\pi}{q}} e^{-qu} - \pi \sqrt{u}(1 - \operatorname{erf}(\sqrt{qu})).$$

The evaluation is elementary, but more complicated than those in the previous section.

We first let $x = u + t^2$ to produce $I = 2e^{-qu}J$, where

$$(4.7.2) \quad J = \int_0^\infty \frac{t^2}{t^2 + u} e^{-qt^2}\, dt.$$

The next step is to write

$$(4.7.3) \quad J = \int_0^\infty e^{-qt^2}\, dt - u \int_0^\infty \frac{e^{-qt^2}}{u + t^2}\, dt.$$

The first integral evaluates as $\sqrt{\pi}/2\sqrt{q}$ and we let

$$(4.7.4) \quad K = \int_0^\infty \frac{e^{-r^2}}{r^2 + qu}\, dr,$$

so that

$$(4.7.5) \quad I = \sqrt{\frac{\pi}{q}} e^{-qu} - 2u\sqrt{q}e^{-qu}K.$$

The change of variables $s = r^2 + qu$ produces, with $v = qu$,

$$(4.7.6) \quad K = \frac{1}{2}e^{qu}\int_v^\infty \frac{e^{-s}\, ds}{s\sqrt{s - v}}.$$

The scaling $s = vy$ reduces the question to the evaluation of

$$(4.7.7) \quad T = \int_1^\infty \frac{e^{-vy}}{y\sqrt{y - 1}}\, dy.$$

Observe that

$$(4.7.8) \qquad \frac{\partial T}{\partial v} = -\int_1^\infty \frac{e^{-vy}}{\sqrt{y-1}}\, dy = -\sqrt{\frac{\pi}{v}} e^{-v},$$

where we have used **3.362.1**. Integrate back to produce

$$T = \sqrt{\pi} \int_v^\infty \frac{e^{-r}}{\sqrt{r}}\, dr$$

$$= \sqrt{\pi} \left(\int_0^\infty \frac{e^{-r}}{\sqrt{r}}\, dr - \int_0^v \frac{e^{-r}}{\sqrt{r}}\, dr \right)$$

$$= \pi(1 - \mathrm{erf}(\sqrt{v})).$$

This gives the stated result.

EXAMPLE 4.7.2. The identity

$$(4.7.9) \qquad \frac{\sqrt{x-u}}{x} = \frac{1}{u}\left(\frac{1}{\sqrt{x-u}} - \frac{\sqrt{x-u}}{x} \right)$$

and the results of **3.362.2** and **3.363.1** give an evaluation of entry **3.363.2**

$$(4.7.10) \qquad \int_u^\infty \frac{e^{-qx}\, dx}{x\sqrt{x-u}} = \frac{\pi}{\sqrt{u}}\left(1 - \mathrm{erf}(\sqrt{qu}) \right).$$

4.8. Differentiation with respect to a parameter

The evaluation of the integral J described in the previous section is an example of a very powerful technique that is illustrated below.

EXAMPLE 4.8.1. The evaluation of **3.466.1**

$$(4.8.1) \qquad \int_0^\infty \frac{e^{-a^2 x^2}\, dx}{x^2 + b^2} = \frac{\pi}{2b}(1 - \mathrm{erf}(ab))e^{a^2 b^2},$$

is simplified first by the scaling $x = bt$. This yields the equivalent form

$$(4.8.2) \qquad \int_0^\infty \frac{e^{-c^2 t^2}\, dt}{1 + t^2} = \frac{\pi}{2}(1 - \mathrm{erf}(c))e^{c^2},$$

with $c = ab$. Introduce the function

$$(4.8.3) \qquad f(c) = \int_0^\infty \frac{e^{-c^2(1+t^2)}}{1 + t^2}\, dt$$

and the identity is equivalent to proving

$$(4.8.4) \qquad f(c) = \frac{\pi}{2}(1 - \mathrm{erf}(c)).$$

Differentiating with respect to c, we get

$$(4.8.5) \qquad f'(c) = -2ce^{-c^2} \int_0^\infty e^{-(ct)^2}\, dt = -\sqrt{\pi} e^{-c^2}.$$

Using the value $f(0) = \frac{\pi}{2}$, we get

$$(4.8.6) \qquad f(c) = \frac{\pi}{2}(1 - \mathrm{erf}(c))$$

as required.

EXAMPLE 4.8.2. The evaluation of entry **3.464**

$$(4.8.7) \qquad \int_0^\infty \left(e^{-\mu x^2} - e^{-\nu x^2} \right) \frac{dx}{x^2} = \sqrt{\pi} \left(\sqrt{\nu} - \sqrt{\mu} \right)$$

is obtained by introducing

$$(4.8.8) \qquad f(\mu) = \int_0^\infty \left(e^{-\mu x^2} - e^{-\nu x^2} \right) \frac{dx}{x^2},$$

and, differentiating with respect to the parameter μ, we obtain

$$(4.8.9) \qquad f'(\mu) = -\int_0^\infty e^{-\mu x^2}\, dx = -\frac{\sqrt{\pi}}{2\sqrt{\mu}}.$$

Integrating back and using $f(\nu) = 0$, we obtain the result.

4.9. A family of Laplace transforms

Several entries in the table [**35**] are special cases of the integral

$$(4.9.1) \qquad L_b(a, q) := \int_0^\infty \frac{e^{-xq}\, dx}{(x + a)^b},$$

where b has the form $n - \frac{1}{2}$ for $n \in \mathbb{N}$. For example, entry **3.362.2** states that

$$(4.9.2) \qquad L_{\frac{1}{2}}(a, q) = \sqrt{\frac{\pi}{q}} e^{aq} \mathrm{erfc}(\sqrt{aq})$$

and entry **3.369** is

$$(4.9.3) \qquad L_{\frac{3}{2}}(a, q) = \frac{2}{\sqrt{a}} - 2\sqrt{\pi q} e^{aq} \mathrm{erfc}(\sqrt{aq}).$$

The function `erfc` is the *complementary error function* defined by

$$(4.9.4) \qquad \mathrm{erfc}(u) := 1 - \mathrm{erf}(u) = \frac{2}{\sqrt{\pi}} \int_u^\infty e^{-x^2}\, dx.$$

These results are special cases of entry **3.382.4**, which gives $L_b(a, u)$ in terms of the incomplete gamma function

$$(4.9.5) \qquad \Gamma(a, c) := \int_c^\infty e^{-t} t^{a-1}\, dt$$

in the form

$$(4.9.6) \qquad L_b(a, u) = e^{au} u^{b-1} \Gamma(1 - b, au).$$

In this paper only the case $b = n - \frac{1}{2}$ is considered. Details for the general situation will be given elsewhere.

EXAMPLE 4.9.1. The integral in **3.362.2**

$$(4.9.7) \qquad \int_0^\infty \frac{e^{-qx}\, dx}{\sqrt{x+a}} = \sqrt{\frac{\pi}{q}} e^{aq} \left(1 - \operatorname{erf}\sqrt{qa}\right)$$

can be established by elementary means. Indeed, the change of variables $x = s^2 - a$ yields

$$(4.9.8) \qquad I = 2e^{qa} \int_{\sqrt{a}}^\infty e^{-qs^2}\, ds,$$

and scaling by $y = \sqrt{q}\, s$ yields

$$(4.9.9) \qquad I = \frac{2e^{qa}}{\sqrt{q}} \int_{\sqrt{qa}}^\infty e^{-y^2}\, dy,$$

which can be written as

$$(4.9.10) \qquad I = \frac{2e^{qa}}{\sqrt{q}} \left(\sqrt{\frac{\pi}{2}} - \int_0^{\sqrt{qa}} e^{-y^2}\, dy \right),$$

and now just write this in terms of the error function to get the stated result.

LEMMA 4.9.2. *Let $m \in \mathbb{N}$ and $a > 0$. Then*

$$(4.9.11) \qquad \int_0^\infty \frac{e^{-x}\, dx}{(x+a)^{m+1/2}} = \frac{(-1)^m 2^m}{(2m-1)!!} \left(\sqrt{\pi} e^a \operatorname{erfc}(\sqrt{a}) - \frac{P_m(a)}{2^{m-1} a^{m-1/2}} \right),$$

where $P_m(a)$ is a polynomial that satisfies the recurrence

$$(4.9.12) \qquad P_m(a) = 2^{m-1} a^{m-1} + 2a \frac{d}{da} P_{m-1}(a) - (2m-3) P_{m-1}(a)$$

and the initial condition $P_0(a) = 0$.

PROOF. The identity (4.9.7) can be expressed as

$$(4.9.13) \qquad \int_0^\infty \frac{e^{-x}\, dx}{\sqrt{x+a}} = \sqrt{\pi} e^a \operatorname{erfc}\sqrt{a}.$$

Now differentiate m times with respect to a and using

$$(4.9.14) \qquad \left(\frac{d}{da}\right)^m \frac{1}{\sqrt{x+a}} = \frac{(-1)^m (2m-1)!!}{2^m (x+a)^{m+1/2}}$$

and the ansatz

$$(4.9.15) \qquad \left(\frac{d}{da}\right)^m \left[\sqrt{\pi} e^a \operatorname{erfc}(\sqrt{a})\right] = \sqrt{\pi} e^a \operatorname{erfc}(\sqrt{a}) - \frac{P_m(a)}{2^{m-1} a^{m-1/2}}$$

give the recurrence for $P_m(a)$.   □

It is possible to obtain a simple expression for the polynomial $P_m(a)$. This is given below.

COROLLARY 4.9.3. *Define*

$$(4.9.16) \qquad R_m(a) := (-1)^{m-1} P_m(-a/2).$$

Then

$$(4.9.17) \qquad R_m(a) = \sum_{j=0}^{m-1} (2j-1)!! a^{m-1-j}.$$

PROOF. The recurrence for $P_m(a)$ gives

$$(4.9.18) \qquad R_m(a) = a^{m-1} - 2a R'_{m-1}(a) + (2m-3) R_{m-1}(a).$$

The claim now follows by induction. $\qquad\qquad\square$

In summary, the integral considered in this section is given in the next theorem.

THEOREM 4.9.4. *Let $m \in \mathbb{N}$ and $a > 0$. Then*

$$\int_0^\infty \frac{e^{-x}\,dx}{(x+a)^{m+1/2}} = \frac{(-1)^m 2^m}{(2m-1)!!} \left(\sqrt{\pi} e^a \, erfc(\sqrt{a}) - \frac{1}{\sqrt{a}} \sum_{j=0}^{m-1} \frac{(-1)^j (2j-1)!!}{(2a)^j} \right).$$

In terms of the original integral, this result gives

$$L_{m+\frac{1}{2}}(a,q) = \int_0^\infty \frac{e^{-qx}\,dx}{(x+a)^{m+1/2}} = \frac{(-1)^m 2^m q^{m-1/2}}{(2m-1)!!}$$

$$\times \left(\sqrt{\pi} e^{aq} erfc(\sqrt{aq}) - \frac{1}{\sqrt{aq}} \sum_{j=0}^{m-1} \frac{(-1)^j (2j-1)!!}{(2aq)^j} \right).$$

4.10. A family involving the complementary error function

The table [35] contains a small number of entries that involve the complementary error function defined in (4.9.4). To study these integrals, introduce the notation

$$(4.10.1) \qquad H_{n,m}(b) := \int_0^\infty x^n erfc^m(x) e^{-bx^2}\,dx.$$

The table [35] contains the values $H_{0,2}(b)$ in **8.258.1**, $H_{1,2}(b)$ in **8.258.2** and **8.258.3** is $H_{3,2}(b)$. The change of variables $x = \sqrt{t}$ yields the form

$$(4.10.2) \qquad H_{n,m}(b) := \frac{1}{2} \int_0^\infty t^{\frac{n-1}{2}} erfc^m(\sqrt{t}) e^{-bt}\,dt.$$

In this format, entry **8.258.4** contains $H_{3,1}(b)$ and $H_{2,1}(b)$ appears as **8.258.5**. This section contains an analysis of this family of integrals.

EXAMPLE 4.10.1. The value

$$(4.10.3) \qquad H_{0,0}(b) = \frac{\sqrt{\pi}}{2\sqrt{b}}$$

is elementary.

The next result presents a recurrence for these integrals.

PROPOSITION 4.10.2. *Assume $n \geq 2$ and $m \geq 1$. The integrals $H_{n,m}(b)$ satisfy*

$$(4.10.4) \qquad H_{n,m}(b) = \frac{n-1}{2b} H_{n-2,m}(b) - \frac{m}{b\sqrt{\pi}} H_{n-1,m-1}(b+1).$$

PROOF. Observe that

$$(4.10.5) \qquad H_{n,m}(b) = -\frac{1}{2b} \int_0^\infty x^{n-1} \left(\operatorname{erfc} x\right)^m \frac{d}{dx} e^{-bx^2}\, dx.$$

Integration by parts gives the result.   □

Note. The family $H_{n,m}(b)$ is determined by the initial conditions $H_{0,m}(b)$, $H_{1,m}(b)$ and $H_{n,0}(b)$. Each of these are analyzed below.

The family $H_{n,0}(b)$ is easy to evaluate.

LEMMA 4.10.3. *The integral $H_{n,0}(b)$ is given by*

$$(4.10.6) \qquad H_{n,0}(b) = \frac{1}{2} b^{-\frac{n+1}{2}} \Gamma\left(\tfrac{n+1}{2}\right).$$

PROOF. This follows from

$$(4.10.7) \qquad H_{n,0}(b) = \int_0^\infty x^n e^{-bx^2}\, dx$$

by the change of variables $s = bx^2$.   □

The family $H_{0,m}(b)$ is considered next.

The first step introduces an auxiliary function.

LEMMA 4.10.4. *Define*

$$(4.10.8) \qquad f_{n,m}(y,b) = \int_0^\infty x^n \left(\int_{xy}^\infty e^{-t^2}\, dt\right)^m e^{-bx^2}\, dx.$$

Then

$$(4.10.9) \qquad \frac{d}{dy} f_{n,m}(y,b) = -m f_{n+1,m-1}(y, b+y^2).$$

PROOF. This follows directly from the definition.   □

LEMMA 4.10.5. *Assume $m, n \in \mathbb{N}$. Then*

$$(4.10.10) \qquad H_{n,m}(b) = H_{n,0}(b) - m \left(\frac{2}{\sqrt{\pi}}\right)^m \int_0^1 f_{n+1,m-1}(y, b+y^2)\, dy.$$

PROOF. Integrating (4.10.9) and using the values

$$(4.10.11) \quad f_{n,m}(1,b) = \left(\frac{\sqrt{\pi}}{2}\right)^m H_{n,m}(b) \text{ and } f_{n,m}(0,b) = \left(\frac{\sqrt{\pi}}{2}\right)^m H_{n,0}(b)$$

gives the result.   □

COROLLARY 4.10.6. *The choice $n = 0$ gives*

$$H_{0,m}(b) = \frac{\sqrt{\pi}}{2\sqrt{b}} - m \left(\frac{2}{\sqrt{\pi}} \right)^m \int_0^1 f_{1,m-1}(y, b + y^2) \, dy.$$

The next examples are obtained by specific choices of the parameter m.

EXAMPLE 4.10.7. The first example deals with $m = 1$. In this case, Corollary 4.10.6 reduces to

$$(4.10.12) \qquad H_{0,1}(b) = \frac{\sqrt{\pi}}{2\sqrt{b}} - \frac{2}{\sqrt{b}} \int_0^1 f_{1,0}(y, b + y^2) \, dy.$$

The value $f_{1,0}(y, r) = \frac{1}{2r}$ produces

$$(4.10.13) \qquad H_{0,1}(b) = \frac{\tan^{-1} \sqrt{b}}{\sqrt{\pi b}}.$$

The computation of $H_{0,2}(b)$ employs alternative expression for the integrand in Corollary 4.10.6.

LEMMA 4.10.8. *The function $f_{1,m-1}$ is given by*
$$(4.10.14)$$
$$f_{1,m-1}(y, b + y^2) = \frac{(\sqrt{\pi}/2)^{m-1}}{2(b + y^2)} - \frac{(m-1)\pi^{m/2-1}}{2^{m-1}(b + y^2)} H_{0,m-2}\left(\frac{b + 2y^2}{y^2} \right).$$

PROOF. Integration by parts gives

$$
\begin{aligned}
f_{1,m-1}(y, b + y^2) &= \int_0^\infty x e^{-(b+y^2)x^2} \left(\int_{xy}^\infty e^{-t^2} \, dt \right)^{m-1} dx \\
&= -\frac{1}{2(b+y^2)} \int_0^\infty \frac{d}{dx}\left(e^{-(b+y^2)x^2} \right) \left(\int_{xy}^\infty e^{-t^2} \, dt \right)^{m-1} dx \\
&= \frac{(\sqrt{\pi}/2)^{m-1}}{2(b+y^2)} - \frac{(m-1)\pi^{m/2-1}}{2^{m-1}} \frac{1}{b+y^2} \\
&\quad \times \int_0^\infty e^{-(b+2y^2)x^2} \left(\int_{xy}^\infty e^{-t^2} \, dt \right)^{m-2} dx.
\end{aligned}
$$

This is the claim. ☐

This expression for the integrand in Corollary 4.10.6 gives the next result.

COROLLARY 4.10.9. *The integral $H_{0,m}(b)$ satifies*

$$H_{0,m}(b) = \frac{\sqrt{\pi}}{2\sqrt{b}} - \frac{m}{\sqrt{\pi b}} \tan^{-1}\left(\frac{1}{\sqrt{b}} \right) + \frac{m(m-1)}{\sqrt{b\pi}} \int_b^\infty \frac{H_{0,m-2}(t+2) \, dt}{t^{1/2}(t+1)}.$$

EXAMPLE 4.10.10. The case $m = 2$ in the previous formula yields the value of $H_{0,2}(b)$. The value of $H_{0,0}(b)$ in (4.10.3) produces

$$H_{0,2}(b) = \frac{\sqrt{\pi}}{2\sqrt{b}} - \frac{2}{\sqrt{\pi b}} \tan^{-1} \frac{1}{\sqrt{b}} + \frac{1}{\sqrt{\pi b}} \int_b^\infty \frac{dt}{(t+1)\sqrt{t}\sqrt{t+2}}.$$

Evaluating the remaining elementary integral gives
(4.10.15)
$$H_{0,2}(b) := \int_0^\infty e^{-bx^2} \operatorname{erfc}^2(x)\, dx = \frac{1}{\sqrt{\pi b}} \left(2\tan^{-1}\sqrt{b} - \cos^{-1}\left(\frac{1}{1+b}\right) \right).$$

This appears as entry **8.258.1** in [**35**].

The family $H_{1,m}(b)$. This is the final piece of the initial conditions.

PROPOSITION 4.10.11. *The integral $H_{1,m}(b)$ satisfies the relation*

(4.10.16)
$$H_{1,m}(b) = \frac{1}{2b}\left(1 - \frac{2m}{\sqrt{\pi}} H_{0,m-1}(b+1) \right).$$

PROOF. Integrate by parts in the representation

(4.10.17)
$$H_{1,m}(b) = -\frac{1}{2b}\int_0^\infty \frac{d}{dx} e^{-bx^2} \times \operatorname{erfc}^m x\, dx,$$

and use $\operatorname{erfc}(0) = 1$. □

EXAMPLE 4.10.12. The relation (4.10.17) in the case $m = 1$ yields

(4.10.18)
$$H_{1,1}(b) = \frac{1}{2b}\left(1 - \frac{1}{\sqrt{b+1}} \right),$$

in view of (4.10.3).

EXAMPLE 4.10.13. The case $m = 2$ gives

(4.10.19)
$$H_{1,2}(b) = \frac{1}{2b}\left(1 - \frac{4}{\sqrt{\pi}} H_{0,1}(b+1) \right).$$

Entry **8.258.2**

(4.10.20)
$$H_{1,2} := \int_0^\infty x\, \operatorname{erfc}^2 x\, e^{-bx^2}\, dx = \frac{1}{2b}\left(1 - \frac{4}{\pi}\frac{\tan^{-1}(\sqrt{1+b})}{\sqrt{1+b}} \right)$$

now follows from (4.10.13).

4.11. A final collection of examples

Sections 6.28 to 6.31 contain many other examples of integrals involving the error function. A selected number of them are established here. A systematic analysis of these sections will be presented elsewhere.

EXAMPLE 4.11.1. Entry **6.281.1** states that

(4.11.1)
$$\int_0^\infty (1 - \operatorname{erf}(px))\, x^{2q-1}\, dx = \frac{\Gamma\left(q + \frac{1}{2}\right)}{2\sqrt{\pi}\, qp^{2q}}.$$

The change of variables $t = px$ shows that this formula is equivalent to the special case $p = 1$. This is an instance of *a fake parameter*.

To show that

(4.11.2)
$$\int_0^\infty (1 - \operatorname{erf}(t))\, t^{2q-1}\, dt = \frac{\Gamma\left(q + \frac{1}{2}\right)}{2\sqrt{\pi}\, q},$$

integrate by parts, with $u = 1 - \operatorname{erf} t$ and $dv = t^{2q-1}$, to obtain

(4.11.3) $$\int_0^\infty (1 - \operatorname{erf}(t))\, t^{2q-1}\, dt = \frac{1}{\sqrt{\pi}\, q} \int_0^\infty t^{2q} e^{-t^2}\, dt.$$

The change of variables $s = t^2$ gives the result.

EXAMPLE 4.11.2. Entry **6.282.1** is

(4.11.4) $$\int_0^\infty \operatorname{erf}(qt)e^{-pt}\, dt = \frac{1}{p}\left[1 - \operatorname{erf}\left(\frac{p}{2q}\right)\right] e^{p^2/4q^2}.$$

The change of variables $x = qt$ and with $a = p/2q$ converts the entry to

(4.11.5) $$\int_0^\infty \operatorname{erf}(x)e^{-2ax}\, dx = \frac{1}{2a}\left[1 - \operatorname{erf} a\right] e^{a^2}.$$

This follows simply by integrating by parts.

EXAMPLE 4.11.3. Entry **6.282.2**, with a minor change from the stated formula in the table, is

$$\int_0^\infty \left[\operatorname{erf}\left(x + \tfrac{1}{2}\right) - \operatorname{erf}\left(\tfrac{1}{2}\right)\right] e^{-\mu x + \frac{1}{4}}\, dx = \frac{1}{\mu}\exp\left(\frac{(\mu+1)^2}{4}\right)\left[1 - \operatorname{erf}\left(\frac{\mu+1}{2}\right)\right].$$

Integration by parts gives

(4.11.6) $$\int_0^\infty \left[\operatorname{erf}\left(x + \tfrac{1}{2}\right) - \operatorname{erf}\left(\tfrac{1}{2}\right)\right] e^{-\mu x + \frac{1}{4}}\, dx = \frac{2}{\mu\sqrt{\pi}} \int_0^\infty e^{-x^2 - (\mu+1)x}\, dx.$$

The result follows by completing the square. This entry in [**35**] has the factor μ in the denominator replaced by $(\mu + 1)(\mu + 2)$. This is incorrect. The formula stated here is the correct one.

EXAMPLE 4.11.4. Entry **6.283.1** states that

(4.11.7) $$\int_0^\infty e^{\beta x}\left[1 - \operatorname{erf}(\sqrt{\alpha x})\right]\, dx = \frac{1}{\beta}\left(\frac{\sqrt{\alpha}}{\sqrt{\alpha - \beta}} - 1\right).$$

The change of variables $t = \sqrt{\alpha x}$ gives

(4.11.8) $$\int_0^\infty e^{\beta x}\left[1 - \operatorname{erf}(\sqrt{\alpha x})\right]\, dx = \frac{2}{\alpha} \int_0^\infty t e^{\beta t^2/\alpha} \operatorname{erfc} t\, dt.$$

This last integral is $H_{1,1}(-\beta/\alpha)$ and it is evaluated in Example 4.10.12.

EXAMPLE 4.11.5. Entry **6.283.2** states that

(4.11.9) $$\int_0^\infty \operatorname{erf}(\sqrt{qt})\, e^{-pt}\, dt = \frac{\sqrt{q}}{p\sqrt{p+q}}.$$

The change of variables $x = \sqrt{qt}$ gives

(4.11.10) $$\int_0^\infty \operatorname{erf}(\sqrt{qt})\, e^{-pt}\, dt = \frac{2}{q} \int_0^\infty x e^{-px^2/q} \operatorname{erf} x\, dx.$$

Integration by parts, with $\operatorname{erf} x$ the term that will be differentiated, gives the result.

Note. The table [**35**] contains many other integrals containing the error function with results involving more advanced special functions. For instance, entry **6.294.1**

$$(4.11.11) \qquad \int_0^\infty [1 - \text{erf}(1/x)] \, e^{-\mu^2 x^2} \, \frac{dx}{x} = -\text{Ei}(-2\mu),$$

where Ei denotes the *exponential integral*. These will be described in a future publication.

CHAPTER 5

Hypergeometric functions

5.1. Introduction

The hypergeometric function defined by

$$(5.1.1) \qquad {}_pF_q\left(a_1, a_2, \cdots, a_p; b_1, b_2, \cdots, b_q; x\right) := \sum_{k=0}^{\infty} \frac{(a_1)_k \cdots (a_p)_k}{(b_1)_k \cdots (b_q)_k} \frac{x^k}{k!}$$

includes, as special cases, many of the elementary special functions. For example,

$$(5.1.2) \qquad \begin{aligned} \log(1+x) &= x \, {}_2F_1\left(1, 1; 2; -x\right) \\ \sin x &= x \, {}_0F_1\left(-; \tfrac{3}{2}; -x^2/4\right) \\ \cosh x &= \lim_{a, b \to \infty} {}_2F_1\left(a, b; \tfrac{1}{2}; x^2/4ab\right). \end{aligned}$$

The binomial theorem, for a real exponent, can also be expressed in hypergeometric form as

$$(5.1.3) \qquad (1-x)^{-a} = {}_1F_0\left(a; -; x\right).$$

The goal of this paper is to verify the integrals in [**35**] that involve this function. Due to the large number of entries in [**35**] that can be related to hypergeometric functions, the list presented here represents the first part of these. More entries will appear in a future publication.

The hypergeometric function satisfies a large number of identities. The reader will find in [**13**] the best introduction to the subject. Some elementary identities are described here in detail. For example, if one of the top parameters (the a_i) agrees with a bottom one (the b_i), the function reduces to one with lower indices. The identity

$$(5.1.4) \qquad {}_2F_1(a, b; a; x) = {}_1F_0(a; -; x)$$

illustrates this point. The binomial theorem identifies the latter as $(1-x)^{-a}$.

5.2. Integrals over $[0, 1]$

The first result is a representation of ${}_2F_1$ in terms of the *beta integral*

$$(5.2.1) \qquad B(a, b) = \int_0^1 t^{a-1}(1-t)^{b-1} \, dt.$$

63

PROPOSITION 5.2.1. *The hypergeometric function $_2F_1$ is given by*

$$(5.2.2) \qquad _2F_1\left(a,\ b;\ c;\ x\right) = \frac{1}{B(b,c-b)} \int_0^1 t^{b-1}(1-t)^{c-b-1}(1-tx)^{-a}\, dt.$$

PROOF. Expand the term $(1-tx)^{-a}$ by the binomial theorem and integrate term by term. $\square$

This representation appears as **3.197.3** in [**35**]. In order to simplify the replacing of parameters, this entry is also written as

$$(5.2.3) \qquad \int_0^1 t^b(1-t)^c(1-tx)^a\, dt = B(b+1,c+1)\, _2F_1\left(-a,\ b+1;\ b+c+2;\ x\right).$$

This is one of the forms in which it will be used here: the integral being the object of primary interest.

EXAMPLE 5.2.2. The special case $a = c = 1$ in (5.2.2) appears as **3.197.10** in [**35**]

$$(5.2.4) \qquad \int_0^1 \frac{t^{b-1}\, dt}{(1-t)^b\,(1+tx)} = \frac{\pi}{\sin \pi b}(1+x)^{-b}.$$

The evaluation is direct. The identity (5.1.4) gives

$$(5.2.5) \qquad _2F_1(1,\ b;\ 1;\ -x) = (1+x)^{-b}$$

and then use $B(b, 1-b) = \Gamma(b)\Gamma(1-b) = \pi/\sin \pi b$ to complete the evaluation.

EXAMPLE 5.2.3. Introduce the index r by $r = a - b$ and take $c = b + r$ in (5.2.2). Then we have

$$(5.2.6) \qquad \int_0^1 t^{b-1}(1-t)^{r-1}(1-tx)^{-b-r}\, dt = B(b,r)\, _2F_1\left(b+r,\ b;\ b+r;\ x\right).$$

The identity (5.1.4) reduces the previous evaluation to

$$(5.2.7) \qquad \int_0^1 t^{b-1}(1-t)^{r-1}(1-tx)^{-b-r}\, dt = B(b,r)\,(1-x)^{-b}.$$

This appears as **3.197.4** in [**35**].

5.3. A linear scaling

In this section we give integrals obtained from the basic representation (5.2.3) by the change of variables $y = tp$. This produces

$$(5.3.1) \qquad \int_0^p y^{b-1}(p-y)^{c-b-1}(p-xy)^{-a}\, dy = p^{c-a-1}B(b,c-b)_2F_1\left(a,\ b;\ c;x\right).$$

EXAMPLE 5.3.1. The special case $c = b + 1$ produces

(5.3.2) $$\int_0^p y^{b-1}(p - xy)^{-a}\, dy = \frac{1}{b} p^{b-a}{}_2F_1\left(a, b; b+1; x\right),$$

where we have used $B(b, 1) = 1/b$. In order to eliminate the factor p^{-a}, we choose $x = -pr$ to obtain

(5.3.3) $$\int_0^p y^{b-1}(1 + ry)^{-a}\, dy = \frac{1}{p} u^p {}_2F_1\left(a, b; b+1; -rp\right),$$

This appears as **3.194.1** in [**35**]. The special case $a = 1$, stating that

(5.3.4) $$\int_0^p \frac{y^{b-1}\, dy}{1 + ry} = \frac{1}{b} p^b {}_2F_1\left(1, b; b+1; -rp\right),$$

appears as **3.194.5** in [**35**].

EXAMPLE 5.3.2. The table [**35**] contains the formula **3.196.1**

(5.3.5) $$\int_0^u (x + b)^\nu (u - x)^{\mu-1}\, dx = \frac{b^\nu u^\mu}{\mu}\, {}_2F_1\left[1, -\nu, 1 + \mu, -\frac{u}{b}\right].$$

We believe that it is a bad idea to have u and μ in the same formula, so we write this as

(5.3.6) $$\int_0^a (x + b)^\nu (a - x)^{\mu-1}\, dx = \frac{b^\nu a^\mu}{\mu}\, {}_2F_1\left[1, -\nu, 1 + \mu, -\frac{a}{b}\right].$$

To prove this, we let $x = at$ to get

(5.3.7) $$\int_0^a (x + b)^\nu (a - x)^{\mu-1}\, dx = b^\nu a^\mu \int_0^1 (1 + at/b)^\nu (1 - t)^{\mu-1}\, dt.$$

The integral representation (5.2.3) now gives the result.

5.4. Powers of linear factors

The hypergeometric function appears in the evaluation of integrals of the form

(5.4.1) $$I = \int_a^b L_1(x)^{\mu-1} L_2(x)^{\nu-1} L_3(x)^{\lambda-1}\, dx$$

where L_j are linear functions and $L_1(a) = L_2(b) = 0$. For example, **3.198**

(5.4.2)
$$\int_0^1 x^{\mu-1}(1 - x)^{\nu-1}\left[ax + b(1 - x) + c\right]^{-(\mu+\nu)}\, dx = (a + c)^{-\mu}(b + c)^{-\nu} B(\mu, \nu)$$

is reduced to the normal form (5.2.3) by writing

(5.4.3) $$I = (b + c)^{-\mu-\nu} \int_0^1 x^{\mu-1}(1 - x)^{\nu-1}(1 - rx)^{-(\mu+\nu)}\, dx$$

with $r = (b - a)/(b + c)$. Then (5.2.3) gives

(5.4.4) $$I = (b + c)^{-\mu-\nu} B(\mu, \nu){}_2F_1\left(\mu + \nu, \mu; \mu + \nu; \frac{b - a}{b + c}\right).$$

To produce the stated answer, simply observe the special value of the hypergeometric function

$$(5.4.5) \qquad\qquad {}_2F_1(a, b; a; z) = (1 - z)^{-b}.$$

Similarly, the evaluation of **3.199**

$$(5.4.6)$$

$$\int_a^b (x-a)^{\mu-1}(b-x)^{\nu-1}(x-c)^{-\mu-\nu}\, dx = (b-a)^{\mu+\nu-1}(b-c)^{-\mu}(a-c)^{-\nu}B(\mu,\nu)$$

is reduced to the interval $[0, 1]$ by $t = (x - a)/(b - a)$ and then the result follows from **3.198**.

The specific form of the answer is sometimes simplified due to a special relation of the parameters μ, ν and λ in (5.4.1), for example, in the evaluation of **3.197.11**

$$(5.4.7)$$

$$\int_0^1 \frac{x^{p-1/2}\, dx}{(1-x)^p\,(1+qx)^p} = \frac{2}{\sqrt{\pi}}\Gamma\left(p + \tfrac{1}{2}\right)\Gamma(1-p)\,\cos^{2p}(\varphi)\frac{\sin((2p-1)\varphi)}{(2p-1)\sin(\varphi)},$$

with $\varphi = \arctan\sqrt{q}$. The standard reduction of the integral to hypergeometric form is easy. Write

$$(5.4.8) \qquad\qquad I = \int_0^1 x^{p-1/2}(1-x)^{-p}(1+qx)^{-p}\, dx$$

and use (5.2.3) to obtain

$$(5.4.9) \qquad\qquad I = B(p + \tfrac{1}{2}, 1 - p)\,{}_2F_1\left(p, p + \tfrac{1}{2}; \tfrac{3}{2}; -q\right).$$

To reduce the answer to the stated form, we employ **9.121.19**

$$ {}_2F_1\left(\frac{n+2}{2}, \frac{n+1}{2}; \frac{3}{2}; -\tan^2 z\right) = \frac{\sin nz\, \cos^{n+1} z}{n \sin z}.$$

The evaluation of **3.197.12**

$$(5.4.10)$$

$$\int_0^1 \frac{x^{p-1/2}\, dx}{(1-x)^p\,(1-qx)^p} = \frac{\Gamma(p + \tfrac{1}{2})\Gamma(1-p)}{\sqrt{\pi}}\frac{\left[(1-\sqrt{q})^{1-2p} - (1+2\sqrt{q})^{1-2q}\right]}{(2p-1)\sqrt{q}}$$

is done in similar form. The reduction to

$$(5.4.11) \qquad\qquad I = B(p + \tfrac{1}{2}, 1 - p)\,{}_2F_1\left(p, p + \tfrac{1}{2}; \tfrac{3}{2}; q\right)$$

is direct from (5.2.3). The stated form now follows from **9.121.4**

$$ {}_2F_1\left(-\tfrac{n-1}{2}, -\tfrac{n}{2} + 1; \tfrac{3}{2}; \tfrac{z^2}{t^2}\right) = \frac{(t+z)^n - (t-z)^n}{2nzt^{n-1}}.$$

5.5. Some quadratic factors

The table [35] contains several entries of the form

$$(5.5.1) \qquad I = \int_a^b Q_1(x)^{\mu-1} L_2(x)^{\nu-1} L_3(x)^{\lambda-1}\, dx$$

where $Q_1(x)$ is a quadratic polynomial and L_j are linear functions. These are discussed in this section.

EXAMPLE 5.5.1. The first entry evaluated here is **3.254.1**

$$\int_0^a x^{\lambda-1}(a-x)^{\mu-1}(x^2+b^2)^{\nu}\, dx = b^{2\nu}a^{\lambda+\mu-1}B(\lambda,\mu)\times$$

$$_3F_2\left(-\nu,\frac{\lambda}{2},\frac{\lambda+1}{2};\frac{\lambda+\mu}{2},\frac{\lambda+\mu+1}{2};-\frac{a^2}{b^2}\right).$$

The conditions given in [35] are $\mathrm{Re}\left(\frac{a}{b}\right) > 0$, $\lambda > 0$, $\mathrm{Re}\,\mu > 0$. This entry appears as entry 186(10) of [28] as an example of the Riemann–Liouville transform

$$(5.5.2) \qquad f(x) \mapsto \frac{1}{\Gamma(\mu)}\int_0^y f(x)(y-x)^{\mu-1}\, dx.$$

It is convenient to scale the formula, by the change of variables $x = at$, to the form

$$\int_0^1 t^{\lambda-1}(1-t)^{\mu-1}(1+c^2t^2)^{\nu}\, dt = B(\lambda,\mu)\,_3F_2$$

$$\times\left(-\nu,\frac{\lambda}{2},\frac{\lambda+1}{2};\frac{\lambda+\mu}{2},\frac{\lambda+\mu+1}{2};-c^2\right),$$

with $c = a/b$. The binomial theorem gives

$$(5.5.3) \qquad (1+c^2t^2)^{\nu} = {}_1F_0(-\nu;-;-c^2t^2) = \sum_{n=0}^{\infty}\frac{(-\nu)_n}{n!}(-1)^n c^{2n}t^{2n},$$

which produces

$$\int_0^1 t^{\lambda-1}(1-t)^{\mu-1}(1+c^2t^2)^{\nu}\, dt = \sum_{n=0}^{\infty}\frac{(-\nu)_n}{n!}(-c^2)^n\int_0^1 t^{\lambda+2n-1}(1-t)^{\mu-1}\, dt$$

$$= \sum_{n=0}^{\infty}\frac{(-\nu)_n}{n!}(-c^2)^n B(\lambda+2n,\mu).$$

Now write the beta term as

$$B(\lambda+2n,\mu) = \frac{\Gamma(\lambda+2n)\,\Gamma(\mu)}{\Gamma(\lambda+2n+\mu)}$$

$$= \Gamma(\mu)\frac{2^{\lambda+2n-1}\Gamma(\frac{\lambda}{2}+n)\Gamma(\frac{\lambda+1}{2}+n)}{2^{\lambda+2n+\mu-1}\Gamma(\frac{\lambda+\mu}{2}+n)\Gamma(\frac{\lambda+\mu+1}{2}+n)}$$

where the duplication formula for the gamma function

$$(5.5.4) \qquad \Gamma(2x) = \frac{2^{2x-1}}{\sqrt{\pi}}\Gamma(x)\Gamma(x + \tfrac{1}{2})$$

has been employed. The relation $\Gamma(x + m) = (x)_m\Gamma(x)$ now yields

$$\int_0^1 t^{\lambda-1}(1 - t)^{\mu-1}(1 + c^2t^2)^\nu \, dt =$$

$$\frac{\Gamma(\mu)\Gamma(\frac{\lambda}{2})\Gamma(\frac{\lambda+1}{2})}{2^\mu\,\Gamma(\frac{\lambda+\mu}{2})\Gamma(\frac{\lambda+\mu+1}{2})}\, {}_3F_2\left(-\nu, \frac{\lambda}{2}, \frac{\lambda+1}{2}; \frac{\lambda+\mu}{2}, \frac{\lambda+\mu+1}{2}; -c^2\right).$$

Now simplify the gamma factors to produce the result.

EXAMPLE 5.5.2. The next entry contains a typo in the 7th edition of [35]. The correct version of **3.254.2** states that

$$(5.5.5) \quad \int_a^\infty x^{-\lambda}(x - a)^{\mu-1}(x^2 + b^2)^\nu \, dx = a^{\mu-\lambda+2\nu} B(\mu, \lambda - \mu - 2\nu)\, {}_3F_2$$

$$\times \left(-\nu, \frac{\lambda-\mu}{2} - \nu, \frac{1+\lambda-\mu}{2} - \nu; \frac{\lambda}{2} - \nu, \frac{1+\lambda}{2} - \nu; -\frac{b^2}{a^2}\right),$$

which follows directly from Example 5.5.1 by the change of variables $y = a^2/x$. It is convenient to scale this entry to the form

$$(5.5.6) \quad \int_1^\infty t^{-\lambda}(t - 1)^{\mu-1}(t^2 + c^2)^\nu \, dt =$$

$$B(\mu, \lambda-\mu-2\nu)\, {}_3F_2\left(-\nu, \frac{\lambda-\nu}{2} - \nu, \frac{1+\lambda-\mu}{2} - \nu; \frac{\lambda}{2} - \nu, \frac{1+\lambda}{2} - \nu; -c^2\right).$$

5.6. A single factor of higher degree

In this section we consider entries in [35] of the

$$(5.6.1) \qquad I = \int_a^b H_1(x)^{\mu-1} L_2(x)^{\nu-1} L_3(x)^{\lambda-1} \, dx$$

where $H_1(x)$ is a polynomial of degree $h \geq 2$ and L_j are linear functions.

EXAMPLE 5.6.1. Entry **3.259.2** of [35] states that

$$\int_0^a x^{\nu-1}(a - x)^{\mu-1}(x^m + b^m)^\lambda \, dx = b^{m\lambda}a^{\mu+\nu-1}B(\mu, \nu)$$

$$\times {}_{m+1}F_m\left(-\lambda, \frac{\nu}{m}, \frac{\nu+1}{m}, \cdots, \frac{\nu+m-1}{m};\right.$$

$$\left.\frac{\mu+\nu}{m}, \frac{\mu+\nu+1}{m}, \cdots, \frac{\mu+\nu+m-1}{m}; -\frac{a^m}{b^m}\right).$$

The scaling $t = x/a$ transforms this entry into

$$\int_0^1 t^{\nu-1}(1-t)^{\mu-1}(1+c^m t^m)^\lambda\, dt = B(\mu,\nu)$$

$$\times\, _{m+1}F_m\left(-\lambda, \frac{\nu}{m}, \frac{\nu+1}{m}, \cdots, \frac{\nu+m-1}{m};\right.$$

$$\left.\frac{\mu+\nu}{m}, \frac{\mu+\nu+1}{m}, \cdots, \frac{\mu+\nu+m-1}{m}; -c^m\right)$$

with $c = a/b$. This is established next using a technique developed by Euler in his proof of the integral representation of $_2F_1$.

Start with

$$I = \int_0^1 t^{\nu-1}(1-t)^{\mu-1}\left(c^m t^m + 1\right)^\lambda\, dt$$

$$= \int_0^1 t^{\nu-1}(1-t)^{\mu-1}\, _1F_0\left(-\lambda; -; -c^m t^m\right)\, dt$$

using the elementary identity (10.3.2). This gives

$$I = \int_0^1 t^{\nu-1}(1-t)^{\mu-1} \sum_{n=0}^\infty \frac{(-\lambda)_n}{n!}\left(-c^m t^m\right)^n\, dt$$

$$= \sum_{n=0}^\infty \frac{(-\lambda)_n}{n!}\left(-c^m\right)^n \int_0^1 t^{\nu+mn-1}(1-t)^{\mu-1}dt.$$

The integral is recognized as a beta function value; therefore,

$$I = \sum_{n=0}^\infty \frac{(-\lambda)_n}{n!}\left(-c^m\right)^n \frac{\Gamma(\nu+mn)\Gamma(\mu)}{\Gamma(\nu+mn+\mu)}$$

$$= \sum_{n=0}^\infty \frac{(-\lambda)_n}{n!}\left(-c^m\right)^n \frac{\Gamma(m(\frac{\nu}{m}+n))\Gamma(\mu)}{\Gamma(m(\frac{\nu+\mu}{m}+n))}$$

$$= \Gamma(\mu)\sum_{n=0}^\infty \frac{(-\lambda)_n(-c^m)^n}{n!}\,\frac{m^{m(\nu/m+n)-1/2}\Gamma(\frac{\nu}{m}+n)\cdots\Gamma(\frac{\nu+m-1}{m}+n)}{m^{m(\frac{\nu+\mu}{m}+m)-1/2}\Gamma(\frac{\nu+\mu}{m}+n)\cdots\Gamma(\frac{\nu+\mu+m-1}{m}+n)}$$

$$= \frac{\Gamma(\mu)}{m^\mu}\,\frac{\Gamma(\frac{\nu}{m})\cdots\Gamma(\frac{\nu+m-1}{m})}{\Gamma(\frac{\nu+\mu}{m})\cdots\Gamma(\frac{\nu+\mu+m-1}{m})} \times \sum_{n=0}^\infty \frac{(-\lambda)_n(\frac{\nu}{m})_n\cdots(\frac{\nu+m-1}{m})_n}{(\frac{\nu+\mu}{m})_n\cdots(\frac{\nu+\mu+m-1}{m})}\frac{(-c^m)^n}{n!}$$

$$= \frac{\Gamma(\mu)}{m^\mu}\,\frac{\Gamma(\frac{\nu}{m})\cdots\Gamma(\frac{\nu+m-1}{m})}{\Gamma(\frac{\nu+\mu}{m})\cdots\Gamma(\frac{\nu+\mu+m-1}{m})} \times$$

$$\times\, _{m+1}F_m(-\lambda, \frac{\nu}{m}, \ldots, \frac{\nu+m-1}{m}; \frac{\nu+\mu}{m}, \ldots, \frac{\nu+\mu+m-1}{m}; -c^m).$$

This is the evaluation presented in entry **3.259.2**.

5.7. Integrals over a half-line

This section considers integrals over a half-line that can be expressed in terms of the hypergeometric function.

EXAMPLE 5.7.1. To write (5.3.3) as an integral over an infinite half-line, make the change of variables $w = 1/y$ to obtain

$$(5.7.1) \qquad \int_{1/u}^{\infty} w^{a-b-1}(1+w/r)^{-a}\, dw = \frac{u^b r^a}{b}\, {}_2F_1\left(a,\, b;\, b+1;\, -ru\right),$$

Now replace u by $1/u$ and r by $1/r$ to produce

$$(5.7.2) \qquad \int_{u}^{\infty} w^{a-b-1}(1+rw)^{-a}\, dw = \frac{1}{bu^b r^a}\, {}_2F_1\left(a,\, b;\, b+1;\, -\frac{1}{ru}\right).$$

Finally, let $b = a - s$ to obtain

$$(5.7.3)$$
$$\int_{u}^{\infty} w^{s-1}(1+rw)^{-a}\, dw = \frac{1}{(a-s)u^{a-s}r^a}\, {}_2F_1\left(a,\, a-s;\, a-s+1;\, -\frac{1}{ru}\right).$$

This appears as **3.194.2** in **[35]**.

EXAMPLE 5.7.2. The change of variable $y = 1/t$ converts (5.2.3) into **3.197.6**

$$(5.7.4) \quad \int_{1}^{\infty} y^{a-c}(y-1)^{c-b-1}(\alpha y - 1)^{-a}\, dy = \alpha^{-a}B(b, c-b)\, {}_2F_1\left(a,\, b;\, c;\, 1/\alpha\right)$$

where we have labelled $\alpha = 1/x$.

EXAMPLE 5.7.3. The change of variables $y = t/(1-t)$ converts (5.2.3) into **3.197.5**:

$$(5.7.5) \qquad \int_{0}^{\infty} y^{b-1}(1+y)^{a-c}(1+\alpha y)^{-a}\, dy = B(b, c-b)\, {}_2F_1\left(a,\, b;\, c;\, 1-\alpha\right)$$

where we have labelled $\alpha = 1 - x$. If we now replace α by $1/\alpha$ we obtain

$$(5.7.6) \quad \int_{0}^{\infty} y^{b-1}(1+y)^{a-c}(y+\alpha)^{-a}\, dy = \alpha^a B(b, c-b)\, {}_2F_1\left(a,\, b;\, c;\, 1-1/\alpha\right).$$

Use the identity

$$(5.7.7) \qquad {}_2F_1\left(a,\, b;\, c;\, 1-1/\alpha\right) = (1-\alpha)^a\, {}_2F_1\left(a,\, c-b;\, c;\, \alpha\right)$$

to produce **3.197.9**:

$$(5.7.8)$$
$$\int_{0}^{\infty} y^{b-1}(1+y)^{a-c}(y+\alpha)^{-a}\, dy = \alpha^a B(b, c-b)\, {}_2F_1\left(a,\, c-b;\, c;\, 1-\alpha\right).$$

EXAMPLE 5.7.4. The change of variables $y = tu$ converts (5.2.3), with $-x$ instead of x, into **3.197.8**:

$$(5.7.9)$$
$$\int_{0}^{u} y^{b-1}(u-y)^{c-b-1}(y+\alpha)^{-a}\, dy = \alpha^{-a}u^{c-1}B(b, c-b)\, {}_2F_1\left(a,\, b;\, c;\, -u/\alpha\right)$$

where we have labelled $\alpha = u/x$.

EXAMPLE 5.7.5. The change of variables $y = st/(1-t)$ converts (5.2.3) into

(5.7.10)
$$\int_0^\infty y^{b-1}(y+s)^{a-c}(y+r)^{-a}\,dy = r^{-a}s^{a+b-c}B(b, c-b)\,_2F_1\left(a, b; c; 1 - \frac{s}{r}\right),$$

where $r = s/(1-x)$. This is **3.197.1** in [**35**]. The special case $a = c-1$ produces **3.227.1**:

(5.7.11)
$$\int_0^\infty \frac{y^{b-1}(y+r)^{1-c}}{y+s}\,dy = r^{1-c}s^{b-1}B(b, c-b)\,_2F_1\left(c-1, b; c; 1 - \frac{s}{r}\right).$$

EXAMPLE 5.7.6. Now shift the lower limit of integration via $x = y + u$ to produce

$$\int_u^\infty (x-u)^{b-1}(x-u+s)^{a-c}(x-u+r)^{-a}\,dx$$
$$= r^{-a}u^{a+b-c}B(b, c-b)\,_2F_1\left(a, b; c; 1 - \frac{s}{r}\right).$$

Choose $s = u$ and introduce the parameter v by $v = r - u$ to get

$$\int_u^\infty x^{a-c}(x-u)^{b-1}(x+v)^{-a}\,dx$$
$$= (v+u)^{-a}u^{a+b-c}B(b, c-b)\,_2F_1\left(a, b; c; \frac{v}{v+u}\right).$$

Introduce new parameters via $a = -p$, keeping b and $c = q - p$. This yields

$$\int_u^\infty x^{-q}(x-u)^{b-1}(x+v)^p\,dx$$
$$= (v+u)^p u^{b-q}B(b, c-b-p)\,_2F_1\left(-p, b; q-p; \frac{v}{v+u}\right)$$
$$= (v+u)^p u^{b-q}B(b, c-b-p)\,_2F_1\left(b, -p; q-p; \frac{v}{v+u}\right)$$

where the symmetry of the hypergeometric function in its two variables has been used.

This result is transformed using **9.131.1**,

(5.7.12) $_2F_1\left(a, b; c; z\right) = (1-z)^{-a}\,_2F_1\left(a, c-b; c; z/(z-1)\right),$

which gives

$$\int_u^\infty x^{-q}(x-u)^{b-1}(x+v)^p\,dx = (v+u)^{b+p}u^{b-q}B(b, q-p-b)\,_2F_1\left(b, q; q-p; -\frac{v}{u}\right).$$

This is the form that is found in **3.197.2**.

5.8. An exponential scale

The change of variables $t = e^{-r}$ in (5.2.3) produces

(5.8.1) $\quad {}_2F_1\left(a,\, b;\, c;\, x\right) = \dfrac{1}{B(b, c-b)} \displaystyle\int_0^\infty e^{-br}(1 - e^{-r})^{c-b-1}(1 - xe^{-r})^{-a}\, dr.$

The parameters are relabeled by $a = \rho,\, b = \mu,\, c = \nu + \mu,\, x = \beta$ to produce **3.312.3**:

(5.8.2) $\quad \displaystyle\int_0^\infty (1 - e^{-x})^{\nu-1}(1 - \beta e^{-x})^{-\rho} e^{-\mu x}\, dx = B(\mu, \nu)\, {}_2F_1\left(\rho,\, \mu;\, \mu + \nu;\, \beta\right).$

5.9. A more challenging example

The evaluation of **3.197.7**

(5.9.1) $\quad \displaystyle\int_0^\infty x^{\mu-1/2}(x+s)^{-\mu}(x+r)^{-\mu}\, dx = \sqrt{\pi}(\sqrt{r} + \sqrt{s})^{1-2\mu}\dfrac{\Gamma(\mu - 1/2)}{\Gamma(\mu)}$

requires some more properties of the hypergeometric function.

The scaling $x = rt$ produces

(5.9.2) $\qquad I = s^{-\mu}\sqrt{r} \displaystyle\int_0^\infty t^{\mu-1/2}(1+t)^{-\mu}(1 + rt/s)^{\mu}\, dt$

and using **3.197.5** we have

(5.9.3) $\qquad I = s^{-\mu}\sqrt{r}B\left(\mu + \tfrac{1}{2}, \mu - \tfrac{1}{2}\right) {}_2F_1\left(\mu, \mu + \tfrac{1}{2}, 2\mu;\, z\right)$

where $z = 1 - r/s$. To simplify this expression we employ the relation

$$
\begin{aligned}
{}_2F_1(\alpha, \beta; \gamma; z) &= \frac{(1-z)^{-\alpha}\,\Gamma(\gamma)\Gamma(\beta - \alpha)}{\Gamma(\beta)\Gamma(\gamma - \alpha)}{}_2F_1\left(\alpha, \gamma - \beta; \alpha - \beta + 1; \frac{1}{1-z}\right) + \\
&+ \frac{(1-z)^{-\beta}\,\Gamma(\gamma)\Gamma(\alpha - \beta)}{\Gamma(\beta)\Gamma(\gamma - \beta)}{}_2F_1\left(\beta, \gamma - \alpha; \beta - \alpha + 1; \frac{1}{1-z}\right)
\end{aligned}
$$

to produce

$$
\begin{aligned}
{}_2F_1\left(\mu, \mu + \tfrac{1}{2}, 2\mu;\, z\right) &= \frac{(1-z)^{-\mu}\Gamma(2\mu)\Gamma(1/2)}{\Gamma(\mu + 1/2)\,\Gamma(\mu)}{}_2F_1\left(\mu, \mu - \tfrac{1}{2}\ \tfrac{1}{2};\ \frac{1}{1-z}\right) \\
&+ \frac{(1-z)^{-\mu-1/2}\Gamma(2\mu)\Gamma(-1/2)}{\Gamma(\mu - 1/2)\,\Gamma(\mu)}{}_2F_1\left(\mu, \mu + \tfrac{1}{2}\ \tfrac{3}{2};\ \frac{1}{1-z}\right).
\end{aligned}
$$

The binomial theorem shows that

(5.9.4) $\qquad {}_2F_1\left(-\dfrac{n}{2}, -\dfrac{n-1}{2}; \dfrac{1}{2}; \dfrac{z^2}{t^2}\right) = \dfrac{1}{2t^n}\left((t+z)^n + (t-z)^n\right),$

which appears as **9.121.2** in [**35**]. Thus

$$
\begin{aligned}
{}_2F_1\left(\mu, \mu - \tfrac{1}{2}; \tfrac{1}{2}; \frac{1}{1-z}\right) &= \frac{1}{2(1-z)^{1/2-\mu}}\left((1 + \sqrt{1-z})^{1-2\mu}\right. \\
&+ \left.(-1 + \sqrt{1-z})^{1-2\mu}\right).
\end{aligned}
$$

Similarly, **9.121.4** states that

$$(5.9.5) \quad {}_2F_1\left(-\frac{n-1}{2}, -\frac{n-2}{2}; \frac{3}{2}; \frac{z^2}{t^2}\right) = \frac{1}{2nzt^{n-1}}\left((t+z)^n - (t-z)^n\right),$$

to produce

$$
{}_2F_1\left(\mu, \mu-\tfrac{1}{2}; \tfrac{3}{2}; \frac{1}{1-z}\right) = \frac{1}{2(1-2\mu)(1-z)^{-\mu}}\left((1+\sqrt{1-z})^{1-2\mu}\right.
$$
$$
\left. - (-1+\sqrt{1-z})^{1-2\mu}\right).
$$

Replacing these values in (5.9.3) produces the result.

5.10. One last example: A combination of algebraic factors and exponentials

Entry **3.389.1** presents an analytic expression for the integral

$$(5.10.1) \qquad I := \int_0^a x^{2\nu-1}(a^2 - x^2)^{\rho-1}e^{\mu x}\,dx.$$

The evaluation begins with an elementary scaling to obtain

$$
I = a^{2(\rho-1)}\int_0^1 x^{2\nu-1}\left(1 - \frac{x^2}{a^2}\right)^{\rho-1}e^{\mu x}\,dx
$$
$$
= \tfrac{1}{2}a^{2\rho-1}\int_0^1 (ay^{1/2})^{2\nu-1}(1-y)^{\rho-1}e^{\mu ay^{1/2}}y^{-1/2}\,dy.
$$

Now use ${}_0F_0(;;x) = e^x$ to obtain

$$
I = \frac{a^{2\rho+2\nu-2}}{2}\int_0^1 y^{\nu-1}(1-y)^{\rho-1}\,{}_0F_0(;;\mu ay^{1/2})\,dy
$$
$$
= \frac{a^{2\rho+2\nu-2}}{2}\int_0^1 y^{\nu-1}(1-y)^{\rho-1}\sum_{n=0}^{\infty}\frac{(\mu ay^{1/2})^n}{n!}\,dy
$$
$$
= \frac{a^{2\rho+2\nu-2}}{2}\sum_{n=0}^{\infty}\frac{(\mu a)^n}{n!}\int_0^1 y^{\nu+n/2-1}(1-y)^{\rho-1}\,dy.
$$

The integral is now recognized as a beta value to conclude that

$$
I = \frac{a^{2\rho+2\nu-2}}{2}\sum_{n=0}^{\infty}\frac{(\mu a)^n}{n!}B(\nu+n/2, \rho)
$$
$$
= \frac{a^{2\rho+2\nu-2}\Gamma(\rho)}{2}\sum_{n=0}^{\infty}\frac{(\mu a)^n}{n!}\frac{\Gamma(\nu+n/2)}{\Gamma(\nu+n/2+\rho)}
$$
$$
= \frac{a^{2\rho+2\nu-2}\Gamma(\rho)\Gamma(\nu)}{2\Gamma(\nu+\rho)}\sum_{k=0}^{\infty}\frac{(\mu a)^{2k}(\nu)_k}{\Gamma(2k+1)(\nu+\rho)_k}
$$
$$
+ \frac{a^{2\rho+2\nu-2}\Gamma(\rho)}{2}\sum_{k=0}^{\infty}\frac{(\mu a)^{2k+1}\Gamma(\nu+k+1/2)}{(2k+1)!\Gamma(\nu+\rho+k+1/2)}
$$

and combining the gamma factors to produce the beta function yields

$$I = \frac{1}{2} a^{2\rho+2\nu-2} B(\rho, \nu) \sum_{k=0}^{\infty} \frac{(\mu^2 a^2)^k (\nu)_k}{(2k) \Gamma(2k)(\nu + \rho)_k} +$$

$$+ \frac{1}{2} a^{2\rho+2\nu-1} \mu \Gamma(\rho) \sum_{k=0}^{\infty} \frac{(\mu a)^{2k}}{\Gamma(2k+2)} \frac{(\nu + 1/2)_k \Gamma(\nu + 1/2)}{(\nu + \rho + 1/2)_k \Gamma(\nu + \rho + 1/2)}.$$

This can be reduced to

$$2I = a^{2\rho+2\nu-2} B(\rho, \nu) \sum_{k=0}^{\infty} \frac{(\nu)_k (\mu^2 a^2)^k}{(\nu + \rho)_k (2k)} \frac{2^{1-2k} \sqrt{\pi}}{\Gamma(k)\Gamma(k+1/2)} +$$

$$+ a^{2\rho+2\nu-1} \mu B(\rho, \nu + 1/2) \sum_{k=0}^{\infty} \frac{(\nu + 1/2)_k}{(\nu + \rho + 1/2)_k} \frac{(\mu^2 a^2)^k 2^{1-2(k+1)} \sqrt{\pi}}{\Gamma(k+1)\Gamma(k+\frac{3}{2})}$$

$$= a^{2\rho+2\nu-2} B(\rho, \nu) \sum_{k=0}^{\infty} \frac{(\nu)_k}{(\nu + \rho)_k (\frac{1}{2})_k k!} \left(\frac{\mu^2 a^2}{4} \right)^k +$$

$$+ a^{2\rho+2\nu-1} \mu B(\rho, \nu + 1/2) \sum_{k=0}^{\infty} \frac{(\nu + 1/2)_k}{(\nu + \rho + 1/2)_k (\frac{3}{2})_k} \left(\frac{\mu^2 a^2}{4} \right)^k$$

$$= a^{2\rho+2\nu-2} B(\rho, \nu) \, {}_1F_2 \left(\nu; \nu + \rho, \frac{1}{2}; \frac{\mu^2 a^2}{4} \right) +$$

$$+ a^{2\rho+2\nu-1} \mu B(\rho, \nu + 1/2) \, {}_1F_2 \left(\nu + 1; \nu + \rho + 1/2, \frac{3}{2}; \frac{\mu^2 a^2}{4} \right).$$

There are many other entries of [**35**] that can be evaluated in terms of hypergeometric functions. A second selection of examples is in preparation.

CHAPTER 6

Hyperbolic functions

6.1. Introduction

The table of integrals [35] contains some entries giving definite integrals where the integrand contains the classical standard *hyperbolic functions*, defined by

$$(6.1.1) \qquad \sinh x = \frac{e^x - e^{-x}}{2} \quad \text{and} \quad \cosh x = \frac{e^x + e^{-x}}{2}.$$

Some of these entries are verified in the present paper.

6.2. Some elementary examples

In the evaluation of **3.511.1** in [35]

$$(6.2.1) \qquad \int_0^\infty \frac{dx}{\cosh ax} = \frac{\pi}{2a}, \quad \text{for } a > 0,$$

the parameter a can be scaled out of the equation. Indeed, the change of variables $t = ax$ yields

$$(6.2.2) \qquad \int_0^\infty \frac{dt}{\cosh t} = \frac{\pi}{2}.$$

This can be reduced to a rational integrand by the change of variables $s = e^t$ to obtain

$$
\begin{aligned}
\int_0^\infty \frac{dt}{\cosh t} &= 2 \int_1^\infty \frac{ds}{s^2 + 1} \\
&= 2 \left(\tan^{-1}(\infty) - \tan^{-1} 1 \right) = \frac{\pi}{2}.
\end{aligned}
$$

Actually, the change of variables $s = e^t$ produces the value of the indefinite integral,

$$(6.2.3) \qquad \int \frac{dt}{\cosh t} = 2 \int \frac{ds}{s^2 + 1},$$

which leads to

$$(6.2.4) \qquad \int \frac{dt}{\cosh t} = 2 \tan^{-1}(e^t).$$

This appears as **2.423.9**.

EXAMPLE 6.2.1. The second elementary example presented here appears as entry **3.527.15**

$$(6.2.5) \qquad \int_0^\infty \frac{\tanh(x/2)\, dx}{\cosh x} = \ln 2.$$

The integral is written as

$$(6.2.6) \qquad \int_0^\infty \frac{\tanh(x/2)\, dx}{\cosh x} = 2 \int_0^\infty \frac{e^x - 1}{e^x + 1} \frac{e^x\, dx}{e^{2x} + 1},$$

and the change of variables $t = e^{-x}$ gives

$$(6.2.7) \qquad \int_0^\infty \frac{\tanh(x/2)\, dx}{\cosh x} = 2 \int_0^1 \frac{1 - t}{(1 + t)(1 + t^2)}\, dt.$$

The result now comes from an elementary partial fraction decomposition.

6.3. An example that is evaluated in terms of the Hurwitz zeta function

Special cases of the evaluation

$$(6.3.1) \qquad \int_0^\infty \frac{x^n\, dx}{\cosh(x^m)} = \frac{\Gamma(p)}{m\, 2^{2p-1}} \left[\zeta\left(p, \tfrac{1}{4}\right) - \zeta\left(p, \tfrac{3}{4}\right) \right]$$

appear in [**35**]. Here $p = \frac{n+1}{m}$ and

$$(6.3.2) \qquad \zeta(z, q) = \sum_{k=1}^\infty \frac{1}{(k + q)^z}$$

is the Hurwitz zeta function. To prove (6.3.1), simply write

$$(6.3.3) \qquad \int_0^\infty \frac{x^n\, dx}{\cosh(x^m)} = 2 \int_0^\infty \frac{x^n e^{-x^m}\, dx}{1 + e^{-2x^m}}$$

and expand the integrand as a geometric series to produce

$$\begin{aligned} I &= 2 \sum_{j=0}^\infty (-1)^j \int_0^\infty x^n e^{-(2j+1)x^m}\, dx \\ &= 2 \sum_{j=0}^\infty \frac{(-1)^j}{(2j + 1)^p} \int_0^\infty t^n e^{-t^m}\, dt. \end{aligned}$$

The change of variables $u = t^m$ shows that

$$\begin{aligned} \int_0^\infty t^n e^{-t^m}\, dt &= \frac{1}{m} \int_0^\infty u^{p-1} e^{-u}\, du \\ &= \frac{1}{m} \Gamma(p). \end{aligned}$$

It follows that

$$(6.3.4) \qquad I = \frac{2\Gamma(p)}{m} \sum_{j=0}^\infty \frac{(-1)^j}{(2j + 1)^p}.$$

Now split the sum according to the parity of j:

$$\sum_{j=0}^{\infty} \frac{(-1)^j}{(2j+1)^p} = \sum_{j=0}^{\infty} \frac{1}{(4j+1)^p} - \sum_{j=0}^{\infty} \frac{1}{(4j+3)^p}$$

$$= 2^{-2p} \left(\zeta(p, \tfrac{1}{4}) - \zeta(p, \tfrac{3}{4}) \right).$$

Thus,
(6.3.5)
$$\int_0^{\infty} \frac{x^n \, dx}{\cosh(x^m)} = \frac{\Gamma(p)}{m \, 2^{2p-1}} \left[\zeta\left(p, \tfrac{1}{4}\right) - \zeta\left(p, \tfrac{3}{4}\right) \right] = \frac{2\Gamma(p)}{m} \sum_{j=0}^{\infty} \frac{(-1)^j}{(2j+1)^p}$$

as claimed.

EXAMPLE 6.3.1. In the case $n = m = 1$, the parameter $p = 2$ and **3.521.2** is obtained

(6.3.6)
$$\int_0^{\infty} \frac{x \, dx}{\cosh x} = 2G,$$

where G is **Catalan's constant** defined by

(6.3.7)
$$G := \sum_{j=0}^{\infty} \frac{(-1)^j}{(2j+1)^2}.$$

The change of variables $u = e^{-t}$ yields **4.231.12**

(6.3.8)
$$\int_0^1 \frac{\ln u \, du}{1 + u^2} = -G$$

EXAMPLE 6.3.2. The case $n = 0$, $m = 2$ yields $p = 1/2$ and **3.511.8**

(6.3.9)
$$\int_0^{\infty} \frac{dx}{\cosh(x^2)} = \sqrt{\pi} \sum_{k=0}^{\infty} \frac{(-1)^k}{\sqrt{2k+1}}$$

follows from $\Gamma(1/2) = \sqrt{\pi}$. This integral has been replaced in the last edition of [**35**] by the elementary entry

(6.3.10)
$$\int_0^{\infty} \frac{dx}{\cosh^2(x)} = 1.$$

EXAMPLE 6.3.3. The case $n = -1/2$, $m = 1$ yields $p = 1/2$ and the evaluation of **3.523.12**

(6.3.11)
$$\int_0^{\infty} \frac{dx}{\sqrt{x} \cosh x} = 2\sqrt{\pi} \sum_{k=0}^{\infty} \frac{(-1)^k}{\sqrt{2k+1}},$$

EXAMPLE 6.3.4. The case $n = 1/2$, $m = 1$ yields $p = 3/2$ and **3.523.11**:

$$(6.3.12) \qquad \int_0^\infty \frac{\sqrt{x}\,dx}{\cosh x} = \sqrt{\pi} \sum_{k=0}^\infty \frac{(-1)^k}{\sqrt{(2k+1)^3}},$$

follows from $\Gamma(3/2) = \sqrt{\pi}/2$.

The evaluation of

$$(6.3.13) \qquad \int_0^\infty \frac{x^n\,dx}{\sinh(x^m)} = \frac{2\Gamma(p)}{m} \sum_{j=0}^\infty \frac{1}{(2j+1)^p}$$

with $p = (n+1)/m$ is done exactly as above. The identity

$$(6.3.14) \qquad \sum_{j=0}^\infty \frac{1}{(2j+1)^p} = \frac{2^p - 1}{2^p} \sum_{j=0}^\infty \frac{1}{j^p}$$

yields

$$(6.3.15) \qquad \int_0^\infty \frac{x^n\,dx}{\sinh(x^m)} = \frac{\Gamma(p)}{m} \frac{2^p - 1}{2^{p-1}} \zeta(p).$$

EXAMPLE 6.3.5. The special case $m = 1$ gives $p = n + 1$ and

$$(6.3.16) \qquad \int_0^\infty \frac{x^n\,dx}{\sinh x} = \Gamma(n+1) \frac{2^{n+1} - 1}{2^n} \zeta(n+1).$$

This appears as **3.523.1** in [**35**]. In particular $n = 1$ gives **3.521.1**

$$(6.3.17) \qquad \int_0^\infty \frac{x\,dx}{\sinh x} = \frac{\pi^2}{4}.$$

This comes in the apparently more general form

$$(6.3.18) \qquad \int_0^\infty \frac{x\,dx}{\sinh ax} = \frac{\pi^2}{4a^2}.$$

But this reduces to the case $a = 1$ by the change of variables $t = ax$.

EXAMPLE 6.3.6. The special case $n = 2k - 1$ gives **3.523.2**

$$(6.3.19) \qquad \int_0^\infty \frac{x^{2k-1}\,dx}{\sinh x} = \frac{2^{2k} - 1}{2k} |B_{2k}| \pi^{2k},$$

using

$$(6.3.20) \qquad \zeta(2k) = \frac{2^{2k-1} |B_{2k}|}{(2k)!} \pi^{2k}.$$

The values $B_4 = -1/30$, $B_6 = 1/42$, and $B_8 = 1/30$ give **3.523.6**

$$(6.3.21) \qquad \int_0^\infty \frac{x^3\,dx}{\sinh x} = \frac{\pi^4}{8},$$

and **3.523.8**

$$(6.3.22) \qquad \int_0^\infty \frac{x^5\,dx}{\sinh x} = \frac{\pi^6}{4},$$

and **3.523.10**:

$$(6.3.23) \qquad \int_0^\infty \frac{x^7\,dx}{\sinh x} = \frac{17\,\pi^8}{16}.$$

6.4. A direct series expansion

Entry **3.523.3** states that

$$(6.4.1) \qquad \int_0^\infty \frac{x^{b-1}\,dx}{\cosh ax} = \frac{2\Gamma(b)}{(2a)^b} \sum_{k=0}^\infty \frac{(-1)^k}{(2k+1)^b}.$$

The change of variables $t = ax$ shows that the entry is equivalent to the special case $a = 1$:

$$(6.4.2) \qquad \int_0^\infty \frac{t^{b-1}\,dt}{\cosh t} = \frac{\Gamma(b)}{2^{b-1}} \sum_{k=0}^\infty \frac{(-1)^k}{(2k+1)^b}.$$

The proof of (6.4.2) is obtained by modifying the integrand and expanding in series

$$(6.4.3) \qquad \int_0^\infty \frac{t^{b-1}e^{-t}\,dt}{1+e^{-2t}} = \sum_{k=0}^\infty (-1)^k \int_0^\infty t^{b-1} e^{-(2k+1)t}\,dt.$$

The result follows via the change of variables $u = (2k+1)t$.

EXAMPLE 6.4.1. In the special case $b = 2n+1$, with $n \in \mathbb{N}$, the evaluation takes the form

$$(6.4.4) \qquad \int_0^\infty \frac{x^{2n}\,dx}{\cosh x} = 2(2n)! \sum_{k=0}^\infty \frac{(-1)^k}{(2k+1)^{2n+1}}.$$

The series is represented in terms of the Euler numbers E_{2n} via the classical expression

$$(6.4.5) \qquad \sum_{k=0}^\infty \frac{(-1)^k}{(2k+1)^{2n+1}} = \frac{\pi^{2n+1}\,|E_{2n}|}{(2n)!\,2^{2n+2}}$$

to obtain **3.523.4**

$$(6.4.6) \qquad \int_0^\infty \frac{x^{2n}\,dx}{\cosh x} = \left(\frac{\pi}{2}\right)^{2n+1} |E_{2n}|.$$

The Euler number can be computed from the exponential generating function

$$(6.4.7) \qquad \frac{1}{\cosh t} = \sum_{n=0}^\infty \frac{E_n}{n!} t^n.$$

The first few values are $E_0 = 1$, $E_2 = -1$, $E_4 = 5$ and $E_6 = 61$. This gives the entries **3.523.5**,

$$(6.4.8) \qquad \int_0^\infty \frac{x^2\,dx}{\cosh x} = \frac{\pi^3}{8},$$

3.523.7,

(6.4.9)
$$\int_0^\infty \frac{x^4 \, dx}{\cosh x} = \frac{5\pi^5}{32},$$

and **3.523.9**,

(6.4.10)
$$\int_0^\infty \frac{x^6 \, dx}{\cosh x} = \frac{61\pi^7}{128}.$$

6.5. An example involving Catalan's constant

Entry **3.527.14** states that

(6.5.1)
$$\int_0^\infty x^2 \frac{\sinh x}{\cosh^2 x} \, dx = 4G,$$

where G is Catalan's constant defined in (6.3.7). The evaluation is obtained by writing the integral as

(6.5.2)
$$\int_0^\infty x^2 \frac{\sinh x}{\cosh^2 x} \, dx = 2 \int_0^\infty \frac{x^2 (e^x - e^{-x}) e^{-2x}}{(1 + e^{-2x})^2} \, dx$$

and expanding in a geometric series to produce

(6.5.3)
$$\int_0^\infty x^2 \frac{\sinh x}{\cosh^2 x} \, dx = -2 \sum_{k=1}^\infty (-1)^k k \int_0^\infty x^2 (e^x - e^{-x}) e^{-2kx} \, dx.$$

Integrate term by term to obtain

(6.5.4)
$$\int_0^\infty x^2 \frac{\sinh x}{\cosh^2 x} \, dx = -4 \sum_{k=1}^\infty (-1)^k k \left[\frac{1}{(2k-1)^3} - \frac{1}{(2k+1)^3} \right].$$

Simple manipulations of the last two series produce the result.

6.6. Quotients of hyperbolic functions

Section 3.5 of [**35**] contains several evaluations where the integrand contains quotients of hyperbolic functions. This section describes a selection of them.

EXAMPLE 6.6.1. Formula **3.511.2** states that

(6.6.1)
$$\int_0^\infty \frac{\sinh ax}{\sinh bx} \, dx = \frac{\pi}{2b} \tan \frac{\pi a}{2b}.$$

To evaluate this entry, start with the change of variables $t = e^{-x}$ to obtain

(6.6.2)
$$\int_0^\infty \frac{\sinh ax}{\sinh bx} \, dx = \int_0^1 \frac{t^{a-b-1} - t^{-a-b-1}}{1 - t^{-2b}} \, dt$$

and continue with $u = t^{2b}$ to produce

(6.6.3)
$$\int_0^\infty \frac{\sinh ax}{\sinh bx} \, dx = \frac{1}{2b} \int_0^1 \frac{u^{-c-1/2} - u^{c-1/2}}{1 - u} \, du.$$

with $c = a/2b$. The evaluation of this last form employs formula **3.231.5** in [**35**]

$$(6.6.4) \qquad \int_0^1 \frac{x^{\mu-1} - x^{\nu-1}}{1 - x} \, dx = \psi(\nu) - \psi(\mu),$$

where $\psi(a) = \frac{d}{da} \ln \Gamma(a)$ is the logarithmic derivative of the gamma function. This formula was established in [**47**]. It follows that

$$(6.6.5) \qquad \int_0^\infty \frac{\sinh ax}{\sinh bx} \, dx = \frac{1}{2b} \left(\psi(c + \tfrac{1}{2}) - \psi(-c + \tfrac{1}{2}) \right).$$

The final form of the evaluation comes from the identity **8.365.9**

$$(6.6.6) \qquad \psi(\tfrac{1}{2} + c) = \psi(\tfrac{1}{2} - c) + \pi \tan \pi c.$$

EXAMPLE 6.6.2. Differentiating (6.6.1) $2m$ times with respect to a yields **3.524.2**

$$(6.6.7) \qquad \int_0^\infty x^{2m} \frac{\sinh ax}{\sinh bx} \, dx = \frac{\pi}{2b} \frac{d^{2m}}{da^{2m}} \left(\tan \frac{\pi a}{2b} \right)$$

with special cases **3.524.9**

$$\int_0^\infty x^2 \frac{\sinh ax}{\sinh bx} \, dx = \frac{\pi^3}{4b^3} \sin \frac{\pi a}{2b} \sec^3 \frac{\pi a}{2b},$$

3.524.10

$$\int_0^\infty x^4 \frac{\sinh ax}{\sinh bx} \, dx = 8 \left(\frac{\pi}{2b} \sec \frac{\pi a}{2b} \right)^5 \cdot \sin \frac{\pi a}{2b} \cdot \left(2 + \sin^2 \frac{\pi a}{2b} \right),$$

and **3.524.11**

$$\int_0^\infty x^6 \frac{\sinh ax}{\sinh bx} \, dx = 16 \left(\frac{\pi}{2b} \sec \frac{\pi a}{2b} \right)^7 \cdot \sin \frac{\pi a}{2b} \cdot \left(45 - 30 \cos^2 \frac{\pi a}{2b} + 2 \cos^4 \frac{\pi a}{2b} \right).$$

An odd number of differentiations of (6.6.1) yields **3.524.8**

$$(6.6.8) \qquad \int_0^\infty x^{2m+1} \frac{\cosh ax}{\sinh bx} \, dx = \frac{\pi}{2b} \frac{d^{2m+1}}{da^{2m+1}} \left(\tan \frac{\pi a}{2b} \right),$$

with special cases **3.524.16**

$$\int_0^\infty x \frac{\cosh ax}{\sinh bx} \, dx = \left(\frac{\pi}{2b} \sec \frac{\pi a}{2b} \right)^2,$$

3.524.17

$$\int_0^\infty x^3 \frac{\cosh ax}{\sinh bx} \, dx = 2 \left(\frac{\pi}{2b} \sec \frac{\pi a}{2b} \right)^4 \left(1 + 2 \sin^2 \frac{\pi a}{2b} \right),$$

3.524.18

$$\int_0^\infty x^5 \frac{\cosh ax}{\sinh bx} \, dx = 8 \left(\frac{\pi}{2b} \sec \frac{\pi a}{2b} \right)^6 \left(15 - 15 \cos^2 \frac{\pi a}{2b} + 2 \cos^4 \frac{\pi a}{2b} \right),$$

and **3.524.19**

$$\int_0^\infty x^7 \frac{\cosh ax}{\sinh bx}\, dx$$

$$= 16\left(\frac{\pi}{2b}\sec\frac{\pi a}{2b}\right)^8\left(315 - 420\cos^2\frac{\pi a}{2b} + 126\cos^4\frac{\pi a}{2b} - 4\cos^6\frac{\pi a}{2b}\right).$$

EXAMPLE 6.6.3. Entry **3.511.4** states that

$$(6.6.9) \qquad \int_0^\infty \frac{\cosh ax}{\cosh bx}\, dx = \frac{\pi}{2b}\sec\frac{\pi a}{2b}.$$

The proof follows the procedure employed in Example 6.6.1. The change of variables $u = e^{-2bx}$ gives

$$(6.6.10) \qquad \int_0^\infty \frac{\cosh ax}{\cosh bx}\, dx = \frac{1}{2b}\int_0^1 \frac{u^{c-1/2} + u^{-c-1/2}}{1+u}\, du.$$

Now employ **3.231.2**

$$(6.6.11) \qquad \int_0^1 \frac{x^{p-1} + x^{-p}}{1+x}\, dx = \frac{\pi}{\sin \pi p}$$

with $p = c + 1/2$. This integral was evaluated in [**20**].

EXAMPLE 6.6.4. Differentiating (6.6.9) an even number of times with respect to the parameter a gives **3.524.6**

$$(6.6.12) \qquad \int_0^\infty x^{2m}\frac{\cosh ax}{\cosh bx}\, dx = \frac{\pi}{2b}\frac{d^{2m}}{da^{2m}}\left(\sec\frac{\pi a}{2b}\right).$$

The special cases **3.524.20**

$$\int_0^\infty x^2\frac{\cosh ax}{\cosh bx}\, dx = \frac{\pi^3}{8b^3}\left(2\sec^3\frac{\pi a}{2b} - \sec\frac{\pi a}{2b}\right),$$

3.524.21

$$\int_0^\infty x^4\frac{\cosh ax}{\cosh bx}\, dx = \left(\frac{\pi}{2b}\sec\frac{\pi a}{2b}\right)^5\left(24 - 20\cos^2\frac{\pi a}{2b} + \cos^4\frac{\pi a}{2b}\right),$$

and **3.524.22**

$$\int_0^\infty x^6\frac{\cosh ax}{\cosh bx}\, dx$$

$$= \left(\frac{\pi}{2b}\sec\frac{\pi a}{2b}\right)^7\left(720 - 840\cos^2\frac{\pi a}{2b} + 184\cos^4\frac{\pi a}{2b} - \cos^6\frac{\pi a}{2b}\right)$$

are obtained by performing the differentiation.

EXAMPLE 6.6.5. Differentiating (6.6.9) an odd number of times with respect to the parameter a gives **3.524.4**

$$(6.6.13) \qquad \int_0^\infty x^{2m+1}\frac{\sinh ax}{\cosh bx}\, dx = \frac{\pi}{2b}\frac{d^{2m+1}}{da^{2m+1}}\left(\sec\frac{\pi a}{2b}\right).$$

The special cases **3.524.12**

$$\int_0^\infty x\frac{\sinh ax}{\cosh bx}\, dx = \frac{\pi^2}{4b^2}\sin\frac{\pi a}{2b}\sec^2\frac{\pi a}{2b},$$

3.524.13

$$\int_0^\infty x^3 \frac{\sinh ax}{\cosh bx} \, dx = \left(\frac{\pi}{2b} \sec \frac{\pi a}{2b}\right)^4 \sin \frac{\pi a}{2b} \left(6 - \cos^2 \frac{\pi a}{2b}\right),$$

3.524.14

$$\int_0^\infty x^5 \frac{\sinh ax}{\cosh bx} \, dx = \left(\frac{\pi}{2b} \sec \frac{\pi a}{2b}\right)^6 \sin \frac{\pi a}{2b} \left(120 - 60 \cos^2 \frac{\pi a}{2b} + \cos^4 \frac{\pi a}{2b}\right),$$

and **3.524.15**

$$\int_0^\infty x^7 \frac{\sinh ax}{\cosh bx} \, dx$$

$$= \left(\frac{\pi}{2b} \sec \frac{\pi a}{2b}\right)^8 \sin \frac{\pi a}{2b} \left(5040 - 4200 \cos^2 \frac{\pi a}{2b} + 546 \cos^4 \frac{\pi a}{2b} - \cos^6 \frac{\pi a}{2b}\right)$$

are obtained as before.

EXAMPLE 6.6.6. Integrating (6.6.9) with respect to the parameter a produces

$$(6.6.14) \qquad \int_0^\infty \frac{\sinh ax}{\cosh bx} \frac{dx}{x} = \ln \tan \left(\frac{\pi a}{4b} + \frac{\pi}{4}\right).$$

This appears as entry **3.524.23** in [**35**]. The evaluation employs the elementary primitive (which appears as entry **2.01.14**)

$$(6.6.15) \qquad \int \sec u \, du = \ln \tan \left(\frac{x}{2} + \frac{\pi}{4}\right).$$

EXAMPLE 6.6.7. Entry **3.527.6** states that

$$(6.6.16) \qquad \int_0^\infty \frac{x^{\mu-1} \sinh ax}{\cosh^2 ax} \, dx = \frac{2\Gamma(\mu)}{a^\mu} \sum_{k=0}^\infty \frac{(-1)^k}{(2k+1)^{\mu-1}},$$

which can be scaled to the case $a = 1$ by $t = ax$,

$$(6.6.17) \qquad \int_0^\infty \frac{t^{\mu-1} \sinh t}{\cosh^2 t} \, dt = 2\Gamma(\mu) \sum_{k=0}^\infty \frac{(-1)^k}{(2k+1)^{\mu-1}}.$$

To evaluate this last form, write the integrand as

$$(6.6.18) \qquad \int_0^\infty \frac{t^{\mu-1} \sinh t}{\cosh^2 t} \, dt = 2 \int_0^\infty t^{\mu-1}(e^t - e^{-t})e^{-2t}\frac{dt}{(1 + e^{-2t})^2}$$

and expand it in a power series and integrate it to obtain
(6.6.19)

$$\int_0^\infty \frac{t^{\mu-1} \sinh t}{\cosh^2 t} \, dt = 2\Gamma(\mu) \left[1 + \sum_{k=1}^\infty \frac{(-1)^k(k+1)}{(2k+1)^\mu} - \sum_{k=1}^\infty \frac{(-1)^{k+1} k}{(2k+1)^\mu}\right].$$

This is the right-hand side of (6.6.17).

The special case $\mu = 2$ and the series

$$(6.6.20) \qquad \sum_{k=0}^\infty \frac{(-1)^k}{2k+1} = \frac{\pi}{4}$$

yield the evaluation of entry **3.527.7**

$$(6.6.21) \qquad \int_0^\infty \frac{x \sinh x}{\cosh^2 x} \, dx = \frac{\pi}{2}.$$

The special case $\mu = 2m + 2$ and the series for the Euler numbers in (6.4.5) produce the evaluation of entry **3.527.8**

$$(6.6.22) \qquad \int_0^\infty \frac{x^{2m+1} \sinh x}{\cosh^2 x} \, dx = (2m + 1) \left(\frac{\pi}{2} \right)^{2m+1} |E_{2m}|.$$

6.7. An evaluation by residues

Entry **3.522.3**

$$(6.7.1) \qquad \int_0^\infty \frac{dx}{(b^2 + x^2) \cosh ax} = \frac{2\pi}{b} \sum_{k=1}^\infty \frac{(-1)^{k-1}}{2ab + (2k - 1)\pi}$$

is now evaluated by the method of residues. The change of variables $t = bx$ shows that it suffices to evaluate this integral for $b = 1$; that is,

$$(6.7.2) \qquad \int_0^\infty \frac{dx}{(1 + x^2) \cosh ax} = 2\pi \sum_{k=1}^\infty \frac{(-1)^{k-1}}{2a + (2k - 1)\pi}.$$

The integrand $f(x)$ is an even function; therefore, the evaluation requested is equivalent to

$$(6.7.3) \qquad \int_{-\infty}^\infty f(x) \, dx = \pi \sum_{k=1}^\infty \frac{(-1)^{k-1}}{2a + (2k - 1)\pi}.$$

The integral is computed by closing the real axis with a semi-circle centered at the origin located in the upper half-plane. An elementary estimate shows that the integral over the circular boundary vanishes as the radius goes to infinity. Therefore,

$$(6.7.4) \qquad \int_{-\infty}^\infty f(x) \, dx = 2\pi i \sum_{p_j} \mathrm{Res}(f; p_j),$$

where p_j is a pole of f in the upper half-plane. The integrand has poles at $z = i$ and $z = \frac{(2k-1)\pi i}{2a}$ for $k \in \mathbb{N}$. The poles are simple, unless $(2k+1)\pi = 2a$ for some k. Aside from this special case, the residues are computed as

$$\mathrm{Res}(f; i) = \frac{1}{2i \cosh(ia)} = \frac{1}{2i \cos a}$$

$$\mathrm{Res}\left(f; \frac{(2k - 1)\pi i}{2a} \right) = \frac{(-1)^{k-1} 4ia}{4a^2 - \pi^2 (2k - 1)^2}.$$

The residue theorem and a partial fraction decomposition give the stated value of the integral.

EXAMPLE 6.7.1. The special case $a = \pi$ and $b = 1$ gives

(6.7.5)
$$\int_0^\infty \frac{dx}{(1+x^2)\cosh \pi x} = 2 \sum_{k=1}^\infty \frac{(-1)^{k-1}}{2k+1}$$

and

(6.7.6)
$$\sum_{k=0}^\infty \frac{(-1)^k}{2k+1} = \frac{\pi}{4}$$

provides entry **3.522.6**

(6.7.7)
$$\int_0^\infty \frac{dx}{(1+x^2)\cosh \pi x} = 2 - \frac{\pi}{2}.$$

EXAMPLE 6.7.2. The special case $a = \pi/2$ and $b = 1$ gives

(6.7.8)
$$\int_0^\infty \frac{dx}{(1+x^2)\cosh \frac{\pi x}{2}} = \sum_{k=1}^\infty \frac{(-1)^{k-1}}{k}.$$

The evaluation

(6.7.9)
$$\sum_{k=1}^\infty \frac{(-1)^{k-1}}{k} = \ln 2$$

yields

(6.7.10)
$$\int_0^\infty \frac{dx}{(1+x^2)\cosh \frac{\pi x}{2}} = \ln 2.$$

This is entry **3.522.8**.

EXAMPLE 6.7.3. The choice $a = \pi/4$ and $b = 1$ gives

(6.7.11)
$$\int_0^\infty \frac{dx}{(1+x^2)\cosh(\pi x/4)} = 4 \sum_{k=1}^\infty \frac{(-1)^{k-1}}{4k-1}.$$

Entry **3.522.10** states that

(6.7.12)
$$\int_0^\infty \frac{dx}{(1+x^2)\cosh(\pi x/4)} = \frac{1}{\sqrt{2}}\left(\pi - 2\ln(\sqrt{2}+1)\right).$$

This is now verified by evaluating the series in (6.7.11). Start by integrating the geometric series

(6.7.13)
$$\sum_{k=1}^\infty (-1)^k x^{4k-2} = \frac{x^2}{1+x^4}$$

to produce

(6.7.14)
$$\sum_{k=1}^\infty \frac{(-1)^{k-1}}{4k-1} = \int_0^1 \frac{x^2\, dx}{1+x^4}.$$

The factorization $x^4 + 1 = (x^2 - \sqrt{2}x + 1)(x^2 + \sqrt{2}x + 1)$ gives the integral by the method of partial fractions.

6.8. An evaluation via differential equations

This section describes a method to evaluate the entries in Section **3.525** by employing differential equations.

EXAMPLE 6.8.1. Entry **3.525.1** states that

$$(6.8.1) \qquad \int_0^\infty \frac{\sinh ax}{\sinh \pi x} \frac{dx}{1+x^2} = -\frac{a}{2} \cos a + \frac{1}{2} \sin a \, \ln[2(1 + \cos a)].$$

To verify this evaluation, define

$$(6.8.2) \qquad y(a) = \int_0^\infty \frac{\sinh ax}{\sinh \pi x} \frac{dx}{1+x^2}.$$

Then

$$(6.8.3) \qquad y''(a) + y(a) = \int_0^\infty \frac{\sinh ax}{\sinh \pi x} dx = \frac{1}{2} \tan \frac{a}{2}$$

according to **3.511.2**. Equation (6.8.3) is solved by the method of variation of parameters. The general solution is of the form

$$(6.8.4) \qquad y(a) = (u_1(a) + A) \cos a + (u_2(a) + B) \sin a$$

where the (unknown) functions u_1, u_2 are determined by solving the system

$$\begin{aligned} u_1' \cos a + u_2' \sin a &= 0 \\ -u_1' \sin a + u_2' \cos a &= \frac{1}{2} \tan \frac{a}{2}. \end{aligned}$$

The solution to this system is

$$(6.8.5) \qquad u_1(a) = \frac{1}{2}(\sin a - a) \text{ and } u_2(a) = \frac{1}{2}(\ln(1 + \cos a) - \cos a).$$

The constants A and B in (6.8.4) are obtained from the values $y(0) = 0$ and

$$(6.8.6) \qquad y(\pi/2) = \int_0^\infty \frac{1}{2\cosh(\pi x/2)} \frac{dx}{1+x^2} = \frac{\ln 2}{2}$$

according to **3.522.8**. This establishes (6.8.1).

Differentiation of (6.8.1) gives **3.525.3**

$$(6.8.7) \qquad \int_0^\infty \frac{\cosh ax}{\sinh \pi x} \frac{x \, dx}{1+x^2} = \frac{1}{2}(a \sin a - 1) + \frac{\cos a}{2} \ln[2(1 + \cos a)].$$

The same procedure gives the remaining integrals in Section **3.525**, namely, **3.525.2**,

$$(6.8.8) \qquad \int_0^\infty \frac{\sinh ax}{\sinh(\pi x/2)} \frac{dx}{1+x^2} = \frac{\pi}{2} \sin a + \frac{\cos a}{2} \ln \frac{1 - \sin a}{1 + \sin a}$$

and its derivative **3.525.4**

$$(6.8.9) \qquad \int_0^\infty \frac{\cosh ax}{\sinh(\pi x/2)} \frac{x \, dx}{1+x^2} = \frac{\pi}{2} \cos a - 1 - \frac{\sin a}{2} \ln \frac{1 + \sin a}{1 - \sin a},$$

as well as **3.525.6**

$$(6.8.10) \qquad \int_0^\infty \frac{\cosh ax}{\cosh \pi x} \frac{dx}{1 + x^2} = 2\cos(a/2) - \frac{\pi}{2}\cos a - \sin a \, \ln \tan \frac{a + \pi}{4}$$

and its derivative **3.525.5**

$$(6.8.11) \qquad \int_0^\infty \frac{\sinh ax}{\cosh \pi x} \frac{x \, dx}{1 + x^2} = -2\sin(a/2) + \frac{\pi}{2}\sin a - \cos a \, \ln \tan \frac{a + \pi}{4}.$$

6.9. Squares in denominators

Section **3.527** contains a collection of integrals where the integrand is a combination of powers of the integration variable and a rational function of hyperbolic functions. The majority of them contain the square of sinh or cosh in the denominator. These integrals are evaluated in this section.

EXAMPLE 6.9.1. Entry **3.527.1** states that

$$(6.9.1) \qquad \int_0^\infty \frac{x^{\mu-1} \, dx}{\sinh^2(ax)} = \frac{4\Gamma(\mu)\,\zeta(\mu - 1)}{(2a)^\mu}.$$

The change of variables $t = ax$ shows that it is sufficient to consider the case $a = 1$. This is

$$(6.9.2) \qquad \int_0^\infty \frac{t^{\mu-1} \, dt}{\sinh^2 t} = 2^{2-\mu}\Gamma(\mu)\zeta(\mu - 1).$$

The integral to be evaluated is

$$\int_0^\infty \frac{t^{\mu-1} \, dt}{\sinh^2 t} = 4 \int_0^\infty \frac{t^{\mu-1} \, dt}{(e^t - e^{-t})^2} = 4 \int_0^\infty \frac{t^{\mu-1} e^{-2t} \, dt}{(1 - e^{-2t})^2}.$$

Expand the integrand into series to obtain

$$(6.9.3) \qquad \int_0^\infty \frac{t^{\mu-1} \, dt}{\sinh^2 t} = 4 \sum_{n=1}^\infty n \int_0^\infty t^{\mu-1} e^{-2nt} \, dt.$$

The change of variables $v = 2nt$ yields

$$(6.9.4) \qquad \int_0^\infty \frac{t^{\mu-1} \, dt}{\sinh^2 t} = 4 \sum_{n=1}^\infty \frac{1}{n^{\mu-1}} \times \frac{1}{2^\mu} \int_0^\infty v^{\mu-1} e^{-v} \, dv.$$

The series gives the Riemann zeta function term $\zeta(\mu - 1)$ and the integral is $\Gamma(\mu)$.

The special case $\mu = 3$ gives

$$(6.9.5) \qquad \int_0^\infty \frac{x^2 \, dx}{\sinh^2 x} = \frac{1}{2}\Gamma(3)\zeta(2).$$

The values $\Gamma(3) = 2$ and $\zeta(2) = \pi^2/6$ give the evaluation of entry **3.527.12**

$$(6.9.6) \qquad \int_{-\infty}^\infty \frac{x^2 \, dx}{\sinh^2 x} = \frac{\pi^2}{3}.$$

The identity

$$(6.9.7) \qquad \zeta(2m) = \frac{(2\pi)^{2m}}{2(2m)!}|B_{2m}|,$$

which provides the values of the Riemann zeta function at even integers in terms of the Bernoulli numbers B_{2m}, gives **3.527.2** (in the scaled form $a = 1$),

$$(6.9.8) \qquad \int_0^\infty \frac{x^{2m}\,dx}{\sinh^2 x} = \pi^{2m}|B_{2m}|.$$

EXAMPLE 6.9.2. Entry **3.527.3** states that

$$(6.9.9) \qquad \int_0^\infty \frac{x^{\mu-1}\,dx}{\cosh^2 x} = 2^{2-\mu}(1 - 2^{2-\mu})\Gamma(\mu)\zeta(\mu - 1)$$

for $\mu \neq 2$ and

$$(6.9.10) \qquad \int_0^\infty \frac{x\,dx}{\cosh^2 x} = \ln 2$$

for the corresponding value for $\mu = 2$. This integral also appears as **3.527.4**. The evaluation proceeds as in the previous example to produce

$$(6.9.11) \qquad \int_0^\infty \frac{x^{\mu-1}\,dx}{\cosh^2 x} = -2^{2-\mu}\Gamma(\mu)\sum_{k=1}^\infty \frac{(-1)^k}{k^{\mu-1}}.$$

The last series can be expressed in terms of the Riemann zeta function by splitting the cases k even and odd to produce the identity

$$(6.9.12) \qquad \sum_{k=1}^\infty \frac{(-1)^k}{k^{\mu-1}} = (2^{2-\mu} - 1)\zeta(\mu - 1)$$

for $\mu > 1$. The case $\mu = 2$ is obtained from the elementary value

$$(6.9.13) \qquad \sum_{k=1}^\infty \frac{(-1)^k}{k} = -\ln 2.$$

As in the previous example, the identity (6.9.7) gives

$$(6.9.14) \qquad \int_0^\infty \frac{x^{2m}\,dx}{\cosh^2 x} = \frac{(2^{2m} - 2)}{2^{2m}\pi^{2m}}|B_{2m}|.$$

This appears as **3.527.5**.

The same procedure provides the evaluation

$$(6.9.15) \qquad \int_0^\infty x^{\mu-1}\frac{\cosh x\,dx}{\sinh^2 x} = 2\Gamma(\mu)\zeta(\mu - 1)(1 - 2^{1-\mu}),$$

which appears as entry **3.527.16**. The special case $\mu = 2m + 2$ appears as entry **3.527.9**

$$(6.9.16) \qquad \int_0^\infty x^{2m+1}\frac{\cosh x}{\sinh^2 x}\,dx = \frac{2^{2m+1} - 1}{2^{2m}}(2m + 1)!\zeta(2m + 1),$$

and $\mu = 2m + 1$ provides entry **3.527.10** in the form

$$(6.9.17) \qquad \int_0^\infty x^{2m} \frac{\cosh x}{\sinh^2 x}\, dx = (2^{2m-1} - 1)\pi^{2m}|B_{2m}|,$$

employing (6.9.7). Entry **3.527.13**

$$(6.9.18) \qquad \int_0^\infty x^2 \frac{\cosh x}{\sinh^2 x}\, dx = \frac{\pi^2}{2}$$

is the special case $\mu = 3$.

6.10. Two integrals giving beta function values

This final section presents the evaluation of the two integrals that constitute Section **3.512**.

EXAMPLE 6.10.1. Entry **3.512.1** states that

$$(6.10.1) \qquad \int_0^\infty \frac{\cosh 2\beta x}{\cosh^{2\nu} ax}\, dx = \frac{4^{\nu-1}}{a} B\left(\nu + \frac{\beta}{a}, \nu - \frac{\beta}{a}\right).$$

The change of variables $t = ax$ and replacing β/a by c provides an equivalent form of the entry:

$$(6.10.2) \qquad \int_0^\infty \frac{\cosh 2ct}{(\cosh t)^{2\nu}}\, dt = 4^{\nu-1} B(\nu + c, \nu - c).$$

The beta function appearing in the answer is defined by its integral representation

$$(6.10.3) \qquad B(x, y) = \int_0^1 t^{x-1}(1 - t)^{y-1}\, dt.$$

To evaluate the left-hand side of (6.10.2), write the integrand in exponential form and let $w = e^{-ct}$ to obtain

$$(6.10.4) \qquad \int_0^\infty \frac{e^{2(c-\nu)t} + e^{-2(c+\nu)t}}{(1 + e^{-2t})^{2\nu}} = \int_0^1 \frac{w^{\nu+c} + w^{\nu-c}}{(1 + w)^{2\nu}}\, dw.$$

The result now comes from the integral representation

$$(6.10.5) \qquad B(x, y) = \int_0^1 \frac{w^{x-1} + w^{y-1}}{(1 + w)^{x+y}}\, dw,$$

which appears as entry **8.380.5** of [35]. An elementary proof of it from (6.10.3) starts with the change of variables $s = t/(1 - t)$ to produce

$$(6.10.6) \qquad B(x, y) = \int_0^\infty \frac{s^{x-1}\, ds}{(1 + s)^{x+y}}$$

given as entry **8.380.3**, and then transforms the integral to $[0, 1]$ by splitting into $[0, 1]$ and $[1, \infty)$ and moving the second integral to $[0, 1]$ by $s_1 = 1/s$.

The special case $\beta = 0$ gives

$$(6.10.7) \qquad \int_0^\infty \frac{dx}{(\cosh x)^{2\mu}} = 4^{\mu-1} B(\mu, \mu)$$

and letting $t = ax$ gives

(6.10.8)
$$\int_0^\infty \frac{dx}{(\cosh at)^{2\mu}} = \frac{4^{\mu-1}}{a} B(\mu, \mu).$$

Differentiate with respect to the parameter a to produce

(6.10.9)
$$\int_0^\infty \frac{x \sinh ax \, dx}{(\cosh ax)^{2\mu+1}} = \frac{2^{2\mu-2}}{\mu a^2} B(\mu, \mu).$$

The duplication formula of the gamma function

(6.10.10)
$$\Gamma(2\mu) = \frac{2^{2\mu-1}}{\sqrt{\pi}} \Gamma(\mu)\Gamma(\mu + \tfrac{1}{2})$$

transforms (6.10.9) into

(6.10.11)
$$\int_0^\infty \frac{x \sinh ax \, dx}{(\cosh ax)^{2\mu+1}} = \frac{\sqrt{\pi}}{4\mu a^2} \frac{\Gamma(\mu)}{\Gamma(\mu + \tfrac{1}{2})}.$$

This appears as entry **3.527.11**.

EXAMPLE 6.10.2. The last entry in Section 3.512 is **3.512.2**

(6.10.12)
$$\int_0^\infty \frac{\sinh^\mu x}{\cosh^\nu x} \, dx = \frac{1}{2} B\left(\frac{\mu+1}{2}, \frac{\nu-\mu}{2}\right).$$

Two proofs of this evaluation are given here. The first one is elementary and the second one enters the realm of hypergeometric functions.

The first proof begins with the change of variables $w = \cosh x$ to obtain

(6.10.13)
$$\int_0^\infty \frac{\sinh^\mu x}{\cosh^\nu x} \, dx = \int_1^\infty (w^2 - 1)^{\frac{\mu-1}{2}} w^{-\nu} \, dw$$

followed by the change of variables $t = w^{-2}$ to produce

(6.10.14)
$$\frac{1}{2} \int_0^1 t^{\frac{\nu-\mu}{2}-1}(1-t)^{\frac{\nu-\mu}{2}-1} \, dt = \frac{1}{2} B\left(\frac{\mu+1}{2}, \frac{\nu-\mu}{2}\right).$$

The second proof begins by writing the integrand as exponentials to obtain

$$\int_0^\infty \frac{\sinh^\mu x}{\cosh^\nu x} \, dx = 2^{\nu-\mu-1} \int_0^1 t^{\nu/2-\mu/2-1}(1-t)^\mu (1+t)^{-\nu} \, dt$$

after the change of variable $t = e^{-2x}$. The integral representation **9.111** states that

$$\int_0^1 t^{b-1}(1-t)^{c-b-1}(1-tz)^{-a} \, dt = B(b, c-b) \, {}_2F_1(a, b; c; z).$$

It follows that

$$\int_0^\infty \frac{\sinh^\mu x}{\cosh^\nu x} \, dx = 2^{\nu-\mu-1} B\left(\frac{\nu-\mu}{2}, 1+\mu\right) {}_2F_1\left(\nu, \frac{\nu-\mu}{2}; 1+\frac{\mu+\nu}{2}; -1\right).$$

Now use **9.131.1** ${}_2F_1(a, b; c; z) = (1-z)^{-a} {}_2F_1\left(a, c-b; c; z/(z-1)\right)$ to transform the integral to the value of a hypergeometric function with $z = 1/2$. The quadratic transformation **9.133** ${}_2F_1\left(2a, 2b; a+b+\tfrac{1}{2}; z\right) = {}_2F_1(a, b; a+b$

$+\frac{1}{2}; 4z(1-z))$ transforms it to the value of a hypergeometric function with $z = 1$. The result now follows from the evaluation

$$_2F_1\left(a, b; c; 1\right) = \frac{\Gamma(c)\,\Gamma(c - a - b)}{\Gamma(c - a)\,\Gamma(c - b)}.$$

6.11. The last two entries of Section 3.525

This section presents a new technique that will produce evaluations of entries **3.525.7** and **3.525.8**. This completes the verification of all entries in this section that started in Section 6.8.

The first step is the computation of a Laplace transform.

LEMMA 6.11.1. *The identity*

$$(6.11.1) \qquad \int_0^\infty \frac{e^{-st}\,dt}{\cosh \lambda t + \cos \lambda p} = \frac{2}{\sin \lambda p} \sum_{n=1}^\infty (-1)^{n-1} \frac{\sin(\lambda pn)}{s + \lambda n}$$

holds.

PROOF. The factorization

$$(6.11.2) \qquad \cosh \lambda t + \cos \lambda p = \frac{e^{\lambda t}}{2}\left(1 + e^{-2\lambda t} + e^{-\lambda t + i\lambda p} + e^{-\lambda t - i\lambda p}\right)$$

gives the decomposition

$$\begin{aligned}
\frac{e^{-st}}{\cosh \lambda t + \cos \lambda p} &= \frac{2e^{-(\lambda + s)t}}{(1 + e^{-\lambda(t - ip)})(1 + e^{-\lambda(t + ip)})} \\
&= \frac{e^{-st}}{\sin \lambda p}\left(\frac{1}{1 + e^{-\lambda(t + ip)}} - \frac{1}{1 + e^{-\lambda(t - ip)}}\right) \\
&= -\frac{2e^{st}}{\sin \lambda p} \sum_{n=0}^\infty (-1)^n e^{-\lambda tn} \sin \lambda pn.
\end{aligned}$$

The result now follows by integration. $\qquad\square$

EXAMPLE 6.11.2. The special case $\lambda = 1$ and $p = \pi - q$ in the lemma gives entry **3.543.2**

$$(6.11.3) \qquad \int_0^\infty \frac{e^{-st}\,dt}{\cosh t - \cos q} = \frac{2}{\sin q} \sum_{n=1}^\infty \frac{\sin(qn)}{s + n}.$$

EXAMPLE 6.11.3. Entry **3.511.5** is established next. Its value is employed in the next example. This entry states

$$\int_0^\infty \frac{\sinh ax \cosh bx}{\sinh cx}\,dx = \frac{\pi}{2c}\left(\frac{\sin \frac{\pi a}{c}}{\cos \frac{\pi a}{c} + \cos \frac{\pi b}{c}}\right).$$

The proof starts by expressing the integrand in exponential form to obtain

$$\int_0^\infty \frac{\sinh ax \cosh bx}{\sinh cx}\,dx = \frac{1}{2}\int_0^\infty \frac{e^{-cx}(e^{ax} - e^{-ax})(e^{bx} + e^{-bx})}{1 - e^{-2cx}}\,dx$$

and using the change of variables $t = e^{-2cx}$ to produce

$$\int_0^\infty \frac{\sinh ax \cosh bx}{\sinh cx} \, dx$$

$$= \frac{1}{4c} \int_0^1 \frac{t^{-A-B-1/2} + t^{-A+B-1/2} - t^{A-B-1/2} - t^{A+B-1/2}}{1-t} \, dt$$

with $A = \frac{a}{2c}$ and $B = \frac{b}{2c}$. Using the formula

$$\int_0^1 \frac{1 - x^{a-1}}{1-x} \, dx = \psi(a) + \gamma$$

given as entry **3.265** (established in [**47**]), it follows that

$$\int_0^\infty \frac{\sinh ax \cosh bx}{\sinh cx} \, dx =$$

$$\frac{1}{4c} \left(\psi \left(\tfrac{1}{2} + A - B \right) - \psi \left(\tfrac{1}{2} - A + B \right) + \psi \left(\tfrac{1}{2} + A + B \right) - \psi \left(\tfrac{1}{2} - A - B \right) \right).$$

The result now follows from the identity

$$\psi(\tfrac{1}{2} + z) - \psi(\tfrac{1}{2} - z) = \pi \tan \pi z.$$

EXAMPLE 6.11.4. Entry **3.525.7** is

$$(6.11.4) \qquad \int_0^\infty \frac{\sinh(ax)}{\sinh(bx)} \frac{s}{s^2 + x^2} \, dx = \pi \sum_{n=1}^\infty \frac{\sin \left(\frac{n(b-a)}{b} \pi \right)}{bs + n\pi}.$$

The evaluation employs the Laplace transform

$$(6.11.5) \qquad \int_0^\infty e^{-st} \cos xt \, dt = \frac{s}{s^2 + x^2}$$

and entry **3.511.5** given in the previous example

$$\int_0^\infty \frac{\sinh(ax)}{\sinh(bx)} \frac{s}{s^2 + x^2} \, dx = \int_0^\infty e^{-st} \left\{ \int_0^\infty \frac{\sinh(ax)}{\sinh(bx)} \cos xt \, dx \right\} dt$$

$$= \int_0^\infty e^{-st} \left\{ \frac{\pi}{2b} \frac{\sin \frac{\pi a}{b}}{\cosh \frac{\pi t}{b} + \cos \frac{\pi a}{b}} \right\} dt$$

$$= \frac{\pi}{2b} \sin \frac{\pi a}{b} \int_0^\infty \frac{e^{-st} \, dt}{\cosh \frac{\pi t}{b} + \cos \frac{\pi a}{b}}.$$

The proof concludes by choosing $\lambda = \pi/b$ and $a = p$ in Lemma 6.11.1 and using $\sin(n(b - a)\pi/b) = (-1)^{n-1} \sin(n\pi a/b)$.

EXAMPLE 6.11.5. Differentiation of entry **3.525.7** with respect to the parameter a gives

$$(6.11.6) \qquad \int_0^\infty \frac{\cosh(ax)}{\sinh(bx)} \frac{x}{s^2 + x^2} \, dx = \frac{\pi}{bs} \sum_{n=1}^\infty \frac{(-1)^{n-1} \pi n}{bs + \pi n} \cos \frac{\pi a n}{b}.$$

To simplify this expression use $\dfrac{\pi n}{bs + \pi n} = 1 - \dfrac{bs}{bs + \pi n}$ and split the series using the Fourier expansion

$$(6.11.7) \qquad \sum_{n=1}^{\infty} (-1)^{n-1} \cos \frac{\pi a n}{b} = \frac{1}{2}.$$

This final result is entry **3.525.8**

$$(6.11.8) \qquad \int_{0}^{\infty} \frac{\cosh ax}{\sinh bx} \frac{x\,dx}{s^2 + x^2} = \frac{\pi}{2bs} + \pi \sum_{n=1}^{\infty} \frac{\cos \frac{n(b-a)}{b}\pi}{bs + n\pi}.$$

The series in (10.2.3) and (6.11.7) are both Abel-convergent. The reader is invited to verify that the series (6.11.8) is convergent and reduces to (6.8.7) when $b = \pi$ and $s = 1$.

REMARK 6.11.6. *Section 4.11 of* [**35**] *contains many analogous formulas to those considered here. For instance, entry* **3.525.1**

$$(6.11.9) \qquad \int_{0}^{\infty} \frac{\sinh ax}{\sinh \pi x} \frac{dx}{1 + x^2} = -\frac{a}{2}\cos a + \frac{1}{2}\sin a \, \ln[2(1 + \cos a)]$$

is related to entry **4.113.3**

$$(6.11.10) \qquad \int_{0}^{\infty} \frac{\sin ax}{\sinh \pi x} \frac{dx}{1 + x^2} = -\frac{a}{2}\cosh a + \frac{1}{2}\sinh a \, \ln[2(1 + \cosh a)].$$

The right-hand side of the last entry appears in [**35**] *in the equivalent form*

$$(6.11.11) \qquad -\frac{a}{2}\cosh a + \sinh a \, \ln[2\cosh a/2].$$

A systematic study of this correspondence and the evaluation of the integrals appearing in Section 4.11 will be presented in a future publication.

CHAPTER 7

Bessel-K functions

7.1. Introduction

This paper is part of the collection initiated in [**49**], aiming to evaluate the entries in [**35**] and to provide some context. This table contains a large variety of entries involving the Bessel functions. The goal of the current work is to evaluate some entries in [**35**] where the integrand is an elementary function and the result involves the so-called modified Bessel function of the second kind, denoted by $K_\nu(x)$. Other types of integrals containing Bessel functions will appear in a future publication. This introduction contains a brief description of the Bessel functions. The reader is referred to [**16, 65, 67, 68**] for more information about this class of functions.

The Bessel differential equation

$$(7.1.1) \qquad x^2 \frac{d^2u}{dx^2} + x\frac{du}{dx} + (x^2 - \nu^2)u = 0$$

arises from the solution of Laplace's equation

$$(7.1.2) \qquad \frac{\partial^2 U}{\partial x^2} + \frac{\partial^2 U}{\partial y^2} + \frac{\partial^2 U}{\partial z^2} = 0$$

in spherical or cylindrical coordinates. The method of Frobenius shows that, for any $\nu \in \mathbb{R}$, the function

$$(7.1.3) \qquad J_\nu(x) = \sum_{k=0}^{\infty} \frac{(-1)^k}{\Gamma(\nu + 1 + k)\, k!} \left(\frac{x}{2}\right)^{\nu+2k}$$

solves (7.1.1). The function $J_\nu(x)$ is called the **Bessel function of the first kind**.

In the case $\nu \notin \mathbb{Z}$, the functions $J_\nu(x)$ and $J_{-\nu}(x)$ are linearly independent, so they form a basis for the space of solutions to (7.1.1). If $\nu = n \in \mathbb{Z}$, the relation $J_{-n}(x) = (-1)^n J_n(x)$ shows that a second function is required. This is usually obtained from

$$(7.1.4) \qquad Y_\nu(x) = \frac{J_\nu(x)\cos \pi\nu - J_{-\nu}(x)}{\sin \pi\nu},$$

and now $\{J_\nu, Y_\nu\}$ is a basis for all $\nu \in \mathbb{R}$. Naturally, when $\nu \in \mathbb{Z}$, the function $Y_\nu(x)$ has to be interpreted as $\lim_{\mu\to\nu} Y_\mu(x)$. The function $Y_\nu(x)$ is called the **Bessel function of the second kind**.

The **modified Bessel equation**

$$(7.1.5) \qquad x^2 \frac{d^2 w}{dx^2} + x \frac{dw}{dx} - (x^2 + \nu^2) w = 0$$

is solved in terms of the **modified Bessel functions**

$$(7.1.6) \qquad I_\nu(x) = \sum_{k=0}^{\infty} \frac{1}{\Gamma(\nu + 1 + k)\, k!} \left(\frac{x}{2} \right)^{\nu + 2k}$$

and

$$(7.1.7) \qquad K_\nu(x) = \frac{\pi}{2} \frac{I_{-\nu}(x) - I_\nu(x)}{\sin \pi \nu}.$$

As before, if $\nu \in \mathbb{Z}$, the function K_ν has to be replaced by its limiting value. The function $I_\nu(x)$ is called of first kind and $K_\nu(x)$ of second kind. The integral repesentation

$$(7.1.8) \qquad I_\nu(z) = \frac{(z/2)^\nu}{\Gamma(\nu + \frac{1}{2}) \Gamma(\frac{1}{2})} \int_{-1}^{1} e^{-zt} (1 - t^2)^{\nu - 1/2} \, dt$$

appears as entry **3.387.1**. A proof may be found in [**65**].

This paper contains entries in [**35**] that involve the function $K_\nu(x)$ in the answers. For instance, entry **3.324.1**, which is a special case of $(7.2.11)$, stating that

$$(7.1.9) \qquad \int_0^\infty \exp\left(-\frac{b}{4x} - ax \right) dx = \sqrt{\frac{b}{a}} K_1(\sqrt{ab}),$$

is an example of the type of problems considered here, but entry **6.512.9**, which is

$$(7.1.10) \qquad \int_0^\infty K_0(ax) J_1(bx) \, dx = \frac{1}{2b} \ln\left(1 + \frac{b^2}{a^2} \right),$$

where the Bessel function appears in the integrand, will be described in a future publication.

Most of the entries presented here appear in the literature. The objective of this paper is to present several techniques that are applicable to this and other integral evaluations. Some typos in the table [**35**] have been corrected. The work presented here employs a variety of techniques. The choice of method used in a specific entry has been determined by pedagogical as well as efficiency reasons.

Many integrals that appear in this article have integrands that are members of the class of hyperexponential expressions. Recall that $f(x)$ is called hyperexponential if $f'(x)/f(x) = r(x)$ is a rational function of x. In other words, $f(x)$ satisfies a first-order linear differential equation with polynomial coefficients, namely, $q(x)f'(x) - p(x)f(x) = 0$, if we write $r(x) = p(x)/q(x)$. A multivariate function is hyperexponential if the above property holds for each single variable. Almkvist and Zeilberger [**2**] developed an algorithm for treating integrals with a hyperexponential integrand in an automatic fashion. The idea is based on the paradigm of creative telescoping: assume one wants to evaluate the integral $\int_a^b f(x, y)\, dx$. Then the goal of the algorithm is to find a differential equation for f of the following, very special, form

$$(7.1.11) \qquad c_m(y)\frac{d^m f}{dy^m} + \cdots + c_1(y)\frac{df}{dy} + c_0(y)f = \frac{d}{dx}\left(q(x, y)f\right),$$

where the $c_i(y)$ are polynomials and $q(x, y)$ is a bivariate rational function. If one integrates this equation and applies the fundamental theorem of calculus, then one obtains a differential equation for the integral. This equation may be used to find a closed form or to prove a certain identity. In many cases, the right-hand side evaluates to zero, yielding a homogeneous o.d.e.; in other cases one may end up with an inhomogeneous one. Care has to be taken that all the integrals that appear do really converge (this may not always be the case). The approach just described will be employed and illustrated in Section 7.7.2.

The Almkvist–Zeilberger algorithm was later extended to general holonomic functions by Chyzak [25]. In this context, a holonomic function is one which satisfies a linear ordinary differential equation with polynomial coefficients for each of its variables (not necessarily of order 1 as in the hyperexponential case). Implementations in Mathematica of these two algorithms are given in the package HolonomicFunctions [43].

7.2. A first integral representation of modified Bessel functions

This section describes the integral representations of the modified Bessel function $K_\nu(z)$. A detailed proof of the first result appears as (9.42) in [65], page 235.

THEOREM 7.2.1. *The function $K_\nu(z)$ admits the integral representation*

$$(7.2.1) \qquad K_\nu(z) = \frac{z^\nu}{2^{\nu+1}}\int_0^\infty t^{-\nu-1}e^{-t-z^2/4t}\, dt.$$

This formula appears as entry **8.432.6** *in* [35].

REMARK 7.2.2. *Several other entries of* [35] *are obtained by elementary manipulations of* (7.2.1). *For instance, it can be written as*

$$(7.2.2) \qquad \int_0^\infty t^{-\nu-1}\, exp\left(-t - \frac{b}{t}\right) dt = \frac{2}{b^{\nu/2}}K_\nu(2\sqrt{b}).$$

EXAMPLE 7.2.3. *Let $b = 1$ in* (7.2.2) *and make the change of variables $t = e^x$ to obtain*

$$(7.2.3) \qquad \int_{-\infty}^\infty exp\left(-\nu x - 2\cosh x\right) dx = 2K_\nu(2).$$

Splitting the integration over $(-\infty, 0)$ and $(0, \infty)$ gives

$$(7.2.4) \qquad \int_0^\infty exp\left(-2\cosh x\right)\cosh \nu x\, dx = K_\nu(2).$$

EXAMPLE 7.2.4. Example 7.2.3 is the special case $\beta = 2$ of entry **3.547.4**

$$(7.2.5) \qquad \int_0^\infty \exp\left(-\beta \cosh x\right) \cosh \nu x \, dx = K_\nu(\beta).$$

The table employs γ instead of ν. This entry also follows directly from (7.2.2). The change of variables $t = \sqrt{b}x$ gives

$$(7.2.6) \qquad \int_0^\infty x^{-\nu-1} \exp\left(-\sqrt{b}\left(x + 1/x\right)\right) dx = 2K_\nu(2\sqrt{b}).$$

The change of variables $y = e^t$ gives an integral over the whole real line. Splitting the integration as in Example 7.2.3 produces the result (7.2.5).

EXAMPLE 7.2.5. Entry **3.395.1** is

$$(7.2.7) \quad \int_0^\infty \left[(\sqrt{x^2-1}+x)^\nu + (\sqrt{x^2-1}+x)^{-\nu} \right] \frac{e^{-\mu x}}{\sqrt{x^2-1}} \, dx = 2K_\nu(\mu).$$

The left-hand side of (7.2.7) transforms as

$$\int_1^\infty \left[(\sinh\theta + \cosh\theta)^\nu + (\sinh\theta + \cosh\theta)^{-\nu} \right] e^{-\mu\cosh\theta} \, d\theta$$

$$= \int_1^\infty \left[e^{\nu\theta} + e^{-\nu\theta} \right] e^{-\mu\cosh\theta} \, d\theta$$

$$= 2 \int_1^\infty \cosh(\nu\theta) e^{-\mu\cosh\theta} \, d\theta$$

and applying (7.2.5) yields (7.2.7).

EXAMPLE 7.2.6. Entry **3.471.12** is

$$(7.2.8) \qquad \int_0^\infty x^{\nu-1} \exp\left(-x - \frac{\mu^2}{4x}\right) dx = 2\left(\frac{\mu}{2}\right)^\nu K_{-\nu}(\mu)$$

and it comes directly from (7.2.2).

EXAMPLE 7.2.7. The change of variables $s = 1/t$ yields

$$(7.2.9) \qquad K_\nu(z) = \frac{z^\nu}{2^{\nu+1}} \int_0^\infty s^{\nu-1} e^{-1/s - z^2 s/4} \, ds,$$

and, followed by $s = w/a$, produces

$$(7.2.10) \qquad K_\nu(z) = \frac{z^\nu}{2^{\nu+1}a^\nu} \int_0^\infty w^{\nu-1} \exp\left(-\frac{a}{w} - \frac{z^2}{4a}w\right) dw.$$

Now introduce the parameter b by the relation $4ab = z^2$ to obtain

$$(7.2.11) \qquad \int_0^\infty w^{\nu-1} \exp\left(-\frac{a}{w} - bw\right) dw = 2\left(\frac{a}{b}\right)^{\nu/2} K_\nu(2\sqrt{ab}).$$

In particular, if $b = 1$, it follows that

$$(7.2.12) \qquad \int_0^\infty w^{\nu-1} \exp\left(-w - \frac{a}{w}\right) dw = 2a^{\nu/2} K_\nu(2\sqrt{a}).$$

Formula (7.2.11) appears as entry **3.471.9** of [**35**]. The special case $\nu = 1$ is entry **3.324.1**, which served as an illustration in (7.1.9).

Now replace a by b and ν by $-\nu$ in (7.2.12) to obtain

$$(7.2.13) \qquad \int_0^\infty w^{-\nu-1} \exp\left(-w - \frac{b}{w}\right) dw = \frac{2}{b^{\nu/2}} K_{-\nu}(2\sqrt{b}).$$

PROPOSITION 7.2.8. *The function K_ν satisfies the symmetry*

$$(7.2.14) \qquad K_\nu(z) = K_{-\nu}(z).$$

PROOF. This symmetry is suggested by the differential equation, as only even powers of ν occur. The actual proof follows directly from (7.1.7). A second proof is obtained by comparing (7.2.2) with (7.2.13). □

EXAMPLE 7.2.9. Entry **3.337.1** is

$$(7.2.15) \qquad \int_{-\infty}^\infty \exp\left(-\alpha x - \beta \cosh x\right) dx = 2K_\alpha(\beta).$$

To establish this identity, make the change of variables $t = \beta e^x / 2$ to produce

$$\int_{-\infty}^\infty \exp\left(-\alpha x - \beta \cosh x\right) dx = \left(\frac{\beta}{2}\right)^\alpha \int_0^\infty t^{-\alpha-1} \exp\left(-t - \frac{\beta^2}{4t}\right) dt.$$

The result (7.2.15) then follows from (7.2.2) and Proposition 7.2.8.

EXAMPLE 7.2.10. The result of Example 7.2.9 is now employed to produce a proof of the evaluation

$$(7.2.16) \qquad \int_0^\infty e^{-2b\sqrt{x^2+1}} dx = K_1(2b).$$

The reader will find the similar looking integral

$$(7.2.17) \qquad \int_0^\infty e^{-2b(x^2+1)^2} dx = 2^{-3/2} e^{-b} K_{1/4}(b)$$

in Section 7.7.

The change of variables $t = \sinh x$ produces

$$\begin{aligned}
\int_0^\infty e^{-2b\sqrt{x^2+1}} dx &= \int_0^\infty \cosh x \exp\left(-2b\cosh x\right) dx \\
&= \frac{1}{2} \int_0^\infty (e^x + e^{-x}) \exp\left(-2b\cosh x\right) dx \\
&= \frac{1}{2} \int_{-\infty}^\infty \exp\left(-x - 2b\cosh x\right) dx.
\end{aligned}$$

The result then follows from (7.2.15).

EXAMPLE 7.2.11. Entry **3.391** is

$$(7.2.18)$$
$$\int_0^\infty [(\sqrt{x+2\beta} + \sqrt{x})^{2\nu} - (\sqrt{x+2\beta} - \sqrt{x})^{2\nu}] e^{-\mu x} dx = 2^{\nu+1} \frac{\nu}{\mu} e^{\beta\mu} K_\nu(\beta\mu).$$

Under the change of variables $x \to 2\beta \sinh^2 x$, the left-hand side becomes

$$(2\beta)^{\nu+1} \int_0^\infty \sinh 2x [(\cosh x + \sinh x)^{2\nu}$$
$$- (\cosh x - \sinh x)^{2\nu}] e^{-\beta\mu(\cosh 2x - 1)} dx$$
$$= (2\beta)^{\nu+1} e^{\beta\mu} \int_0^\infty [e^{2\nu x} - e^{-2\nu x}] e^{-2\beta\mu \cosh 2x} \sinh 2x \, dx$$
$$= (2\beta)^{\nu+1} e^{\beta\mu} \int_0^\infty [\cosh(\nu+1)x] - \cosh(\nu-1)x] e^{-\beta\mu \cosh x} dx$$
$$= \frac{1}{2}(2\beta)^{\nu+1} e^{\beta\mu} \int_{-\infty}^\infty \{\exp[(\nu+1)x - \beta\mu \cosh x]$$
$$- \exp[-(\nu-1)x - \beta\mu \cosh x]\} dx$$
$$= (2\beta)^{\nu+1} e^{\beta\mu} [K_{\nu-1}(\beta\mu) - K_{\nu+1}(\beta\mu)],$$

where, in the last step, Example 7.2.9 was used. Finally, by the recursion formula for the modified Bessel functions, this reduces, as claimed, to the right-hand side of (7.2.18).

EXAMPLE 7.2.12. Entry **3.547.2**, given by

$$(7.2.19) \qquad \int_0^\infty \exp(-\beta \cosh x) \sinh(\gamma x) \sinh x \, dx = \frac{\gamma}{\beta} K_\gamma(\beta),$$

follows by rewriting the integral as

$$2e^{-\beta} \int_0^\infty \exp(-\beta(\cosh 2x - 1)) \sinh(2\gamma x) \sinh 2x \, dx$$
$$= e^\beta \int_0^\infty \exp(-2\beta \sinh^2 x) \left(e^{2\gamma x} - e^{-2\gamma x}\right) \sinh 2x \, dx$$
$$= e^\beta \int_0^\infty \exp(-2\beta \sinh^2 x) \left[(\cosh x + \sinh x)^{2\gamma} - (\cosh x - \sinh x)^{2\gamma}\right] d(\sinh^2 x)$$
$$= e^\beta \int_0^\infty e^{-2\beta u} \left[(\sqrt{u^2 + 1} + \sqrt{u})^{2\gamma} - (\sqrt{u^2 + 1} - \sqrt{u})^{2\gamma}\right] du$$

and applying (7.2.18).

EXAMPLE 7.2.13. Entry **3.478.4** is

$$(7.2.20) \qquad \int_0^\infty x^{\nu-1} \exp\left(-\beta x^p - \gamma x^{-p}\right) dx = \frac{2}{p} \left(\frac{\gamma}{\beta}\right)^{\frac{\nu}{2p}} K_{\nu/p}(2\sqrt{\beta\gamma}).$$

To evaluate this entry, let $y = \beta x^p$ to obtain

$$(7.2.21) \qquad \int_0^\infty x^{\nu-1} \exp\left(-\beta x^p - \gamma x^{-p}\right) dx = \frac{1}{p\beta^{\nu/p}} \int_0^\infty y^{\nu/p-1} e^{-y - \beta\gamma/y} dy.$$

The value of this last integral is obtained from (7.2.1).

7.3. A second integral representation of modified Bessel functions

The next integral representation of the modified Bessel function appears as Entry **3.387.3** of [**35**] and it can also be found as (9.43) in [**65**], page 236. In order to make this paper as self-contained as possible, a proof is presented here.

THEOREM 7.3.1. *The modified Bessel function K_ν satisfies*

$$(7.3.1) \qquad \int_1^\infty (x^2 - 1)^{\alpha - 1/2} e^{-\mu x}\, dx = \frac{1}{\sqrt{\pi}} \left(\frac{2}{\mu}\right)^\alpha \Gamma(\alpha + \tfrac{1}{2}) K_\alpha(\mu).$$

PROOF. Let C be the contour starting at ∞, running along, and just above, the positive real axis to go into a counterclockwise circle of radius larger than 1 about the origin and then back to ∞ along, and just below, the positive real axis. Then

$$
\begin{aligned}
(7.3.2)\oint_C e^{-zt}(t^2 - 1)^{\nu - 1/2}dt &= \oint_C e^{-zt}t^{2\nu - 1}(1 - t^{-2})^{\nu - 1/2}dt \\
&= \sum_{k=0}^\infty \frac{\Gamma(\tfrac{1}{2} - \nu + k)}{k!\Gamma(\tfrac{1}{2} - \nu)} \oint_C t^{2\nu - 1 - 2k} e^{-zt}dt.
\end{aligned}
$$

The last integral in (7.3.2) is Hankel's integral representation for the gamma function, so

$$
\begin{aligned}
(7.3.3) \oint_C e^{-zt}(t^2 - 1)^{\nu - 1/2}dt &= \frac{2\pi i}{\Gamma(\tfrac{1}{2} - \nu)} \sum_{k=0}^\infty \frac{\Gamma(\tfrac{1}{2} - \nu + k)z^{2k - 2\nu}}{k!\Gamma(2k - 2\nu + 1)} \\
&= \frac{2^{\nu + 1}\pi i e^{-i\nu\pi}\Gamma(1/2)}{\Gamma(\tfrac{1}{2} - \nu)} \frac{J_{-\nu}(iz)}{(iz)^\nu}.
\end{aligned}
$$

Thus

$$(7.3.4) \qquad I_{-\nu} = \frac{\Gamma(\tfrac{1}{2} - \nu)e^{2\pi\nu i}(z/2)^\nu}{2\pi i\Gamma(1/2)} \oint_C e^{-zt}(t^2 - 1)^{\nu - 1/2}dt.$$

Since C encloses ± 1, branch points of the integrand at which it vanishes, we can collapse C to the real axis from -1 to ∞ (the branch cut runs from -1 to 1). We have, integrating over the two segments above ($t - 1 = (1 - t)e^{i\pi}$) and below ($t - 1 = (1 - t)e^{-i\pi}$) the positive real axis,

$$(7.3.5) \quad I_{-\nu}(z) = \frac{\Gamma(\tfrac{1}{2} - \nu)e^{2\pi\nu i}(z/2)^\nu}{2\pi i\Gamma(1/2)} \times$$

$$\{(1 - e^{-4\pi\nu i})\int_1^\infty e^{-zt}(t^2 - 1)^{\nu - 1/2}dt + i(e^{-\pi\nu i} + e^{-3\pi\nu i})\int_{-1}^1 e^{-zt}(1 - t^2)^{\nu - 1/2}dt\}.$$

Therefore, from (7.1.8) and (7.3.5),

$$(7.3.6) \qquad \frac{I_{-\nu}(z) - I_\nu(z)}{\sin \pi\nu} = \frac{\Gamma(\tfrac{1}{2} - \nu)}{\pi\Gamma(\tfrac{1}{2})} \left(\frac{z}{2}\right)^\nu \int_1^\infty e^{-zt}(t^2 - 1)^{\nu - 1/2}dt.$$

Consequently, by (7.1.7),

$$(7.3.7) \qquad \int_1^\infty e^{-zt}(t^2-1)^{\nu-1/2}dt = \frac{\Gamma(\nu+\frac{1}{2})}{\Gamma(\frac{1}{2})}\left(\frac{2}{z}\right)^\nu K_\nu(z).$$

This completes the proof.   □

Several entries of [35] are now obtained by simple manipulations of (7.3.1).

EXAMPLE 7.3.2. The scaled version

$$(7.3.8) \qquad \int_a^\infty (x^2-a^2)^{\nu-1}e^{-\mu x}\,dx = \frac{1}{\sqrt{\pi}}\left(\frac{2a}{\mu}\right)^{\nu-\frac{1}{2}}\Gamma(\nu)K_{\nu-\frac{1}{2}}(a\mu)$$

appears as entry **3.387.6** in [35]. To establish this formula, let $t = ax$ to obtain

$$(7.3.9) \qquad \int_a^\infty (x^2-a^2)^{\nu-1}e^{-\mu x}\,dx = a^\nu \int_1^\infty (t^2-1)^{\nu-1}e^{-\mu at}\,dt.$$

Now use (7.3.1) with $\alpha = \nu - \frac{1}{2}$ and μa instead of μ.

EXAMPLE 7.3.3. The change of variables $x \to \cosh x$ in (7.3.1) yields entry **3.547.9**

$$(7.3.10) \qquad \int_0^\infty \exp(-\beta\cosh x)\sinh^{2\nu}x\,dx = \frac{1}{\sqrt{\pi}}\left(\frac{2}{\beta}\right)^\nu \Gamma\left(\nu+\frac{1}{2}\right)K_\nu(\beta).$$

EXAMPLE 7.3.4. Entry **3.479.1**, given by

$$(7.3.11) \qquad \int_0^\infty \frac{x^{\mu-1}\exp\left(-\beta\sqrt{1+x}\right)}{\sqrt{1+x}}\,dx = \frac{2}{\sqrt{\pi}}\left(\frac{\beta}{2}\right)^{1/2-\nu}\Gamma(\nu)K_{\frac{1}{2}-\nu}(\beta),$$

comes from (7.3.1) by the change of variables $t = \sqrt{1+x}$ and the symmetry of K_ν with respect to the order ν.

EXAMPLE 7.3.5. Entry **3.462.25** states that

$$(7.3.12) \qquad \int_0^\infty \frac{\exp\left(-px^2\right)}{\sqrt{a^2+x^2}}\,dx = \frac{1}{2}\exp\left(\frac{a^2p}{2}\right)K_0\left(\frac{a^2p}{2}\right).$$

To evaluate this example, let $x = at$ to produce

$$(7.3.13) \qquad \int_0^\infty \frac{\exp\left(-px^2\right)}{\sqrt{a^2+x^2}}\,dx = \int_0^\infty \frac{\exp\left(-bt^2\right)}{\sqrt{t^2+1}}\,dt,$$

with $b = pa^2$. The change of variables $y = t^2 + 1$ then gives

$$(7.3.14) \qquad \int_0^\infty \frac{\exp\left(-bt^2\right)}{\sqrt{t^2+1}}\,dt = \frac{e^b}{2}\int_1^\infty \frac{e^{-by}}{\sqrt{y^2-y}}\,dy.$$

Now complete the square to write $y^2-y = (y-1/2)^2-1/4$ and let $y-1/2 = w/2$ to obtain

$$(7.3.15) \qquad \int_0^\infty \frac{\exp\left(-px^2\right)}{\sqrt{a^2+x^2}}\,dx = \frac{1}{2}e^{b/2}\int_1^\infty (w^2-1)^{-1/2}e^{-bw/2}\,dw.$$

This is evaluated by taking $\alpha = 0$ and $\mu = b/2$ in (7.3.1).

EXAMPLE 7.3.6. After replacing a by $2a$ in the original formulation in [35], entry **3.364.3** is given by

$$(7.3.16) \qquad \int_0^\infty \frac{e^{-px}\,dx}{\sqrt{x(x+2a)}} = e^{ap}K_0(ap).$$

To verify this formula, complete the square and define a new variable of integration by $x + \frac{a}{2} = \frac{1}{2}at$. This yields

$$(7.3.17) \qquad \int_0^\infty \frac{e^{-px}\,dx}{\sqrt{x(x+2a)}} = e^{ap}\int_1^\infty (t^2-1)^{-1/2}e^{-pat}\,dt.$$

The result now follows from Theorem 7.3.1.

EXAMPLE 7.3.7. Entry **3.383.8** of [35] is

$$(7.3.18) \qquad \int_0^\infty x^{\nu-1}(x+2a)^{\nu-1}e^{-\mu x}\,dx = \frac{1}{\sqrt{\pi}}\left(\frac{2a}{\mu}\right)^{\nu-\frac{1}{2}}e^{\mu a}\Gamma(\nu)K_{\frac{1}{2}-\nu}(a\mu),$$

where we have replaced the original parameter β in [35] by $2a$ to simplify the form of the result. To establish this formula, let $t = x + a$ to obtain

$$(7.3.19) \qquad \int_0^\infty x^{\nu-1}(x+2a)^{\nu-1}e^{-\mu x}\,dx = e^{\mu a}\int_a^\infty (t^2-a^2)^{\nu-1}e^{-\mu t}\,dt.$$

The result again follows from Theorem 7.3.1.

EXAMPLE 7.3.8. The special case $a = 1$ and $\nu = n + \frac{1}{2}$ and replacing the parameter μ by p in Example 7.3.7 gives

$$(7.3.20) \qquad \int_0^\infty x^{n-1/2}(x+2)^{n-1/2}e^{-px}\,dx = \frac{1}{\sqrt{\pi}}\left(\frac{2}{p}\right)^n e^p\Gamma(n+\tfrac{1}{2})K_{-n}(p).$$

The result is brought to the form

$$(7.3.21) \qquad \int_0^\infty x^{n-1/2}(2+x)^{n-1/2}e^{-px}\,dx = \frac{(2n-1)!!}{p^n}e^pK_n(p),$$

given in entry **3.372** of [35], by using the fact that K is an even function of its order and employing the identity

$$(7.3.22) \qquad (2n-1)!! = \frac{2^n}{\sqrt{\pi}}\Gamma(n+\tfrac{1}{2}).$$

This reduction of the double-factorials appears as entry **8.339.2**.

EXAMPLE 7.3.9. Entry **3.383.3** is

$$(7.3.23) \qquad \int_a^\infty x^{\mu-1}(x-a)^{\mu-1}e^{-2bx}\,dx = \frac{1}{\sqrt{\pi}}\left(\frac{a}{2b}\right)^{\mu-\frac{1}{2}}\Gamma(\mu)e^{-ab}K_{\mu-\frac{1}{2}}(ab),$$

where we have replaced u by a and β by $2b$ to simplify the answer and avoid confusion between u and μ. To prove this, let $t = x - a$ to convert the requested identity into

$$(7.3.24) \qquad \int_0^\infty t^{\mu-1}(t+a)^{\mu-1}e^{-2bt}\,dt = \frac{1}{\sqrt{\pi}}\left(\frac{a}{2b}\right)^{\mu-\frac{1}{2}}\Gamma(\mu)e^{ab}K_{\mu-\frac{1}{2}}(ab).$$

This comes directly from Example 7.3.7 and the symmetry of $K_\alpha(z)$ with respect to α.

EXAMPLE 7.3.10. Entry **3.388.2** is

$$(7.3.25) \qquad \int_0^\infty (2\beta x + x^2)^{\nu-1} e^{-\mu x}\, dx = \frac{1}{\sqrt{\pi}} \left(\frac{2\beta}{\mu}\right)^{\nu-\frac{1}{2}} e^{\beta\mu}\Gamma(\nu) K_{\nu-\frac{1}{2}}(\beta\mu).$$

This comes directly from Example 7.3.7.

EXAMPLE 7.3.11. Entry **3.471.4** states that
(7.3.26)
$$I = \int_0^a x^{-2\mu}(a-x)^{\mu-1} e^{-\beta/x}\, dx = \frac{1}{\sqrt{\pi a}} \beta^{1/2-\mu} e^{-\beta/2a}\Gamma(\mu) K_{\mu-1/2}\left(\frac{\beta}{2a}\right),$$

where we have replaced u by a to avoid confusion. To evaluate this integral, let $t = a/x - 1$ to produce

$$(7.3.27) \qquad I = \frac{e^{-\beta/a}}{a^\mu} \int_0^\infty t^{\mu-1}(t+1)^{\mu-1} e^{-\beta t/a}\, dt.$$

The formula established in Example 7.3.7 now gives the result.

EXAMPLE 7.3.12. The proof of entry **3.471.8**
(7.3.28)
$$\int_0^a x^{-2\mu}(a^2 - x^2)^{\mu-1} e^{-\beta/x}\, dx = \frac{1}{\sqrt{\pi}} \left(\frac{2}{\beta}\right)^{\mu-1/2} a^{\mu-3/2}\Gamma(\mu) K_{\mu-1/2}\left(\frac{\beta}{a}\right)$$

is obtained employing the same change of variables as in Example 7.3.11.

7.4. A family with typos

Section **3.462** of [**35**] contains five incorrect entries involving the modified Bessel function. There are some typos in both the form of the integrand as well as the value of the integral.

EXAMPLE 7.4.1. The first entry analyzed here is **3.462.24**: it appears incorrectly as

$$(7.4.1) \qquad \int_0^\infty \frac{x^{2n} \exp\left(-a\sqrt{x+b^2}\right)}{\sqrt{x^2+b^2}}\, dx = (2n-1)!! \left(\frac{b}{a}\right)^n K_n(ab),$$

with the correct version being

$$(7.4.2) \qquad \int_0^\infty \frac{x^{2n} \exp\left(-a\sqrt{x^2+b^2}\right)}{\sqrt{x^2+b^2}}\, dx = \frac{2^n}{\sqrt{\pi}}\Gamma\left(n+\tfrac{1}{2}\right) \left(\frac{b}{a}\right)^n K_n(ab).$$

The argument of the exponential appears incorrectly as $-a\sqrt{x+b^2}$. The presentation in [**35**] also employs the relation (7.3.22). This becomes inconvenient for $n = 0$.

To confirm (7.4.1), make the change of variables $t = \sqrt{x^2+b^2}$ to obtain

$$(7.4.3) \qquad \int_0^\infty \frac{x^{2n} \exp\left(-a\sqrt{x^2+b^2}\right)}{\sqrt{x^2+b^2}}\, dx = \int_b^\infty (t^2-b^2)^{n-1/2} e^{-at}\, dt.$$

The result then follows from (7.3.8).

EXAMPLE 7.4.2. Entry **3.462.20** states incorrectly that

$$(7.4.4) \qquad \int_0^\infty \frac{\exp\left(-a\sqrt{x+b^2}\right)}{\sqrt{x^2+b^2}}\, dx = K_0(ab).$$

This should be written as

$$(7.4.5) \qquad \int_0^\infty \frac{\exp\left(-a\sqrt{x^2+b^2}\right)}{\sqrt{x^2+b^2}}\, dx = K_0(ab)$$

and follows from (7.4.2) with $n = 0$.

EXAMPLE 7.4.3. Entries **3.462.21**, **3.462.22**, *and* **3.462.23** are the special cases of (7.4.2) with $n = 1, 2, 3$. Each one of these entries has the term $\sqrt{x+b^2}$ instead of the correct $\sqrt{x^2+b^2}$. Entry **3.462.22** has an additional typo in the answer: it has $K_1(ab)$ instead of $K_2(ab)$.

7.5. The Mellin transform method

The *Mellin transform* of a locally integrable function $f : (0,\infty) \to \mathbb{C}$ is defined by

$$(7.5.1) \qquad M[f; s] = \tilde{f}(z) = \int_0^\infty t^{s-1} f(t)\, dt$$

whenever the integral converges. Suppose the integral (7.5.1) converges in a strip $a < \Re s < b$. A function $f(t)$ may be recovered from its Mellin transform $\tilde{f}(s)$ via the *inversion formula*

$$f(t) = \frac{1}{2\pi i} \int_{c-i\infty}^{c+i\infty} t^{-s} \tilde{f}(s)\, ds$$

where $a < c < b$.

EXAMPLE 7.5.1. The Mellin transform of the exponential function $e^{-\mu x}$ is $\mu^{-s}\Gamma(s)$. By the inversion formula, we have, for $s > 0$,

$$(7.5.2) \qquad e^{-\mu x} = \frac{1}{2\pi i} \int_{c-i\infty}^{c+i\infty} x^{-s}\mu^{-s}\Gamma(s)\, ds.$$

LEMMA 7.5.2. *The Mellin transform of $K_\nu(t)$ evaluates as*

$$(7.5.3) \qquad \int_0^\infty t^{s-1} K_\nu(t)\, dt = 2^{s-2}\Gamma\left(\frac{s}{2} + \frac{\nu}{2}\right)\Gamma\left(\frac{s}{2} - \frac{\nu}{2}\right).$$

PROOF. Example 7.3.12 gives

$$\int_0^\infty t^{s-1} K_\nu(t)\, dt = \frac{\sqrt{\pi}}{2^\nu \Gamma(\nu + 1/2)} \int_0^1 x^{-2\nu-1}(1-x^2)^{\nu-1/2} \int_0^\infty t^{\nu+s-1} e^{-t/x}\, dt\, dx$$

$$= \frac{\sqrt{\pi}\,\Gamma(\nu + s)}{2^\nu \Gamma(\nu + 1/2)} \int_0^1 x^{s-\nu-1}(1-x^2)^{\nu-1/2}\, dx$$

$$= \frac{\sqrt{\pi}\,\Gamma(\nu + s)}{2^{\nu+1} \Gamma(\nu + 1/2)} \int_0^1 u^{(s-\nu)/2-1}(1-u)^{\nu+1/2-1}\, du$$

$$= \frac{\sqrt{\pi}\,\Gamma(\nu + s)\,\Gamma(\frac{s-\nu}{2})\Gamma(\nu + 1/2)}{2^{\nu+1}\Gamma(\nu + 1/2)\Gamma(\frac{s+\nu+1}{2})}$$

$$= 2^{s-2}\Gamma\left(\frac{s+\nu}{2}\right)\Gamma\left(\frac{s-\nu}{2}\right).$$

□

An alternative proof is offered next.

PROOF. Since $K_\nu = K_{-\nu}$, we may assume that $\nu \geq 0$. By the Mellin inversion formula, the evaluation (7.5.3) is equivalent to

$$(7.5.4) \qquad K_\nu(ax) = \frac{1}{2\pi i} \int_{c-i\infty}^{c+i\infty} 2^{s-2} a^{-s}\Gamma\left(\frac{s}{2} + \frac{\nu}{2}\right)\Gamma\left(\frac{s}{2} - \frac{\nu}{2}\right) x^{-s}\, dx$$

where $c > \nu$. The integrand has poles at $s = \pm\nu - 2n$ for $n = 0, 1, 2, \ldots$. Assuming that $\nu \notin \mathbb{Z}$, all poles are of first order and the residue at $s = \pm\nu - 2n$ is $2(-1)^n/n!$. Closing the contour of (7.5.4) to the left and collecting the residues yields

$$\frac{1}{2}\sum_{n=0}^\infty \frac{(-1)^n}{n!}\left[\Gamma(\nu - n)\left(\frac{ax}{2}\right)^{-\nu+2n} + \Gamma(-\nu - n)\left(\frac{ax}{2}\right)^{\nu+2n}\right].$$

Using Euler's reflection formula in the form

$$\Gamma(\mu - n) = \frac{(-1)^n}{\Gamma(1 - \mu + n)}\frac{\pi}{\sin(\pi\mu)},$$

this becomes

$$\frac{\pi}{2\sin(\pi\nu)}\sum_{n=0}^\infty \frac{1}{n!}\left[\frac{1}{\Gamma(1 - \nu + n)}\left(\frac{ax}{2}\right)^{-\nu+2n} - \frac{1}{\Gamma(1 + \nu + n)}\left(\frac{ax}{2}\right)^{\nu+2n}\right].$$

The definitions (7.1.6) and (7.1.7) show that this last term is $K_\nu(ax)$, as claimed.

□

EXAMPLE 7.5.3. Entry **3.389.4** of [**35**] is

$$(7.5.5) \qquad \int_a^\infty x(x^2 - a^2)^{\nu-1} e^{-\mu x}\, dx = \frac{2^{\nu-1/2}}{\sqrt{\pi}}\mu^{1/2-\nu} a^{\nu+1/2}\Gamma(\nu) K_{\nu+1/2}(a\mu),$$

where we have replaced the original parameter u in [**35**] by a in order to avoid confusion with the parameter μ. This identity is now verified.

Use the formula (7.5.2) to replace the term $e^{-\mu x}$ and reverse the order of integration to obtain

$$\int_a^\infty x(x^2-a^2)^{\nu-1}e^{-\mu x}\,dx = \frac{1}{2\pi i}\int_{c-i\infty}^{c+i\infty}\mu^{-s}\Gamma(s)\left(\int_a^\infty x^{1-s}(x^2-a^2)^{\nu-1}\,dx\right)ds.$$

LEMMA 7.5.4. *The inner integral is given by*

(7.5.6)
$$\int_a^\infty x^{1-s}(x^2-a^2)^{\nu-1}\,dx = \frac{1}{\sqrt{\pi}\Gamma(s)}a^{2\nu-s}\Gamma\left(\frac{s}{2}-\nu\right)\Gamma(\nu)2^{s-2}\Gamma\left(\frac{s}{2}+\frac{1}{2}\right).$$

PROOF. Let $x=at$ and $t=y^{-1/2}$ to produce

$$\int_a^\infty x^{1-s}(x^2-a^2)^{\nu-1}\,dx = a^{2\nu-s}\int_1^\infty t^{1-s}(t^2-1)^{\nu-1}\,dt$$

$$= \frac{1}{2}a^{2\nu-s}\int_0^1 y^{s/2-\nu-1}(1-y)^{\nu-1}\,dy$$

$$= \frac{1}{2}a^{2\nu-s}B\left(s/2-\nu,\nu\right)$$

$$= \frac{a^{2\nu-s}\Gamma(s/2-\nu)\Gamma(\nu)}{2\Gamma(s/2)}.$$

Now employ the duplication formula for the gamma function

(7.5.7)
$$\Gamma(2s) = \frac{2^{2s-1}}{\sqrt{\pi}}\Gamma(s)\Gamma(s+\tfrac{1}{2})$$

to obtain the result.   □

This produces

$$\int_a^\infty x(x^2-a^2)^{\nu-1}e^{-\mu x}\,dx = \frac{a^{2\nu}\Gamma(\nu)}{8\pi^{3/2}i}\int_{c-i\infty}^{c+i\infty}\left(\frac{a\mu}{2}\right)^{-s}\Gamma\left(\frac{s}{2}+\frac{1}{2}\right)\Gamma\left(\frac{s}{2}-\nu\right)ds.$$

The parameter ν is assumed to be real. Now shift the contour of integration by $z=s-\nu+\frac{1}{2}$ to obtain, with $c'=c-\nu+\frac{1}{2}$,

$$\int_a^\infty x(x^2-a^2)^{\nu-1}e^{-\mu x}\,dx =$$

$$\frac{\Gamma(\nu)}{\sqrt{\pi}}\left(\frac{2}{\mu}\right)^{\nu-1/2}a^{\nu+1/2}\int_{c'-i\infty}^{c'+i\infty}\left(\frac{a\mu}{2}\right)^{-z}\frac{1}{4}\Gamma\left(\frac{z}{2}+\frac{\nu+1/2}{2}\right)\Gamma\left(\frac{z}{2}-\frac{\nu+1/2}{2}\right)dz.$$

The result now follows from Lemma 7.5.2.

EXAMPLE 7.5.5. The special case $\nu=\frac{1}{2}$ of Example 7.5.3 is

(7.5.8)
$$\int_a^\infty \frac{xe^{-\mu x}\,dx}{\sqrt{x^2-a^2}} = aK_1(a\mu).$$

This appears as entry **3.365.2** of [**35**].

EXAMPLE 7.5.6. Entry **3.366.2** is

$$(7.5.9) \qquad \int_0^\infty \frac{(x + \beta)\, e^{-\mu x}\, dx}{\sqrt{x^2 + 2\beta x}} = \beta e^{\beta \mu} K_1(\beta \mu).$$

To evaluate this result, let $t = x + \beta$ and use Example 7.5.5.

7.6. A family of integrals and a recurrence

Section **3.461** of [35] contains four entries that are part of the family

$$(7.6.1) \qquad f_n(a, b) := \int_0^\infty x^{2n} \exp\left(-a\sqrt{x^2 + b^2}\right)\, dx.$$

The evaluation of this family is discussed in this section.

The change of variables $t = \sqrt{x^2 + b^2}$ produces

$$(7.6.2) \qquad f_n(a, b) = \int_b^\infty t(t^2 - b^2)^{n - \frac{1}{2}} e^{-at}\, dt.$$

This integral was evaluated in Example 7.5.3 as

$$(7.6.3) \qquad f_n(a, b) = \frac{b}{\sqrt{\pi}} \Gamma(n + \tfrac{1}{2}) \left(\frac{2b}{a}\right)^n K_{n+1}(ab).$$

The example $n = 0$ appears as entry **3.461.6** in the form

$$(7.6.4) \qquad \int_0^\infty \exp\left(-a\sqrt{x^2 + b^2}\right)\, dx = bK_1(ab).$$

The remaining examples of the stated family are simplified using the recurrence

$$(7.6.5) \qquad K_\nu(z) = \frac{2(\nu - 1)}{z} K_{\nu-1}(z) + K_{\nu-2}(z).$$

EXAMPLE 7.6.1. Entry **3.461.7** states that

$$(7.6.6) \quad f_1(a, b) = \int_0^\infty x^2 \exp\left(-a\sqrt{x^2 + b^2}\right)\, dx = \frac{2b}{a^2} K_1(ab) + \frac{b^2}{a} K_0(ab).$$

The form given in (7.6.3) is

$$(7.6.7) \qquad f_1(a, b) = \frac{b^2}{a} K_2(ab).$$

The recurrence (7.6.5) gives

$$(7.6.8) \qquad K_2(ab) = \frac{2}{ab} K_1(ab) + K_0(ab),$$

which produces the result.

EXAMPLE 7.6.2. The same procedure used in Example 7.6.1 gives the evaluation of entry **3.461.8** as

$$(7.6.9) \quad f_2(a, b) = \int_0^\infty x^4 \exp\left(-a\sqrt{x^2 + b^2}\right)\, dx = \frac{12b^2}{a^3} K_2(ab) + \frac{3b^3}{a^2} K_1(ab)$$

and entry **3.461.9** as
(7.6.10)
$$f_3(a, b) = \int_0^\infty x^6 \exp\left(-a\sqrt{x^2 + b^2}\right) dx = \frac{90b^3}{a^4} K_3(ab) + \frac{15b^4}{a^3} K_2(ab).$$

REMARK 7.6.3. *The recurrence* (7.6.5) *converts the evaluation of* $f_n(a, b)$ *into an expression depending only upon* $K_0(ab)$ *and* $K_1(ab)$. *For instance,*

(7.6.11) $$f_2(a, b) = \frac{12b^2}{a^3} K_0(ab) + \left(\frac{24b}{a^4} + \frac{3b^3}{a^2}\right) K_1(ab)$$

and

(7.6.12) $$f_3(a, b) = \left(\frac{360b^2}{a^5} + \frac{15b^4}{a^3}\right) K_0(ab) + \left(\frac{720b}{a^6} + \frac{120b^3}{a^4}\right) K_1(ab).$$

Experimentally we discovered that introducing the scaling

(7.6.13) $$g_n(c) = \frac{a^{2n} 2^n n!}{b\,(2n)!} f_n(a, b)$$

and label $c = ab$ *and* $x = K_0(c)$, $y = K_1(c)$, *the expressions for the integrals simplify. The first few polynomials are*

$$
\begin{aligned}
g_3(c) &= c(c^2 + 24)x + 8(c^2 + 6)y \\
g_4(c) &= 12c(c^2 + 16)x + (c^4 + 72c^2 + 384)y \\
g_5(c) &= c(c^4 + 144c^2 + 1920)x + 6(3c4 + 128c^2 + 640)y \\
g_6(c) &= 24c(c^4 + 80c^2 + 960)x + (c^6 + 288c^4 + 9600c^2 + 46080)y.
\end{aligned}
$$

Properties of the polynomials appearing in the coefficients will be reported elsewhere. For example, the function $g_n(c)$ *satisfies the differential equation*

(7.6.14) $$b^2 g_n''(b) - (2n - 1)b g_n'(b) - \left((ab)^2 + 2n + 1\right) g_n(b) = 0,$$

and also the recurrence

(7.6.15) $$g_{n+2}(b) - 2(n + 2)g_{n+1}(b) - (ab)^2 g_n(b) = 0.$$

7.7. A hyperexponential example

This section discusses several evaluations of entry **3.323.3**

(7.7.1) $$\int_0^\infty \exp\left(-\beta^2 x^4 - 2\gamma^2 x^2\right) dx = 2^{-3/2} \frac{\gamma}{\beta} e^{\gamma^4/2\beta^2} K_{1/4}\left(\frac{\gamma^4}{2\beta^2}\right).$$

This example also appears as entry **3.469.1** in the form

(7.7.2) $$\int_0^\infty \exp\left(-\mu x^4 - 2\nu x^2\right) dx = \frac{1}{4}\sqrt{\frac{2\nu}{\mu}} \exp\left(\frac{\nu^2}{2\mu}\right) K_{1/4}\left(\frac{\nu^2}{2\mu}\right).$$

The change of variables $x = \gamma t/\beta$ converts (7.7.1) into the form

(7.7.3) $$\int_0^\infty e^{-2b(t^2+1)^2} dt = 2^{-3/2} e^{-b} K_{1/4}(b),$$

with $b = \gamma^4/2\beta^2$. A similar change of variables converts (7.7.2) to (7.7.3).

7.7.1. A traditional proof. Recall that K_ν is defined in terms of I_ν. The definition of I_ν as the series (7.1.7) is equivalent to the hypergeometric representation

$$(7.7.4) \qquad \Gamma(\nu+1)I_\nu(x) = \left(\frac{x}{2}\right)^\nu {}_0F_1\left({{-}\atop{\nu+1}}\bigg|\frac{x^2}{4}\right).$$

Applying Kummer's second transformation (see for instance [13, Section 4.1]), to (7.7.4), one obtains

$$(7.7.5) \qquad \Gamma(\nu+1)I_\nu(x) = \left(\frac{x}{2}\right)^\nu e^{-x} {}_1F_1\left({{\nu+\frac{1}{2}}\atop{2\nu+1}}\bigg|2x\right).$$

Consider the integral in (7.7.3). The change of variables $x = t^2$ followed by a series expansion and the further change of variables $s = x^2$ gives

$$\int_0^\infty e^{-2b(t^2+1)^2}\,dt = \frac{1}{2}e^{-2b}\int_0^\infty x^{-1/2}e^{-2bx^2-4bx}\,dx$$

$$= \frac{1}{2}e^{-2b}\sum_{k=0}^\infty \frac{(-4b)^k}{k!}\int_0^\infty x^{k-1/2}e^{-2bx^2}\,dx$$

$$= \frac{1}{4}e^{-2b}\sum_{k=0}^\infty \frac{(-4b)^k}{k!}\int_0^\infty s^{k/2-3/4}e^{-2bs}\,ds$$

$$= \frac{1}{4}e^{-2b}\sum_{k=0}^\infty \frac{(-4b)^k}{k!}\frac{\Gamma(1/4+k/2)}{(2b)^{k/2+1/4}}$$

$$= \frac{e^{-2b}}{4(2b)^{1/4}}\sum_{k=0}^\infty \frac{(-2\sqrt{2b})^k}{k!}\Gamma\left(\frac{k}{2}+\frac{1}{4}\right).$$

Writing the terms according to the parity of the index k produces

$$\int_0^\infty e^{-2b(t^2+1)^2}\,dt$$

$$= \frac{e^{-2b}}{4(2b)^{1/4}}\left[\sum_{k=0}^\infty \frac{(8b)^k}{(2k)!}\Gamma\left(k+\frac{1}{4}\right) - 2\sqrt{2b}\sum_{k=0}^\infty \frac{(8b)^k}{(2k+1)!}\Gamma\left(k+\frac{3}{4}\right)\right].$$

Now use the definition of the Pochhammer symbol

$$(7.7.6) \qquad (a)_k = \frac{\Gamma(a+k)}{\Gamma(a)}$$

to write

$$(7.7.7) \qquad \Gamma\left(k+\tfrac{1}{4}\right) = \left(\tfrac{1}{4}\right)_k \Gamma\left(\tfrac{1}{4}\right), \quad \Gamma\left(k+\tfrac{3}{4}\right) = \left(\tfrac{3}{4}\right)_k \Gamma\left(\tfrac{3}{4}\right),$$

and

$$(7.7.8) \qquad (2k)! = 2^{2k}\left(\tfrac{1}{2}\right)_k (1)_k, \quad (2k+1)! = 2^{2k}\left(\tfrac{3}{2}\right)_k (1)_k$$

to produce

$$\int_0^\infty e^{-2b(t^2+1)^2}\, dt$$

$$= \frac{e^{-2b}}{4(2b)^{1/4}} \left[\Gamma\left(\frac{1}{4}\right) \sum_{k=0}^\infty \frac{(2b)^k}{k!} \frac{(1/4)_k}{(1/2)_k} - 2\sqrt{2b}\, \Gamma\left(\frac{3}{4}\right) \sum_{k=0}^\infty \frac{(2b)^k}{k!} \frac{(3/4)_k}{(3/2)_k} \right]$$

$$= \frac{e^{-2b}}{4(2b)^{1/4}} \left\{ \Gamma\left(\frac{1}{4}\right) {}_1F_1\left(\begin{matrix}1/4\\1/2\end{matrix}\Big|2b\right) - 2\sqrt{2b}\, \Gamma\left(\frac{3}{4}\right) {}_1F_1\left(\begin{matrix}3/4\\3/2\end{matrix}\Big|2b\right) \right\}.$$

Applying the representation (7.7.5) of I_ν gives

$$(7.7.9) \qquad \int_0^\infty e^{-2b(t^2+1)^2}\, dt = \frac{\pi}{4} e^{-b} \left(I_{-1/4}(b) - I_{1/4}(b) \right).$$

This completes the traditional proof.

7.7.2. An automatic proof. This second proof of (7.7.1) is computer generated. The reader will find in [44] a selection of examples from [35] where similar computer-generated proofs are described.

The condition $\mathrm{Re}\,\beta^2 > 0$, stated below, ensures convergence of the integral. Observe that the left-hand side of (7.7.10) is analytic in both γ and β, while the right-hand side needs to be interpreted such that it shares this analyticity. In order to not worry about taking the right branch-cuts on the right-hand side, we restrict to $\gamma \geq 0$ and $\beta > 0$. These conditions can then be removed at the end of the argument by analytic continuation.

THEOREM 7.7.1. *For complex γ, β such that $\mathrm{Re}\left(\beta^2\right) > 0$, we have*
(7.7.10)
$$F(\gamma) := \int_0^\infty \exp\left(-\beta^2 x^4 - 2\gamma^2 x^2\right) dx = 2^{-3/2} \frac{\gamma}{\beta} \exp\left(\frac{\gamma^4}{2\beta^2}\right) K_{1/4}\left(\frac{\gamma^4}{2\beta^2}\right).$$

PROOF. Since the integrand is hyperexponential, we can apply the Almkvist-Zeilberger algorithm [2], which is a differential analog to Zeilberger's celebrated summation algorithm for hypergeometric summands. These algorithms sometimes are also subsumed under the name WZ theory. In the following we denote the integrand by $f(x,\gamma) := \exp\left(-\beta^2 x^4 - 2\gamma^2 x^2\right)$. Using creative telescoping, one finds that

$$(7.7.11) \qquad \left(A + D_x \cdot 4\gamma^3 x\right) \cdot f(x,\gamma) = 0$$

where $A := \beta^2 \gamma D_\gamma^2 - \left(4\gamma^4 + \beta^2\right) D_\gamma - 4\gamma^3$ and $D_x = \frac{d}{dx}$, $D_\gamma = \frac{d}{d\gamma}$. Hence it follows that

$$A \cdot \int_0^T f(x,\gamma)\, dx = \int_0^T A \cdot f(x,\gamma)\, dx$$

$$= -\int_0^T D_x \cdot 4\gamma^3 x \cdot f(x,\gamma)\, dx$$

$$= -4\gamma^3 T \cdot f(T,\gamma).$$

In the limit $T \to \infty$, we therefore have

$$A \cdot \int_0^\infty f(x, \gamma) \, dx = 0.$$

Let $G(\gamma)$ be the right-hand side of (7.7.10). In light of the differential equation (7.1.5) satisfied by the modified Bessel function $K_{1/4}$, a direct calculation shows that $G(\gamma)$ is also annihilated by A, that is,

$$A \cdot G(\gamma) = A \cdot 2^{-3/2} \frac{\gamma}{\beta} \exp\left(\frac{\gamma^4}{2\beta^2}\right) K_{1/4}\left(\frac{\gamma^2}{2\beta^2}\right) = 0.$$

Thus the claim follows by checking that $F(0) = G(0)$ and $F'(0) = G'(0)$. The explicit evaluations

$$F(0) = \int_0^\infty \exp\left(-\beta^2 x^4\right) \, dx = \frac{\Gamma(1/4)}{4\sqrt{\beta}}$$

$$F'(0) = \left[-4\gamma \int_0^\infty x^2 \exp\left(-\beta^2 x^4 - 2\gamma^2 x^2\right) \, dx\right]_{\gamma=0} = 0$$

confirm that these values agree with $G(0)$ and $G'(0)$. $\qquad\square$

REMARK 7.7.2. *It remains to explain how the relation (7.7.11) can be found using the Mathematica package* HolonomicFunctions [**43**]. *After loading the package, one just has to type*

In[69]:= **CreativeTelescoping[Exp[−b^2 * x^4 − 2 * c^2 * x^2], Der[x], Der[c]]**

Out[69]= $\{\{b^2 c D_c^2 + (-b^2 - 4c^4) D_c - 4c^3\}, \{4c^3 x\}\}$

REMARK 7.7.3. *Instead of to (7.7.1), the creative telescoping approach can also be applied to (7.7.3). However, in that case, the task of comparing initial values is not so simple, as the integral (7.7.3) does not converge for $b = 0$. As a solution, one could compute the initial values at $b = 1$ but the resulting integrals are not trivial themselves.*

7.7.3. An evaluation by the method of brackets. This method was developed by I. Gonzalez and I. Schmidt in [**33**] in the context of definite integrals coming from Feynman diagrams. The complete operational rules are described in [**31, 32**]. Even though this is a formal method for integration, some of the rules have been made rigorous in [**5**]. A code has been produced in [**40**].

The basic idea is to associate a *bracket* to the divergent integral

$$(7.7.12) \qquad \langle a \rangle = \int_0^\infty x^{a-1} \, dx.$$

This extends to the integral of a function expanded in power series: let f be a formal power series

$$(7.7.13) \qquad f(x) = \sum_{n=0}^\infty a_n x^{\alpha n + \beta - 1}.$$

The symbol

$$(7.7.14) \qquad \int_0^\infty f(x)\, dx \doteq \sum_n a_n \langle \alpha n + \beta \rangle$$

represents a *bracket series* assignement to the integral on the left. Rule 2 describes how to evaluate this series.

The symbol

$$(7.7.15) \qquad \phi_n = \frac{(-1)^n}{\Gamma(n+1)}$$

will be called the *indicator of n* it gives a simpler form for the bracket series associated to an integral. For example,

$$(7.7.16) \qquad \int_0^\infty x^{a-1} e^{-x}\, dx \doteq \sum_n \phi_n \langle n + a \rangle.$$

The integral is the gamma function $\Gamma(a)$ and the right-hand side its bracket expansion.

RULE 1. For $\alpha \in \mathbb{C}$, the expression

$$(7.7.17) \qquad (a_1 + a_2 + \cdots + a_r)^\alpha$$

is assigned the bracket series

$$(7.7.18) \qquad \sum_{m_1,\cdots,m_r} \phi_{1,2,\cdots,r}\, a_1^{m_1} \cdots a_r^{m_r} \frac{\langle -\alpha + m_1 + \cdots + m_r \rangle}{\Gamma(-\alpha)},$$

where $\phi_{1,2,\cdots,r}$ is a shorthand notation for the product $\phi_{m_1} \phi_{m_2} \cdots \phi_{m_r}$.

RULE 2. The series of brackets

$$(7.7.19) \qquad \sum_n \phi_n f(n) \langle an + b \rangle$$

is given the *value*

$$(7.7.20) \qquad \frac{1}{a} f(n^*) \Gamma(-n^*)$$

where n^* solves the equation $an + b = 0$.

RULE 3. A two-dimensional series of brackets

$$(7.7.21) \qquad \sum_{n_1,n_2} \phi_{n_1,n_2} f(n_1, n_2) \langle a_{11}n_1 + a_{12}n_2 + c_1 \rangle \langle a_{21}n_1 + a_{22}n_2 + c_2 \rangle$$

is assigned the *value*

$$(7.7.22) \qquad \frac{1}{|a_{11}a_{22} - a_{12}a_{21}|} f(n_1^*, n_2^*) \Gamma(-n_1^*) \Gamma(-n_2^*)$$

where (n_1^*, n_2^*) is the unique solution to the linear system

$$(7.7.23) \qquad \begin{array}{rcl} a_{11}n_1 + a_{12}n_2 + c_1 &=& 0, \\ a_{21}n_1 + a_{22}n_2 + c_2 &=& 0, \end{array}$$

obtained by the vanishing of the expressions in the brackets. A similar rule applies to higher dimensional series, that is,

$$\sum_{n_1} \cdots \sum_{n_r} \phi_{1,\cdots,r} f(n_1, \cdots, n_r) \langle a_{11} n_1 + \cdots a_{1r} n_r + c_1 \rangle \cdots \langle a_{r1} n_1 + \cdots a_{rr} n_r + c_r \rangle$$

is assigned the value

(7.7.24) $$\frac{1}{|\det(A)|} f(n_1^*, \cdots, n_r^*) \Gamma(-n_1^*) \cdots \Gamma(-n_r^*),$$

where A is the matrix of coefficients (a_{ij}) and $\{n_i^*\}$ is the solution of the linear system obtained by the vanishing of the brackets. The value is not defined if the matrix A is not invertible.

RULE 4. In the case where the assignment leaves free parameters, any divergent series in these parameters is discarded. In case several choices of free parameters are available, the series that converge in a common region are added to contribute to the integral.

The method of brackets is now employed to verify (7.7.1) in its original form

$$\int_0^\infty \exp\left(-\beta^2 x^4 - 2\gamma^2 x^2\right) \, dx = 2^{-\frac{3}{2}} \frac{\gamma}{\beta} e^{\frac{\gamma^4}{2\beta^2}} K_{1/4}\left(\frac{\gamma^4}{2\beta^2}\right).$$

Start with the bracket series

$$\int_0^\infty e^{-(\beta^2 x^4 + 2\gamma^2 x^2)} \, dx = \int_0^\infty \sum_{n_1} \phi_{n_1} (\beta^2 x^4 + 2\gamma^2 x^2)^{n_1} \, dx$$

$$= \int_0^\infty \sum_{n_1} \phi_{n_1} x^{2n_1} (\beta^2 x^2 + 2\gamma^2)^{n_1} \, dx$$

and expand the term $(\beta^2 x^2 + 2\gamma^2)^{n_1}$ in a double bracket series to obtain

$$\int_0^\infty e^{-(\beta^2 x^4 + 2\gamma^2 x^2)} \, dx$$

$$= \int_0^\infty \sum_{n_1} \phi_{n_1} x^{2n_1} \left(\sum_{n_2} \sum_{n_3} \phi_{n_2} \phi_{n_3} (\beta^2 x^2)^{n_2} (2\gamma^2)^{n_3} \frac{\langle -n_1 + n_2 + n_3 \rangle}{\Gamma(-n_1)} \right) dx$$

$$= \sum_{n_1} \sum_{n_2} \sum_{n_3} \phi_{n_1} \phi_{n_2} \phi_{n_3} \frac{2^{n_3} \beta^{2n_2} \gamma^{2n_3}}{\Gamma(-n_1)} \langle 2n_1 + 2n_2 + 1 \rangle \langle -n_1 + n_2 + n_3 \rangle.$$

The result is a three-dimensional sum with two brackets. The rules state that the integral is now expressed as a single sum in the free parameter coming from solving the system

$$2n_1 + 2n_2 + 1 = 0$$
$$-n_1 + n_2 + n_3 = 0.$$

The system is of rank 2, so there are three cases to consider according to the choice of the free parameter.

Case 1: n_1 free: the resulting system is

$$2n_2 = -2n_1 - 1$$
$$n_2 + n_3 = n_1,$$

and the corresponding matrix has $\det(A) = -2$. The solutions are $n_3^* = 2n_1 + \frac{1}{2}$ and $n_2^* = -n_1 - \frac{1}{2}$. The resulting sum is

$$\sum_{n_1} \frac{(-1)^{n_1} 2^{2n_1 - 1/2} \beta^{-2n_1 - 1} \gamma^{4n_1 + 1} \Gamma(-2n_1 - 1/2) \Gamma(n_1 + 1/2)}{\Gamma(n_1 + 1) \Gamma(-n_1)}$$

and it vanishes due to the presence of $\Gamma(-n_1)$ in the denominator.

Case 2: n_2 free: in this case the matrix of coefficients satisfies $\det(A) = 2$ and the solutions are $n_1^* = -n_2 - \frac{1}{2}$ and $n_3^* = -2n_2 - \frac{1}{2}$. The resulting sum

$$\sum_{n_2} \frac{(-1)^{n_2} 2^{-2n_2 - 3/2} \beta^{2n_2} \gamma^{-4n_2 - 1} \Gamma(2n_2 + \frac{1}{2})}{\Gamma(n_2 + 1)}$$

is divergent, so it is discarded.

Case 3: n_3 free: then $\det(A) = 4$ and $n_1^* = \frac{1}{2}n_3 - \frac{1}{4}$ and $n_2^* = -\frac{1}{2}n_3 - \frac{1}{4}$. The corresponding series is

$$\sum_{n_3} \frac{(-1)^{n_3} 2^{n_3 - 2} \beta^{-n_3 - 1/2} \gamma^{2n_3} \Gamma(n_3/2 + 1/4)}{\Gamma(n_3 + 1)} = \frac{1}{4\sqrt{\beta}} \sum_{n_3} (-1)^{n_3} \delta^{n_3} \frac{\Gamma(\frac{1}{2}n_3 + \frac{1}{4})}{\Gamma(n_3 + 1)},$$

with $\delta = 2\gamma^2/\beta$. In order to simplify the result, split the sum according to the parity of n_3 to produce

$$S := \frac{1}{4\sqrt{\beta}} \sum_{n=0}^{\infty} \delta^{2n} \frac{\Gamma(n + \frac{1}{4})}{\Gamma(2n + 1)} - \frac{1}{4\sqrt{\beta}} \sum_{n=0}^{\infty} \delta^{2n+1} \frac{\Gamma(n + \frac{3}{4})}{\Gamma(2n + 2)}.$$

Now use (7.7.7) and (7.7.8) to produce

$$S = \frac{1}{4\sqrt{\beta}} \left\{ \Gamma\left(\frac{1}{4}\right) {}_1F_1\left(\frac{1/4}{1/2}\Big|\frac{\delta^2}{4}\right) - \delta\Gamma\left(\frac{3}{4}\right) {}_1F_1\left(\frac{3/4}{3/2}\Big|\frac{\delta^2}{4}\right) \right\}.$$

The claim is thus seen to be equivalent to the identity

$$\Gamma\left(\frac{1}{4}\right) {}_1F_1\left(\frac{1/4}{1/2}\Big| b\right) - 2\sqrt{b}\,\Gamma\left(\frac{3}{4}\right) {}_1F_1\left(\frac{3/4}{3/2}\Big| b\right) = \sqrt{2}b^{1/4}e^{b/2}K_{1/4}\left(\frac{b}{2}\right),$$

where $b = \delta^2/4$. The identity to be established is now expressed in terms of the Bessel function I_ν using (7.1.7). The result is

$$\Gamma\left(\frac{1}{4}\right) {}_1F_1\left(\frac{1/4}{1/2}\Big| b\right) - 2\sqrt{b}\,\Gamma\left(\frac{3}{4}\right) {}_1F_1\left(\frac{3/4}{3/2}\Big| b\right)$$
$$= \pi b^{1/4} e^{b/2} \left(I_{-1/4}\left(\frac{b}{2}\right) - I_{1/4}\left(\frac{b}{2}\right) \right).$$

Using the expansion (7.1.6) shows that the right-hand side of the previous expression is $\pi e^{b/2}$ times the series

$$\sum_{k=0}^{\infty} \frac{1}{\Gamma(k+3/4)k!} \frac{b^{2k}}{2^{4k-1/2}} - \sum_{k=0}^{\infty} \frac{1}{\Gamma(k+5/4)k!} \frac{b^{2k+1/2}}{2^{4k+1/2}}.$$

Each of these series can be simplified. Introduce $c = b^2/16$ and write

$$\sum_{k=0}^{\infty} \frac{1}{\Gamma(k+3/4)k!} \frac{b^{2k}}{2^{4k-1/2}} = \frac{\sqrt{2}}{\Gamma(3/4)} \sum_{k=0}^{\infty} \frac{1}{(3/4)_k} \frac{c^k}{k!} = \frac{\sqrt{2}}{\Gamma(3/4)} {}_0F_1 \left(\begin{array}{c} - \\ 3/4 \end{array} \bigg| c \right)$$

and

$$\sum_{k=0}^{\infty} \frac{1}{\Gamma(k+5/4)k!} \frac{b^{2k+1/2}}{2^{4k+1/2}} = \frac{\sqrt{b}}{\sqrt{2}\Gamma(5/4)} \sum_{k=0}^{\infty} \frac{1}{(5/4)_k} \frac{c^k}{k!}$$

$$= \frac{\sqrt{b}}{\sqrt{2}\Gamma(5/4)} {}_0F_1 \left(\begin{array}{c} - \\ 5/4 \end{array} \bigg| c \right).$$

The proof of the main identity (7.7.1) by the method of brackets is now reduced to verifying

$$(7.7.25) \quad \Gamma\left(\frac{1}{4}\right) {}_1F_1 \left(\begin{array}{c} 1/4 \\ 1/2 \end{array} \bigg| b \right) - 2\sqrt{b}\,\Gamma\left(\frac{3}{4}\right) {}_1F_1 \left(\begin{array}{c} 3/4 \\ 3/2 \end{array} \bigg| b \right) =$$

$$\pi e^{b/2} \left\{ \frac{\sqrt{2}}{\Gamma(3/4)} {}_0F_1 \left(\begin{array}{c} - \\ 3/4 \end{array} \bigg| c \right) - {}_0F_1 \left(\begin{array}{c} - \\ 5/4 \end{array} \bigg| c \right) \right\}.$$

The exponents appearing in the series above are either integers or $\frac{1}{2}$ plus an integer. Matching these two types separately shows that the main evaluation follows from the identities

$${}_1F_1 \left(\begin{array}{c} 1/4 \\ 1/2 \end{array} \bigg| b \right) = e^{b/2} {}_0F_1 \left(\begin{array}{c} - \\ 3/4 \end{array} \bigg| \frac{b^2}{16} \right) \text{ and } {}_1F_1 \left(\begin{array}{c} 3/4 \\ 3/2 \end{array} \bigg| b \right) = e^{b/2} {}_0F_1 \left(\begin{array}{c} - \\ 5/4 \end{array} \bigg| \frac{b^2}{16} \right).$$

These are special cases of Kummer's second transformation, which is exhibited in the equivalence of (7.7.4) and (7.7.5). This completes the proof of Example 7.7.1.

CHAPTER 8

Combination of logarithms and rational functions

8.1. Introduction

The table of integrals [35] contains many entries of the form

$$(8.1.1) \qquad \int_a^b R_1(x) \ln R_2(x)\, dx$$

where R_1 and R_2 are rational functions. Some of these examples have appeared in previous papers: entry **4.291.1**

$$(8.1.2) \qquad \int_0^1 \frac{\ln(1+x)}{x}\, dx = \frac{\pi^2}{12}$$

as well as entry **4.291.2**

$$(8.1.3) \qquad \int_0^1 \frac{\ln(1-x)}{x}\, dx = -\frac{\pi^2}{6}$$

have been established in [10], entry **4.212.7**

$$(8.1.4) \qquad \int_1^e \frac{\ln x\, dx}{(1+\ln x)^2} = \frac{e}{2} - 1$$

appears in [8] and entry **4.231.11**

$$(8.1.5) \qquad \int_0^a \frac{\ln x\, dx}{x^2 + a^2} = \frac{\pi \ln a}{4a} - \frac{G}{a},$$

where

$$(8.1.6) \qquad G = \sum_{k=0}^{\infty} \frac{(-1)^k}{(2k+1)^2}$$

is the Catalan constant, has appeared in [20]. The value of entry **4.233.1**

$$(8.1.7) \qquad \int_0^1 \frac{\ln x\, dx}{x^2 + x + 1} = \frac{2}{9}\left[\frac{2\pi^2}{3} - \psi'\left(\frac{1}{3}\right)\right],$$

where $\psi(x) = \Gamma'(x)/\Gamma(x)$ is the digamma function, was established in [54].

A standard trick employed in the evaluations of integrals over $[0, \infty)$ is to transform the interval $[1, \infty)$ back to $[0,1]$ via $t = 1/x$. This gives

$$(8.1.8) \qquad \int_0^{\infty} R(x) \ln x\, dx = \int_0^1 \left[R(x) - \frac{1}{x^2} R\left(\frac{1}{x}\right)\right] dx.$$

In particular, if the rational function satisfies

$$(8.1.9) \qquad R\left(\frac{1}{x}\right) = x^2 R(x),$$

then

$$(8.1.10) \qquad \int_0^\infty R(x) \ln x \, dx = 0.$$

This is the case for $R(x) = \dfrac{1 + x^2}{(1 - x^2)^2}$ and (8.1.10) appears as entry **4.234.3** in [**35**].

The goal of this paper is to present a systematic evaluation of the entries in [**35**] of the form (8.1.1).

8.2. Combinations of logarithms and linear rational functions

EXAMPLE 8.2.1. Entry **4.291.3** states that

$$(8.2.1) \qquad \int_0^{1/2} \frac{\ln(1-x)}{x} \, dx = \frac{\ln^2 2}{2} - \frac{\pi^2}{12}.$$

To evaluate this integral, let $t = -\ln(1-x)$ to produce

$$(8.2.2) \qquad \int_0^{1/2} \frac{\ln(1-x)}{x} \, dx = -\int_0^{\ln 2} \frac{te^{-t} \, dt}{1 - e^{-t}}.$$

This last integral can be written as

$$(8.2.3) \qquad \int_0^{\ln 2} t \, dt - \int_0^{\ln 2} \frac{t \, dt}{1 - e^{-t}}.$$

The first integral is elementary and has value $\frac{1}{2} \ln^2 2$. The second integral was evaluated as $\pi^2/12$ in [**9**].

EXAMPLE 8.2.2. The change of variables $t = x/2$ converts (8.2.1) to

$$(8.2.4) \qquad \int_0^{1/2} \ln\left(1 - \frac{t}{2}\right) \frac{dt}{t} = \frac{\ln^2 2}{2} - \frac{\pi^2}{12}.$$

This is entry **4.291.4** of [**35**].

EXAMPLE 8.2.3. Entry **4.291.5** states that

$$(8.2.5) \qquad \int_0^1 \ln\left(\frac{1+x}{2}\right) \frac{dx}{1-x} = \frac{\ln^2 2}{2} - \frac{\pi^2}{12}.$$

To evaluate this entry, let $u = (1-x)/2$ to reduce it to (8.2.1).

EXAMPLE 8.2.4. Differentiating

$$(8.2.6) \qquad \int_0^1 (1+x)^{-a} \, dx = \frac{2^{-a}(2^a - 2)}{a - 1}$$

with respect to a gives

$$(8.2.7) \quad \int_0^1 (1+x)^{-a} \ln(1+x)\, dx = \frac{1}{(a-1)^2} \left(2^{-a}(-2 + 2^a + 2\ln 2 - 2a\ln 2) \right).$$

Now let $a \to 1$ to obtain

$$(8.2.8) \quad \int_0^1 \frac{\ln(1+x)}{1+x}\, dx = \frac{1}{2}\ln^2 2.$$

This is entry **4.291.6**.

EXAMPLE 8.2.5. The partial fraction decomposition

$$(8.2.9) \quad \frac{1}{x(1+x)} = \frac{1}{x} - \frac{1}{1+x}$$

gives

$$(8.2.10) \quad \int_0^1 \frac{\ln(1+x)}{x(1+x)}\, dx = \int_0^1 \frac{\ln(1+x)}{x}\, dx - \int_0^1 \frac{\ln(1+x)}{1+x}.$$

The first integral is entry **4.291.1** and it has value $\pi^2/12$, as shown in [**10**]. The second integral is $\frac{1}{2}\ln^2 2$, as established in Example 8.2.4. This gives entry **4.291.12**

$$(8.2.11) \quad \int_0^1 \frac{\ln(1+x)}{x(1+x)}\, dx = \frac{\pi^2}{12} - \frac{1}{2}\ln^2 2.$$

EXAMPLE 8.2.6. Entry **4.291.13** is

$$(8.2.12) \quad \int_0^\infty \frac{\ln(1+x)\, dx}{x(1+x)} = \frac{\pi^2}{6}.$$

Split the integral over $[0,1]$ and $[1,\infty)$ and make the change of variables $t = 1/x$ in the second part. This gives

$$(8.2.13) \quad \int_0^\infty \frac{\ln(1+x)\, dx}{x(1+x)} = \int_0^1 \frac{\ln(1+x)\, dx}{x(1+x)} + \int_0^1 \frac{\ln(1+t) - \ln t}{1+t}\, dt.$$

Expand the first integral in partial fractions to obtain

$$(8.2.14) \quad \int_0^\infty \frac{\ln(1+x)\, dx}{x(1+x)} = \int_0^1 \frac{\ln(1+x)}{x}\, dx - \int_0^1 \frac{\ln x}{1+x}\, dx.$$

Integrate by parts the second integral to obtain

$$(8.2.15) \quad \int_0^\infty \frac{\ln(1+x)\, dx}{x(1+x)} = 2\int_0^1 \frac{\ln(1+x)}{x}\, dx.$$

The evaluation

$$(8.2.16) \quad \int_0^1 \frac{\ln(1+x)}{x}\, dx = \frac{\pi^2}{12}$$

that appears as entry **4.291.1** has been established in [**10**].

8.3. Combinations of logarithms and rational functions with denominators that are squares of linear terms

This section evaluates integrals of the form

$$(8.3.1) \qquad \int_a^b R_2(x) \ln R_1(x)\, dx$$

where R_1, R_2 are rational functions and the denominator of R_2 is a quadratic polynomial of the form $(cx + d)^2$.

EXAMPLE 8.3.1. Entry **4.291.14** is

$$(8.3.2) \qquad \int_0^1 \frac{\ln(1+x)}{(ax+b)^2}\, dx = \frac{1}{a(a-b)} \ln \frac{a+b}{b} + \frac{2\ln 2}{b^2 - a^2}$$

and

$$(8.3.3) \qquad \int_0^1 \frac{\ln(1+x)\, dx}{(x+1)^2} = \frac{1 - \ln 2}{2}$$

gives the value when $a = b$, after scaling.

To evaluate the first case, integrate by parts to get

$$(8.3.4) \qquad \int_0^1 \frac{\ln(1+x)}{(ax+b)^2}\, dx = -\frac{\ln 2}{a(a+b)} + \frac{1}{a} \int_0^1 \frac{dx}{(1+x)(ax+b)}.$$

The result now follows by expanding the second integrand in partial fractions.

The case $a = b$ is obtained by a direct integration by parts:

$$(8.3.5) \qquad \int_0^1 \frac{\ln(1+x)}{(1+x)^2}\, dx = -\frac{\ln 2}{2} + \int_0^1 \frac{dx}{(1+x)^2}.$$

This last integral is $1/2$ and the result has been established.

The same procedure gives entry **4.291.20**

$$(8.3.6) \qquad \int_0^1 \frac{\ln(ax+b)}{(1+x)^2}\, dx = \frac{1}{2(a-b)} \left[(a+b)\ln(a+b) - 2b\ln b - 2a\ln 2 \right]$$

for $a \neq b$.

EXAMPLE 8.3.2. The partial fraction decomposition

$$(8.3.7) \qquad \frac{1 - x^2}{(ax+b)^2\,(bx+a)^2} = \frac{1}{a^2 - b^2} \left[\frac{1}{(ax+b)^2} - \frac{1}{(bx+a)^2} \right]$$

and Example 8.3.1 gives the evaluation of entry **4.291.25**

$$\int_0^1 \frac{(1-x^2)\ln(1+x)\, dx}{(ax+b)^2\,(bx+a)^2} = \frac{1}{(a^2 - b^2)(a-b)} \left[\frac{a+b}{ab} \ln(a+b) - \frac{\ln b}{a} - \frac{\ln a}{b} \right]$$
$$- \frac{4\ln 2}{(a^2 - b^2)^2}.$$

The answer may be written in the more compact form

$$(8.3.8) \qquad \frac{-a^2 \ln a - b\left[b\ln b + a\ln(16ab) \right] + (a+b)^2 \ln(a+b)}{ab(a^2 - b^2)^2},$$

but this form hides the symmetry of the integral.

EXAMPLE 8.3.3. Entry **4.291.15** is

$$(8.3.9) \qquad \int_0^\infty \frac{\ln(1+x)\,dx}{(ax+b)^2} = \frac{\ln a - \ln b}{a(a-b)}$$

for $a \neq b$. In the case $a = b$, the integral scales to

$$(8.3.10) \qquad \int_0^\infty \frac{\ln(1+x)\,dx}{(1+x)^2} = 1.$$

To evaluate this entry, integrate by parts to obtain

$$(8.3.11) \qquad \int_0^\infty \frac{\ln(1+x)\,dx}{(ax+b)^2} = \frac{1}{a} \int_0^\infty \frac{dx}{(1+x)(ax+b)}.$$

This last integral is evaluated by using the partial fraction decomposition

$$(8.3.12) \qquad \frac{1}{(1+x)(ax+b)} = \frac{1}{b-a}\left(\frac{1}{1+x} - \frac{a}{ax+b}\right).$$

Integration by parts in the case $a = b$ (taken to be 1 by scaling) gives

$$(8.3.13) \qquad \int_0^\infty \frac{\ln(1+x)\,dx}{(1+x)^2} = \int_0^\infty \frac{dx}{(1+x)^2} = 1.$$

The same procedure gives entry **4.291.21**

$$(8.3.14) \qquad \int_0^\infty \frac{\ln(ax+b)\,dx}{(1+x)^2} = \frac{a\ln a - b\ln b}{a-b}$$

for $a \neq b$. The value of entry **4.291.17**

$$(8.3.15) \qquad \int_0^\infty \frac{\ln(a+x)}{(b+x)^2}\,dx = \frac{a\ln a - b\ln b}{b(a-b)}$$

is obtained from (8.3.14) by the change of variables $x = bt$.

EXAMPLE 8.3.4. The partial fraction decomposition (8.3.7) given in Example 8.3.2 produces the value of entry **4.291.26**

$$(8.3.16) \qquad \int_0^\infty \frac{(1-x^2)\ln(1+x)\,dx}{(ax+b)^2\,(bx+a)^2} = \frac{\ln b - \ln a}{ab(a^2-b^2)}$$

from Example 8.3.3.

8.4. Combinations of logarithms and rational functions with quadratic denominators

This section considers integrals of the form (8.1.1) where the denominator of $R_2(x)$ is a polynomial of degree 2 with non-real roots.

EXAMPLE 8.4.1. Entry **4.291.8** states that

$$(8.4.1) \qquad \int_0^1 \frac{\ln(1+x)\,dx}{1+x^2} = \frac{\pi}{8}\ln 2.$$

The proof of this evaluation is based on some entries of [**35**] that have been established in [**10**]. The reader is invited to provide a direct proof.

The change of variables $x = \tan \varphi$ gives

$$\int_0^1 \frac{\ln(1+x)\,dx}{1+x^2} = \int_0^{\pi/4} \ln(1+\tan\varphi)d\varphi$$

$$= \int_0^{\pi/4} \ln(\sin\varphi + \cos\varphi)d\varphi - \int_0^{\pi/4} \ln\cos\varphi\,d\varphi.$$

The value

$$\int_0^{\pi/4} \ln(\sin\varphi + \cos\varphi)d\varphi = -\frac{\pi}{8}\ln 2 + \frac{G}{2}$$

is entry **4.225.2** and

$$\int_0^{\pi/4} \ln\cos\varphi\,d\varphi = -\frac{\pi}{4}\ln 2 + \frac{G}{2}$$

is entry **4.224.5**. Both examples are evaluated in [**10**]. This gives the result.

The same technique gives entry **4.291.10**

$$(8.4.2) \qquad \int_0^1 \frac{\ln(1-x)\,dx}{1+x^2} = \frac{\pi}{8}\ln 2 - G.$$

This time, entry **4.225.1**

$$\int_0^{\pi/4} \ln(\cos\varphi - \sin\varphi)d\varphi = -\frac{\pi}{8}\ln 2 - \frac{G}{2}$$

is employed.

EXAMPLE 8.4.2. Entry **4.291.9**

$$(8.4.3) \qquad \int_0^\infty \frac{\ln(1+x)\,dx}{1+x^2} = \frac{\pi}{4}\ln 2 + G$$

is equivalent, via $x = \tan\varphi$, to the identity

$$(8.4.4) \qquad \int_0^{\pi/2} \ln(\sin\varphi + \cos\varphi)d\varphi - \int_0^{\pi/2} \ln\cos\varphi\,d\varphi = \frac{\pi}{4}\ln 2 + G.$$

The first integral is entry **4.225.2** and it has the value $-\frac{1}{4}\pi\ln 2 + G$; the second integral is entry **4.224.6** with value $-\frac{1}{2}\pi\ln 2$. Both of these examples have been established in [**10**]

EXAMPLE 8.4.3. The change of variables $t = 1/x$ gives

$$(8.4.5) \qquad \int_1^\infty \frac{\ln(x-1)\,dx}{1+x^2} = \int_0^1 \frac{\ln(1-t)\,dt}{1+t^2} - \int_0^1 \frac{\ln t\,dt}{1+t^2}.$$

The first integral has the value $\frac{1}{8}\pi\ln 2 - G$ and it appears as entry **4.291.10** (it has been established as (8.4.2)). The second integral is the special case $a = 1$ of (8.1.5). This gives the value of entry **4.291.11**:

$$(8.4.6) \qquad \int_1^\infty \frac{\ln(x-1)\,dx}{1+x^2} = \frac{\pi}{8}\ln 2.$$

EXAMPLE 8.4.4. A small number of entries in [**35**] can be evaluated from entry **4.231.9**

$$(8.4.7) \qquad \int_0^\infty \frac{\ln x \, dx}{x^2 + q^2} = \frac{\pi}{2} \frac{\ln q}{q},$$

evaluated in [**10**]. Expanding in partial fractions gives the identity

$$(8.4.8) \qquad \int_0^\infty \frac{\ln x \, dx}{(x^2 + a^2)(x^2 + b^2)} = \frac{\pi}{2(b^2 - a^2)} \left(\frac{\ln a}{a} - \frac{\ln b}{b} \right).$$

This provides the evaluation of entry **4.234.6**

$$(8.4.9) \qquad \int_0^\infty \frac{\ln x \, dx}{(a^2 + b^2 x^2)(1 + x^2)} = \frac{\pi b}{2a(b^2 - a^2)} \ln \frac{a}{b}$$

via the relation

$$(8.4.10) \qquad \int_0^\infty \frac{\ln x \, dx}{(a^2 + b^2 x^2)(1 + x^2)} = \frac{1}{b^2} \int_0^\infty \frac{\ln x \, dx}{(x^2 + a^2/b^2)(x^2 + 1)},$$

entry **4.234.7**

$$(8.4.11) \qquad \int_0^\infty \frac{\ln x \, dx}{(x^2 + a^2)(1 + b^2 x^2)} = \frac{\pi}{2(1 - a^2 b^2)} \left(\frac{\ln a}{a} + b \ln b \right)$$

via the relation

$$(8.4.12) \qquad \int_0^\infty \frac{\ln x \, dx}{(x^2 + a^2)(1 + b^2 x^2)} = \frac{1}{b^2} \int_0^\infty \frac{\ln x \, dx}{(x^2 + a^2)(x^2 + 1/b^2)},$$

and finally, entry **4.234.8**

$$(8.4.13) \qquad \int_0^\infty \frac{x^2 \ln x \, dx}{(a^2 + b^2 x^2)(1 + x^2)} = \frac{\pi a}{2b(b^2 - a^2)} \ln \frac{b}{a}$$

using the partial fraction decomposition

$$(8.4.14) \qquad \frac{x^2}{(a^2 + b^2 x^2)(1 + x^2)} = \frac{1}{(b^2 - a^2)} \frac{1}{x^2 + 1} - \frac{a^2}{b^2(b^2 - a^2)} \frac{1}{x^2 + a^2/b^2}.$$

The details are left to the reader.

8.5. An example via recurrences

The integral

$$(8.5.1) \qquad F_n(s) = \int_0^1 x^n (1 + x)^s \, dx$$

for $n \in \mathbb{N}$ and $s \in \mathbb{R}$ is integrated by parts (with $u = x^n (x + 1)$ and $dv = (x + 1)^{s-1} \, dx$) to produce the recurrence

$$(8.5.2) \qquad F_n(s) = \frac{2^{s+1}}{n + s + 1} - \frac{n}{n + s + 1} F_{n-1}(s).$$

The initial condition is

$$(8.5.3) \qquad F_0(s) = \int_0^1 (x + 1)^s \, dx = \frac{2^{s+1} - 1}{s + 1}.$$

The recurrence permits the evaluation of $F_n(s)$, for any fixed $n \in \mathbb{N}$. For instance,

$$F_1(s) = \frac{s2^{s+1} + 1}{(s+1)(s+2)}$$

$$F_2(s) = \frac{2\left[2^s(s^2 + s + 2) - 1\right]}{(s+1)(s+2)(s+3)}$$

$$F_3(s) = \frac{2\left[2^s(s^3 + 3s^2 + 8s) + 3\right]}{(s+1)(s+2)(s+3)(s+4)}.$$

Differentiating (8.5.2) produces a recurrence for

$$(8.5.4) \qquad G_n(s) = \int_0^1 \frac{x^n \ln(1+x)}{(1+x)^s} \, dx$$

in the form

$$(8.5.5) \quad G_n(s) = -\frac{2^{1-s}}{(n+1-s)^2} + \frac{2^{1-s} \ln 2}{n+1-s}$$

$$+ \frac{n}{(n-s+1)^2} F_n(-s) - \frac{n}{n-s+1} G_{n-1}(s).$$

This produces the value of $G_n(s)$, starting from

$$(8.5.6) \qquad G_0(s) = \int_0^1 \frac{\ln(1+x)}{(1+x)^s} \, dx = \frac{2^{1-s} \ln 2}{1-s} - \frac{2^{1-s} - 1}{(1-s)^2}.$$

For example,

$$(8.5.7) \quad G_1(s) = \frac{2^s(2s-3) - 2\ln 2 s^3 + 2(3\ln 2 - 1)s^2 - 4\ln 2 s + 4}{2^s(s-1)^2(s-2)^2}.$$

EXAMPLE 8.5.1. Entry **4.291.23** in **[35]** states that

$$(8.5.8) \qquad \int_0^1 \ln(1+x) \frac{1+x^2}{(1+x)^4} \, dx = -\frac{\ln 2}{3} + \frac{23}{72}.$$

This corresponds to the value $G_0(4) + G_2(4)$. The recurrence (8.5.5) gives the required data to verify this entry.

8.6. An elementary example

Integrals of the form

$$(8.6.1) \qquad \int_a^b \ln R_1(x) \frac{d}{dx} R_2(x) \, dx$$

for rational functions R_1, R_2 can be reduced to the integration of a rational function. Indeed, integration by parts yields

$$(8.6.2) \qquad \int_a^b \ln R_1(x) \frac{d}{dx} R_2(x) \, dx = \text{boundary terms} - \int_a^b R_3(x) \, dx$$

with $R_3 = R_1' R_2 / R_1$.

EXAMPLE 8.6.1. Entry **4.291.27** states that

(8.6.3)
$$\int_0^1 \ln(1+ax)\,\frac{1-x^2}{(1+x^2)^2}\,dx = \frac{(1+a)^2\ln(1+a)}{1+a^2}\,\frac{\ln(1+a)}{2} - \frac{\ln 2}{2}\,\frac{a}{1+a^2} - \frac{\pi}{4}\,\frac{a^2}{1+a^2}.$$

This example fits the pattern described above, since

(8.6.4)
$$\frac{1-x^2}{(1+x^2)^2} = \frac{d}{dx}\frac{x}{1+x^2}.$$

Therefore

$$\begin{aligned}
\int_0^1 \ln(1+ax)\,\frac{1-x^2}{(1+x^2)^2}\,dx &= \int_0^1 \ln(1+ax)\,\frac{d}{dx}\frac{x}{1+x^2}\,dx \\
&= \frac{\ln(1+a)}{2} - a\int_0^1 \frac{x\,dx}{(1+x^2)(1+ax)}.
\end{aligned}$$

The partial fraction decomposition

$$\frac{x}{(1+x^2)(1+ax)} = -\frac{a}{1+a^2}\frac{1}{1+ax} + \frac{a}{1+a^2}\frac{1}{1+x^2} + \frac{1}{1+a^2}\frac{x}{1+x^2}$$

and the evaluation of the remaining elementary integrals completes the solution to this problem.

EXAMPLE 8.6.2. Entry **4.291.28**

(8.6.5)
$$\int_0^\infty \ln(a+x)\,\frac{b^2-x^2}{(b^2+x^2)^2}\,dx = \frac{1}{a^2+b^2}\left(a\ln\frac{b}{a} - \frac{\pi b}{2}\right)$$

also fits the pattern in this section since

(8.6.6)
$$\frac{d}{dx}\frac{x}{x^2+b^2} = \frac{b^2-x^2}{(b^2+x^2)^2}.$$

Integrating by parts and checking that the boundary terms vanish produces

(8.6.7)
$$\int_0^\infty \ln(a+x)\,\frac{b^2-x^2}{(b^2+x^2)^2}\,dx = -\int_0^\infty \frac{x\,dx}{(x^2+b^2)(x+a)}.$$

It is convenient to introduce the scaling $x = bt$ to transform the last integral to

(8.6.8)
$$\int_0^\infty \frac{x\,dx}{(x^2+b^2)(x+a)} = \frac{1}{b}\int_0^\infty \frac{t\,dt}{(1+t^2)(t+c)}$$

with $c = a/b$. The evaluation is completed using the partial fraction decomposition

$$\frac{t}{(t^2+1)(t+c)} = -\frac{c}{c^2+1}\frac{1}{t+c} + \frac{1}{1+c^2}\frac{1}{t^2+1} + \frac{c}{c^2+1}\frac{t}{t^2+1}$$

and integrating from $t=0$ to $t=N$ and taking the limit as $N\to\infty$. The reader will easily check that the divergent pieces coming from $1/(t+c)$ and $t/(t^2+1)$ cancel out.

EXAMPLE 8.6.3. Entry **4.291.29** appears as

$$(8.6.9) \qquad \int_0^\infty \ln^2(a-x) \frac{b^2 - x^2}{(b^2 + x^2)^2} \, dx = \frac{2}{a^2 + b^2} \left(a \ln \frac{a}{b} - \frac{\pi b}{2} \right)$$

but it should be written as

$$(8.6.10) \qquad \int_0^\infty \ln\left[(a-x)^2\right] \frac{b^2 - x^2}{(b^2 + x^2)^2} \, dx = \frac{2}{a^2 + b^2} \left(a \ln \frac{a}{b} - \frac{\pi b}{2} \right).$$

This is a singular integral and the value should be interpreted as a Cauchy principal value

$$\int_0^\infty \ln\left[(a-x)^2\right] \frac{b^2 - x^2}{(b^2 + x^2)^2} \, dx =$$

$$\lim_{\varepsilon \to 0} \int_0^{a-\varepsilon} \ln\left[(a-x)^2\right] \frac{b^2 - x^2}{(b^2 + x^2)^2} \, dx + \int_{a+\varepsilon}^\infty \ln\left[(a-x)^2\right] \frac{b^2 - x^2}{(b^2 + x^2)^2} \, dx.$$

The first integral is

$$\int_0^{a-\varepsilon} \ln\left[(a-x)^2\right] \frac{b^2 - x^2}{(b^2 + x^2)^2} \, dx = \int_0^{a-\varepsilon} 2\ln(a-x) \frac{d}{dx} \frac{x}{x^2 + b^2} \, dx$$

$$= \frac{2(a-\varepsilon)}{(a-\varepsilon)^2 + b^2} \ln\varepsilon + \int_0^{a-\varepsilon} \frac{2x \, dx}{(a-x)(x^2 + b^2)},$$

after integration by parts. The second integral produces

$$\int_{a+\varepsilon}^\infty \ln\left[(a-x)^2\right] \frac{b^2 - x^2}{(b^2 + x^2)^2} \, dx = \int_{a+\varepsilon}^\infty 2\ln(x-a) \frac{d}{dx} \frac{x}{x^2 + b^2} \, dx$$

$$= -\frac{2(a+\varepsilon)}{(a+\varepsilon)^2 + b^2} \ln\varepsilon + \int_{a+\varepsilon}^\infty \frac{2x \, dx}{(x-a)(x^2 + b^2)}.$$

The reader will check that the boundary terms vanish as $\varepsilon \to 0$. This produces

$$(8.6.11) \qquad \int_0^\infty \ln\left[(a-x)^2\right] \frac{b^2 - x^2}{(b^2 + x^2)^2} \, dx =$$

$$\lim_{\varepsilon \to 0} \int_0^{a-\varepsilon} \frac{2x \, dx}{(a-x)(x^2 + b^2)} + \int_{a+\varepsilon}^\infty \frac{2x \, dx}{(a-x)(x^2 + b^2)}.$$

The partial fraction decomposition
(8.6.12)

$$\frac{2x}{(a-x)(x^2 + b^2)} = -\frac{2a}{a^2 + b^2} \frac{1}{x-a} - \frac{2b}{a^2 + b^2} \frac{b}{x^2 + b^2} + \frac{a}{a^2 + b^2} \frac{2x}{x^2 + b^2}$$

gives

$$\int_0^{a-\varepsilon} \frac{2x \, dx}{(a-x)(x^2 + b^2)} =$$

$$\frac{2a}{a^2 + b^2} \left[\ln a - \ln\varepsilon\right] - \frac{2b}{a^2 + b^2} \tan^{-1} \frac{a-\varepsilon}{b} + \frac{a}{a^2 + b^2} \left[\ln[(a-\varepsilon)^2 + b^2] - 2\ln b\right].$$

A similar computation yields

$$\int_{a+\varepsilon}^{N} \frac{2x\,dx}{(a-x)(x^2+b^2)} =$$

$$\frac{a}{a^2+b^2}\left\{\ln(N^2+b^2)-2\ln(N-a)+2\ln\varepsilon-\ln\left[(a+\varepsilon)^2+b^2\right]\right\}$$

$$+\frac{2b}{a^2+b^2}\left[\tan^{-1}\left(\frac{a+\varepsilon}{b}\right)-\tan^{-1}\left(\frac{N}{b}\right)\right].$$

Now let $N \to \infty$ and use $\ln(N^2+b^2)-2\ln(N-a) \to 0$ to obtain

$$\int_{a+\varepsilon}^{\infty} \frac{2x\,dx}{(a-x)(x^2+b^2)} =$$

$$\frac{a}{a^2+b^2}\left\{2\ln\varepsilon-\ln\left[(a+\varepsilon)^2+b^2\right]\right\}+\frac{2b}{a^2+b^2}\left[\tan^{-1}\left(\frac{a+\varepsilon}{b}\right)-\frac{\pi}{2}\right].$$

Observe that the singular terms in (8.6.11), namely, those containing the factor $\ln\varepsilon$, cancel out. The remaining terms produce the stated answer as $\varepsilon \to 0$. This completes the evaluation.

EXAMPLE 8.6.4. Entry **4.291.30** written as

$$(8.6.13) \qquad \int_{0}^{\infty} \ln\left[(a-x)^2\right] \frac{x\,dx}{(b^2+x^2)^2} = \frac{1}{a^2+b^2}\left(\ln b - \frac{\pi a}{2b} + \frac{a^2}{b^2}\ln a\right)$$

is evaluated as Example 8.6.3. Start with the identity

$$(8.6.14) \qquad \frac{d}{dx}\left(-\frac{1}{2(x^2+b^2)}\right) = \frac{x}{(x^2+b^2)^2}$$

and then proceed as before. The details are elementary and they are left to the reader.

8.7. Some parametric examples

This section considers some entries of [**35**] that depend on a parameter.

EXAMPLE 8.7.1. Entry **4.291.18** states that

$$(8.7.1) \qquad \int_{0}^{a} \frac{\ln(1+ax)\,dx}{1+x^2} = \frac{1}{2}\tan^{-1}a\,\ln(1+a^2).$$

Differentiating the left-hand side with respect to a gives

$$(8.7.2) \qquad \frac{\ln(1+a^2)}{1+a^2} + \int_{0}^{a} \frac{x\,dx}{(1+ax)(1+x^2)}.$$

The verification of this entry will start with the evaluation of the rational integral

$$(8.7.3) \qquad R(a) := \int_{0}^{a} \frac{x\,dx}{(1+ax)(1+x^2)}.$$

The partial fraction decomposition
(8.7.4)
$$\frac{x}{(1+ax)(1+x^2)} = -\frac{1}{1+a^2}\frac{a}{1+ax} + \frac{a}{1+a^2}\frac{1}{1+x^2} + \frac{1}{2(1+a^2)}\frac{2x}{1+x^2}$$

gives

(8.7.5) $$R(a) = -\frac{\ln(1+a^2)}{1+a^2} + \frac{a}{1+a^2}\tan^{-1}a + \frac{\ln(1+a^2)}{2(1+a^2)}.$$

Motivated by the expression in the entry being evaluated, observe that

(8.7.6) $$\int_0^a \frac{x\,dx}{(1+ax)(1+x^2)} + \frac{\ln(1+a^2)}{1+a^2} = \frac{1}{2}\frac{d}{da}\left[\tan^{-1}a\,\ln(1+a^2)\right].$$

Now integrate this identity from 0 to a to obtain

(8.7.7) $$\int_0^a \left[\int_0^b \frac{x\,dx}{(1+bx)(1+x^2)} + \frac{\ln(1+b^2)}{1+b^2}\right]db+$$
$$\int_0^a \frac{\ln(1+b^2)}{1+b^2}\,db = \frac{1}{2}\tan^{-1}a\,\ln(1+a^2).$$

Exchange the order of integration to produce

$$\int_0^a\int_0^b \frac{x\,dx}{(1+bx)(1+x^2)}\,db = \int_0^a \frac{x}{1+x^2}\int_x^a \frac{db}{1+bx}\,dx$$
$$= \int_0^a \frac{1}{1+x^2}\left[\ln(1+ax) - \ln(1+x^2)\right]\,dx.$$

The result now follows from (8.7.7).

EXAMPLE 8.7.2. Entry **4.291.16** states that

(8.7.8) $$\int_0^1 \frac{\ln(a+x)\,dx}{a+x^2} = \frac{1}{2\sqrt{a}}\cot^{-1}\sqrt{a}\,\ln[a(1+a)].$$

The change of variables $x = \sqrt{a}t$ gives
(8.7.9)
$$\int_0^1 \frac{\ln(a+x)\,dx}{a+x^2} = \frac{1}{\sqrt{a}}\left[\ln a\int_0^{1/\sqrt{a}} \frac{dt}{1+t^2} + \int_0^{1/\sqrt{a}} \frac{\ln(1+t/\sqrt{a})}{1+t^2}\,dt\right].$$

The first integral is elementary and the second one corresponds to (8.7.1).

EXAMPLE 8.7.3. Entry **4.291.19** states that

(8.7.10) $$\int_0^1 \frac{\ln(1+ax)\,dx}{1+ax^2} = \frac{1}{2\sqrt{a}}\tan^{-1}\sqrt{a}\,\ln(1+a).$$

This follows directly from (8.7.1) by the change of variables $x = t/\sqrt{a}$ and replacing a by $\sqrt{a}$.

EXAMPLE 8.7.4. Entry **4.291.7** is the identity

$$(8.7.11) \qquad \int_0^\infty \frac{\ln(1+ax)\,dx}{1+x^2} = \frac{\pi}{4}\ln(1+a^2) - \int_0^a \frac{\ln u\,du}{1+u^2}.$$

Differentiating the left-hand side gives

$$\frac{d}{da}\int_0^\infty \frac{\ln(1+ax)\,dx}{1+x^2} = \int_0^\infty \frac{x\,dx}{(1+ax)(1+x^2)}$$

$$= \frac{\pi}{2}\frac{a}{1+a^2} - \frac{\ln a}{1+a^2},$$

where the last evaluation is established by partial fractions. The result now follows by integrating back with respect to a.

REMARK 8.7.5. *The current version of* Mathematica *gives*

$$\int_0^a \frac{\ln x\,dx}{1+x^2} = \tan^{-1} a\,\ln a - \frac{i}{2}PolyLog[2,-ia] + \frac{i}{2}PolyLog[2,ia]$$

but is unable to provide an analytic expression for the integral

$$\int_0^\infty \frac{\ln(1+ax)\,dx}{1+x^2}.$$

Entries of [**35**] *that can be evaluated in terms of polylogarithms will be described in a future publication.*

EXAMPLE 8.7.6. Entry **4.291.24** states that

$$\int_0^1 \frac{(1+x^2)\ln(1+x)}{(a^2+x^2)(1+a^2x^2)}\,dx = \frac{1}{2a(1+a^2)}\left[\frac{\pi}{2}\ln(1+a^2) - 2\tan^{-1} a\,\ln a\right].$$

The evaluation of this entry starts with the partial fraction decomposition

$$(8.7.12) \qquad \frac{1+x^2}{(a^2+x^2)(1+a^2x^2)} = \frac{1}{1+a^2}\left[\frac{1}{x^2+a^2} + \frac{1}{1+a^2x^2}\right],$$

which yields the identity

$$\int_0^1 \frac{(1+x^2)\ln(1+x)}{(a^2+x^2)(1+a^2x^2)}\,dx = \frac{1}{1+a^2}\left[\int_0^1 \frac{\ln(1+x)\,dx}{x^2+a^2} + \int_0^1 \frac{\ln(1+x)\,dx}{1+a^2x^2}\right],$$

and the change of variables $t = 1/x$ then produces

$$\int_0^1 \frac{\ln(1+x)\,dx}{1+a^2x^2} = \int_1^\infty \frac{\ln(1+t)\,dt}{t^2+a^2} - \int_1^\infty \frac{\ln t\,dt}{t^2+a^2}.$$

Therefore

$$\int_0^1 \frac{(1+x^2)\ln(1+x)}{(a^2+x^2)(1+a^2x^2)}\,dx = \frac{1}{1+a^2}\left[\int_0^\infty \frac{\ln(1+x)\,dx}{x^2+a^2} - \int_1^\infty \frac{\ln x\,dx}{x^2+a^2}\right].$$

The change of variables $x = at$ and Example 8.7.4 give

$$\int_0^\infty \frac{\ln(1+x)\,dx}{x^2+a^2} = \frac{1}{a}\int_0^\infty \frac{\ln(1+at)\,dt}{1+t^2}$$

$$= \frac{\pi}{4a}\ln(1+a^2) - \frac{1}{a}\int_0^a \frac{\ln t\,dt}{1+t^2}.$$

Therefore

$$\int_0^1 \frac{(1+x^2)\ln(1+x)}{(a^2+x^2)(1+a^2x^2)}\,dx$$

$$(8.7.13) \qquad = \frac{1}{1+a^2}\left[\frac{\pi}{4a}\ln(1+a^2) - \frac{1}{a}\int_0^a \frac{\ln x\,dx}{1+x^2} - \int_1^\infty \frac{\ln x\,dx}{x^2+a^2}\right].$$

The change of variables $x = at$ gives

$$\int_1^\infty \frac{\ln x\,dx}{x^2+a^2} = \frac{\ln a}{a}\int_{1/a}^\infty \frac{dt}{1+t^2} + \frac{1}{a}\int_{1/a}^\infty \frac{\ln t\,dt}{1+t^2}$$

$$= \frac{\ln a}{a}\int_{1/a}^\infty \frac{dt}{1+t^2} - \frac{1}{a}\int_0^a \frac{\ln u\,du}{1+u^2},$$

after the change of variables $u = 1/t$ in the last integral. Replacing in (8.7.13) gives the result.

EXAMPLE 8.7.7. The last entry of [35] discussed here is **4.291.22**

$$\int_0^\infty \frac{x\ln(a+x)}{(b^2+x^2)^2}\,dx = \frac{1}{2(a^2+b^2)}\left(\ln b + \frac{\pi a}{2b} + \frac{a^2}{b^2}\ln a\right).$$

As before, start with the identity

$$(8.7.14) \qquad \frac{x}{(x^2+b^2)^2} = -\frac{d}{dx}\frac{1}{2(x^2+b^2)}$$

and integrate by parts to produce

$$\int_0^\infty \frac{x\ln(a+x)}{(b^2+x^2)^2}\,dx = \frac{\ln a}{2b^2} + \frac{1}{2}\int_0^\infty \frac{dx}{(x+a)(x^2+b^2)}.$$

This last integral is evaluated by the method of partial fractions to obtain the result.

Summary. The examples presented here complete the evaluation of every entry in Section 4.291 of the table [35]. The entries not appearing here have been presented in [10, 20, 50].

8.8. Integrals yielding partial sums of the zeta function

Some entries of [35] contain as the integrand the product of $\ln x$ and a rational function coming from manipulations of a geometric series. This section presents the evaluation of some of these examples. These evaluations can be written in terms of the Riemann zeta function

$$(8.8.1) \qquad \zeta(s) = \sum_{k=1}^\infty \frac{1}{n^s}$$

and the generalized harmonic numbers

$$(8.8.2) \qquad H_{n,m} = \sum_{k=1}^n \frac{1}{k^m}.$$

EXAMPLE 8.8.1. Entry **4.231.18** states that

$$(8.8.3) \qquad \int_0^1 \frac{1 - x^{n+1}}{(1-x)^2} \ln x \, dx = -\frac{(n+1)\pi^2}{6} + \sum_{k=1}^n \frac{n-k+1}{k^2}.$$

This can be expressed as

$$(8.8.4) \qquad \int_0^1 \frac{1 - x^{n+1}}{(1-x)^2} \ln x \, dx = -(n+1)\zeta(2) + (n+1)H_{n,2} - H_{n,1}.$$

The evaluation begins with the identity

$$(8.8.5) \qquad \frac{1}{(1-x)^2} = \sum_{k=0}^\infty (k+1)x^k$$

and its shift

$$(8.8.6) \qquad \frac{1 - x^{n+1}}{(1-x)^2} = \sum_{k=0}^n (k+1)x^k + (n+1)\sum_{k=n+1}^\infty x^k.$$

Integrate term by term and use the value

$$(8.8.7) \qquad \int_0^1 x^k \ln x \, dx = -\frac{1}{(k+1)^2}$$

to obtain

$$(8.8.8) \qquad \int_0^1 \frac{1 - x^{n+1}}{(1-x)^2} \ln x \, dx = -\sum_{k=0}^n \frac{1}{k+1} - (n+1)\sum_{k=n+1}^\infty \frac{1}{(k+1)^2}.$$

This can now be transformed to the form stated in [**35**].

EXAMPLE 8.8.2. Entry **4.262.7**

$$(8.8.9) \qquad \int_0^1 \frac{1 - x^{n+1}}{(1-x)^2} (\ln x)^3 \, dx = -\frac{(n+1)\pi^4}{15} + 6\sum_{k=1}^n \frac{n-k+1}{k^4}$$

is obtained by using (8.8.6), the identity

$$(8.8.10) \qquad \int_0^1 (\ln x)^3 x^k \, dx = -\frac{6}{(k+1)^4},$$

and the value

$$(8.8.11) \qquad \sum_{k=1}^\infty \frac{1}{k^4} = \zeta(4) = \frac{\pi^4}{90}.$$

EXAMPLE 8.8.3. Replacing x by x^2 in (8.8.6) gives

$$(8.8.12) \qquad \frac{1 - x^{2n+2}}{(1-x^2)^2} = \sum_{k=0}^n (k+1)x^{2k} + (n+1)\sum_{k=n+1}^\infty x^{2k}.$$

This gives

$$\int_0^1 \frac{1 - x^{2n+2}}{(1 - x^2)^2} \ln x \, dx = \sum_{k=0}^n (k+1) \int_0^1 x^{2k} \ln x \, dx$$

$$+ (n+1) \sum_{k=n+1}^{\infty} \int_0^1 x^{2k} \ln x \, dx$$

$$= - \sum_{k=0}^n \frac{k+1}{(2k+1)^2} - (n+1) \sum_{k=n+1}^{\infty} \frac{1}{(2k+1)^2}.$$

The value

$$(8.8.13) \qquad \sum_{k=1}^{\infty} \frac{1}{(2k-1)^2} = \frac{\pi^2}{8} = \frac{3}{4}\zeta(2)$$

is obtained by separating the terms forming the series for $\zeta(2)$ into even and odd indices. Now write

$$(8.8.14) \qquad \sum_{k=n+1}^{\infty} \frac{1}{(2k+1)^2} = \frac{3}{4}\zeta(2) - \sum_{k=1}^{n+1} \frac{1}{(2k-1)^2}$$

to obtain, after some elementary algebraic manipulatons, the evaluation

$$(8.8.15) \qquad \int_0^1 \frac{1 - x^{2n+2}}{(1 - x^2)^2} \ln x \, dx = -\frac{3}{4}(n+1)\zeta(2) + \sum_{k=1}^n \frac{n-k+1}{(2k-1)^2}.$$

This is entry **4.231.16**.

EXAMPLE 8.8.4. The alternating geometric series

$$(8.8.16) \qquad \frac{1}{1+x} = \sum_{k=0}^{\infty} (-1)^k x^k$$

is used as before to derive the identity

$$(8.8.17) \qquad \frac{1 + (-1)^n x^{n+1}}{(1+x)^2} = (n+1) \sum_{k=0}^{\infty} (-1)^k x^k - \sum_{k=0}^n (-1)^k (n-k) x^k.$$

Integrating yields
(8.8.18)

$$\int_0^1 \frac{1 + (-1)^n x^{n+1}}{(1+x)^2} \ln x \, dx = -(n+1) \sum_{k=0}^{\infty} \frac{(-1)^k}{(k+1)^2} - \sum_{k=1}^n \frac{(-1)^k(n-k+1)}{k^2}.$$

This is entry **4.231.17**, written in the form

$$(8.8.19) \qquad \int_0^1 \frac{1 + (-1)^n x^{n+1}}{(1+x)^2} \ln x \, dx = -\frac{(n+1)\pi^2}{12} - \sum_{k=1}^n \frac{(-1)^k(n-k+1)}{k^2},$$

using the value

$$(8.8.20) \qquad \sum_{k=0}^{\infty} \frac{(-1)^k}{(k+1)^2} = \frac{\pi^2}{12}.$$

EXAMPLE 8.8.5. Entry **4.262.8**
(8.8.21)
$$\int_0^1 \frac{1 + (-1)^n x^{n+1}}{(1+x)^2} (\ln x)^3 \, dx = -\frac{7(n+1)\pi^4}{120} + 6 \sum_{k=1}^n (-1)^{k-1} \frac{n-k+1}{k^4}$$

is obtained by using (8.8.17) and the identities employed in Example 8.8.2. The procedure employed in Example 8.8.3 now gives entry **4.262.9**

$$(8.8.22) \qquad \int_0^1 \frac{1 - x^{2n+2}}{(1-x^2)^2} (\ln x)^3 \, dx = -\frac{(n+1)\pi^4}{16} + 6 \sum_{k=1}^n \frac{n-k+1}{(2k-1)^4}.$$

8.9. A singular integral

The last evaluation presented here is entry **4.231.10**

$$(8.9.1) \qquad \int_0^\infty \frac{\ln x \, dx}{a^2 - b^2 x^2} = -\frac{\pi^2}{4ab}.$$

The parameters a, b have the same sign, so it may be assumed that $a, b > 0$. Observe that this is a singular integral, since the integrand is discontinuous at $x = a/b$.

The change of variables $t = bx/a$ gives

$$(8.9.2) \qquad \int_0^\infty \frac{\ln x \, dx}{a^2 - b^2 x^2} = \frac{1}{ab} \left[\ln \frac{a}{b} \int_0^\infty \frac{dt}{1 - t^2} + \int_0^\infty \frac{\ln t \, dt}{1 - t^2} \right].$$

The first integral is singular and is computed as the limit as $\varepsilon \to 0$ of
(8.9.3)
$$\int_0^{1-\varepsilon} \frac{dt}{1 - t^2} + \int_{1+\varepsilon}^\infty \frac{dt}{1 - t^2} = \frac{1}{2} \ln \left(\frac{2-\varepsilon}{\varepsilon} \right) + \frac{1}{2} \ln \left(\frac{\varepsilon}{2+\varepsilon} \right) = \frac{1}{2} \ln \left(\frac{2-\varepsilon}{2+\varepsilon} \right)$$

obtained by the method of partial fractions. Therefore this singular integral has value 0. The second integral is

$$(8.9.4) \qquad \int_0^\infty \frac{\ln t \, dt}{1 - t^2} = 2 \int_0^1 \frac{\ln t \, dt}{1 - t^2},$$

because the integral over $[1, \infty)$ is the same as over $[0, 1]$. The method of partial fractions and the values

$$(8.9.5) \qquad \int_0^1 \frac{\ln x \, dx}{1 - x} = -\frac{\pi^2}{6} \quad \text{and} \quad \int_0^1 \frac{\ln x \, dx}{1 + x} = -\frac{\pi^2}{12},$$

which appear as entries **4.231.2** and **4.231.1**, respectively, give the final result. These last two entries were evaluated in [4].

The change of variables $t = \ln x$ converts this integral into entry **3.417.2**

$$(8.9.6) \qquad \int_{-\infty}^\infty \frac{t \, dt}{a^2 e^t - b^2 e^{-t}} = \frac{\pi^2}{4ab}.$$

The same change of variables gives the evaluation of entry **3.417.1**

$$(8.9.7) \qquad \int_{-\infty}^\infty \frac{t \, dt}{a^2 e^t + b^2 e^{-t}} = \frac{\pi}{2ab} \ln \frac{b}{a}.$$

from entry **4.231.8**

(8.9.8)
$$\int_0^\infty \frac{\ln x \, dx}{a^2 + b^2 x^2} = -\frac{\pi}{2ab} \ln \frac{b}{a}$$

evaluated in [10].

Summary. The examples presented here complete the evaluation of every entry in Section 4.231 of the table [35]. The entries not appearing here have been presented in [10, 20, 50].

CHAPTER 9

Polylogarithm functions

9.1. Introduction

The table of integrals [35] contains many entries that are expressible in terms of the *polylogarithm function*

$$(9.1.1) \qquad \text{Li}_s(z) := \sum_{k=1}^{\infty} \frac{z^k}{k^s}.$$

In this paper we describe the evaluation of some of them. The series $(9.1.1)$ converges for $|z| < 1$ and $\text{Re}\, s > 1$. The integral representation

$$(9.1.2) \qquad \text{Li}_s(z) = \frac{z}{\Gamma(s)} \int_0^\infty \frac{x^{s-1}\, dx}{e^x - z}$$

provides an analytic extension to $\mathbb{C}$. Here $\Gamma(s)$ is the classical *gamma function* defined by

$$(9.1.3) \qquad \Gamma(s) := \int_0^\infty x^{s-1} e^{-x}\, dx.$$

The polylogarithm function is a generalization of the *Riemann zeta function*

$$(9.1.4) \qquad \zeta(s) := \sum_{k=1}^{\infty} \frac{1}{k^s} = \text{Li}_s(1).$$

A second special value is given by

$$(9.1.5) \qquad \text{Li}_s(-1) = \sum_{k=1}^{\infty} \frac{(-1)^k}{k^s} = -(1 - 2^{1-s})\zeta(s),$$

the last equality being obtained by splitting the sum according to the parity of the summation index.

The first result is an identity between an integral and a series coming from the evaluation of the polylogarithm at two values on the unit circle. Many of the entries presented here are special cases. This is a classical result; the proof is presented here in order to keep the paper as self-contained as possible.

THEOREM 9.1.1. *Let $\nu \in \mathbb{C}$ with $\text{Re}\, \nu > 0$ and $0 < t < \pi$. Then*

$$(9.1.6) \qquad \int_0^\infty \frac{x^{\nu-1}\, dx}{\cosh x - \cos t} = \frac{2\Gamma(\nu)}{\sin t} \sum_{k=1}^{\infty} \frac{\sin kt}{k^\nu}.$$

PROOF. The integral representation (9.1.2) gives

$$i\left[\mathrm{Li}_s(e^{-it}) - \mathrm{Li}_s(e^{it})\right] = \frac{i}{\Gamma(s)} \int_0^\infty x^{s-1} \left[\frac{1}{e^{x+it}-1} - \frac{1}{e^{x-it}-1}\right] dx$$

$$= \frac{\sin t}{\Gamma(s)} \int_0^\infty \frac{x^{s-1}\, dx}{\cosh x - \cos t}.$$

The series representation (9.1.1) gives

$$i\left[\mathrm{Li}_s(e^{-it}) - \mathrm{Li}_s(e^{it})\right] = 2\sum_{k=1}^\infty \frac{e^{ikt} - e^{-ikt}}{2ik^s}$$

$$= 2\sum_{k=1}^\infty \frac{\sin kt}{k^s}.$$

This proves the result.   □

COROLLARY 9.1.2. *Let $\nu \in \mathbb{C}$ with $\mathrm{Re}\,\nu > 0$ and $0 < t < \pi$. Then*

$$(9.1.7) \qquad \int_0^\infty \frac{x^{\nu-1}\, dx}{\cosh x + \cos t} = \frac{2\Gamma(\nu)}{\sin t} \sum_{k=1}^\infty (-1)^{k-1} \frac{\sin kt}{k^\nu}.$$

PROOF. Replace t by $\pi - t$ in the statement of Theorem 9.1.1.   □

This corollary appears as entry **3.531.7** in [**35**].

REMARK 9.1.3. *In the special case $\nu = 2$, the series in the corollary appears in the expansion of the Lobachevsky function*

$$(9.1.8)\quad L(t) := -\int_0^t \ln \cos s\, ds = t\,\ln 2 - \frac{1}{2}\sum_{k=1}^\infty (-1)^{k-1}\frac{\sin 2kt}{k^2}, \quad 0 < t < \frac{\pi}{2}.$$

This special case of the corollary can be stated as

$$(9.1.9) \qquad \int_0^\infty \frac{x\, dx}{\cosh x + \cos 2t} = \frac{4\,(t\,\ln 2 - L(t))}{\sin 2t}, \quad 0 < t < \frac{\pi}{2}.$$

This is entry **3.531.2** *of* [**35**]. *Observe that this is written as*

$$(9.1.10) \qquad \int_0^\infty \frac{x\, dx}{\cosh 2x + \cos 2t} = \frac{t\,\ln 2 - L(t)}{\sin 2t}, \quad 0 < t < \frac{\pi}{2}.$$

The fact that this is the only entry in Section 3.531 with $\cosh 2x$ instead of $\cosh x$ can lead to confusion.

9.2. Some examples from the table by Gradshteyn and Ryzhik

This section presents the evaluation of some entries from the table [**35**] by making specific choices for the parameters ν and t in Theorem 9.1.1 and Corollary 9.1.2. Naturally a closed form for the integral is obtained in those cases for which the series can be evaluated.

EXAMPLE 9.2.1. Take $\nu = 2$ and $t = \pi/3$. Theorem 9.1.1 gives

$$\int_0^\infty \frac{x\,dx}{\cosh x - \frac{1}{2}} = \frac{2\Gamma(2)}{\sin \pi/3} \sum_{k=1}^\infty \frac{\sin(k\pi/3)}{k^2}$$

$$= \frac{4}{\sqrt 3} \sum_{k=1}^\infty \frac{\sin(k\pi/3)}{k^2}.$$

The function $\sin(\pi k/3)$ is periodic, with period 6, and repeating values

$$\frac{\sqrt 3}{2}, \frac{\sqrt 3}{2}, 0, -\frac{\sqrt 3}{2}, -\frac{\sqrt 3}{2}, 0.$$

Therefore

$$\sum_{k=1}^\infty \frac{\sin\left(\frac{\pi k}{3}\right)}{k^2} = \frac{\sqrt 3}{2}\left(\sum_{k=0}^\infty \frac{1}{(6k+1)^2} + \sum_{k=0}^\infty \frac{1}{(6k+2)^2} - \sum_{k=0}^\infty \frac{1}{(6k+4)^2} \right.$$

$$\left. - \sum_{k=0}^\infty \frac{1}{(6k+5)^2} \right).$$

To evaluate this series, recall the series representation of the *digamma* function $\psi(x) = \Gamma'(x)/\Gamma(x)$, given by

(9.2.1) $$\psi(x) = -\gamma - \frac{1}{x} + \sum_{k=1}^\infty \frac{x}{k(x+k)}, \qquad \text{for } x > 0.$$

Differentiation yields

(9.2.2) $$\psi'(x) = \sum_{k=0}^\infty \frac{1}{(x+k)^2}, \qquad \text{for } x > 0,$$

and we obtain

$$\sum_{k=0}^\infty \frac{1}{(6k+j)^2} = \frac{1}{36} \sum_{k=0}^\infty \frac{1}{(k+\frac{j}{6})^2} = \frac{1}{36} \psi'\left(\frac{j}{6}\right).$$

This provides the expression

(9.2.3) $$\sum_{k=1}^\infty \frac{\sin\left(\frac{\pi k}{3}\right)}{k^2} = \frac{\sqrt 3}{72}\left(\psi'\left(\frac{1}{6}\right) + \psi'\left(\frac{2}{6}\right) - \psi'\left(\frac{4}{6}\right) - \psi'\left(\frac{5}{6}\right) \right).$$

The identities

(9.2.4) $$\psi(1-x) = \psi(x) + \pi \cot \pi x, \qquad \text{for } 0 < x < 1,$$

and

(9.2.5) $$\psi(2x) = \frac{1}{2}\left(\psi(x) + \psi(x + \tfrac{1}{2})\right) + \ln 2$$

produce

$$\psi'\left(\frac{1}{6}\right) = 5\psi'\left(\frac{1}{3}\right) - \frac{4\pi^2}{3}, \ \psi'\left(\frac{2}{3}\right) = -\psi'\left(\frac{1}{3}\right) + \frac{4\pi^2}{3}, \ \psi'\left(\frac{5}{6}\right)$$

$$= -5\psi'\left(\frac{1}{3}\right) + \frac{16\pi^2}{3}.$$

It follows that

(9.2.6)
$$\int_0^\infty \frac{x\,dx}{\cosh x - \frac{1}{2}} = \frac{2}{3}\psi'\left(\frac{1}{3}\right) - \frac{4\pi^2}{9}.$$

This example appears as entry **3.531.1**. The value stated there is given in terms of the Lobachevsky function using (9.1.9)

(9.2.7)
$$\int_0^\infty \frac{x\,dx}{\cosh x - \frac{1}{2}} = \frac{8}{\sqrt{3}}\left(\frac{\pi}{3}\ln 2 - L\left(\frac{\pi}{3}\right)\right).$$

Comparing these two evaluations gives

(9.2.8)
$$L\left(\frac{\pi}{3}\right) = -\frac{1}{4\sqrt{3}}\psi'\left(\frac{1}{3}\right) + \frac{\pi^2}{6\sqrt{3}} + \frac{\pi}{3}\ln 2.$$

This example also appears as entry **3.418.1** in the form

(9.2.9)
$$\int_0^\infty \frac{x\,dx}{e^x + e^{-x} - 1} = \frac{1}{3}\left[\psi'\left(\frac{1}{3}\right) - \frac{2\pi^2}{3}\right].$$

EXAMPLE 9.2.2. Entry **3.514.1** in [**35**] is

(9.2.10)
$$\int_0^\infty \frac{dx}{\cosh ax + \cos t} = \frac{t}{a\sin t}, \quad \text{for } 0 < t < \pi, \, a > 0.$$

The case of arbitrary $a > 0$ is equivalent to the special case $a = 1$. This follows from the change of variables $ax \mapsto x$. The integral

(9.2.11)
$$\int_0^\infty \frac{dx}{\cosh x + \cos t} = \frac{t}{\sin t}, \quad \text{for } 0 < t < \pi,$$

is now evaluated by elementary methods.

The next sequence of identities gives the result

$$\int_0^\infty \frac{dx}{\cosh x + \cos t} = 2\int_0^\infty \frac{e^x\,dx}{e^{2x} + 2e^x \cos t + 1}$$

$$= 2\int_1^\infty \frac{dr}{r^2 + 2r\cos t + 1}$$

$$= 2\int_{1+\cos t}^\infty \frac{du}{u^2 + \sin^2 t}$$

$$= \frac{2}{\sin t}\int_{\cot(t/2)}^\infty \frac{dv}{v^2 + 1}$$

$$= \frac{t}{\sin t}.$$

EXAMPLE 9.2.3. The exponential generating function for the Bernoulli polynomials $B_n(x)$ is $te^{xt}/(e^t - 1)$, so for real x and t with $0 < |t| < 2\pi$,

$$(9.2.12) \qquad \frac{te^{xt}}{e^t - 1} = \sum_{n=0}^{\infty} B_n(x)\frac{t^n}{n!}.$$

For $n = 2m + 1$ an odd integer, these polynomials have a Fourier sine series given by

$$(9.2.13)$$
$$\frac{2^{2m}\pi^{2m+1}(-1)^m}{(2m+1)!}B_{2m+1}\left(\frac{t}{2\pi} + \frac{1}{2}\right) = \sum_{k=1}^{\infty} \frac{(-1)^{k-1}}{k^{2m+1}} \sin kt, \text{ for } |t| < \pi.$$

For example, $n = 3$ gives

$$(9.2.14) \qquad \frac{t(\pi^2 - t^2)}{12} = \sum_{k=1}^{\infty}(-1)^{k-1}\frac{\sin kt}{k^3}, \text{ for } |t| < \pi,$$

and $n = 5$ gives

$$(9.2.15) \qquad \frac{t(\pi^2 - t^2)(7\pi^2 - 3t^2)}{720} = \sum_{k=1}^{\infty}(-1)^{k-1}\frac{\sin kt}{k^5}, \text{ for } |t| < \pi.$$

These representations and Corollary 9.1.2 give the evaluations

$$(9.2.16) \qquad \int_0^{\infty} \frac{x^2\, dx}{\cosh x + \cos t} = \frac{t(\pi^2 - t^2)}{3\sin t}, \qquad \text{for } 0 < t < \pi,$$

and

$$(9.2.17) \qquad \int_0^{\infty} \frac{x^4\, dx}{\cosh x + \cos t} = \frac{t(\pi^2 - t^2)(7\pi^2 - 3t^2)}{15\sin t}, \qquad \text{for } 0 < t < \pi.$$

These integrals appear as entries **3.531.3** and **3.531.4**, respectively. The Fourier sine series

$$(9.2.18) \qquad \frac{t}{2} = \sum_{k=1}^{\infty}(-1)^{k-1}\frac{\sin kt}{k}, \text{ for } |t| < \pi,$$

shows that the evaluation given in Example 9.2.2 is also part of this family.

EXAMPLE 9.2.4. The limiting case $t \to 0$ in Corollary 9.1.2 gives, for $\nu \neq 2$, the evaluation

$$(9.2.19) \qquad \int_0^{\infty} \frac{x^{\nu-1}\, dx}{\cosh x + 1} = 2(1 - 2^{2-\nu})\Gamma(\nu)\zeta(\nu - 1).$$

The proof uses the elementary limit $\sin kt/\sin t \to k$ as $t \to 0$ and (9.1.5). The identity (9.2.19) is part of entry **3.531.6**. An alternative direct proof is presented next.

The integral representation

$$(9.2.20) \qquad \int_0^{\infty} \frac{x^{s-1}\, dx}{e^{px} + 1} = \frac{(1 - 2^{1-s})}{p^s}\Gamma(s)\zeta(s)$$

appears as entry **9.513.1** in [35] and it is established in [66] and in [4].

Differentiating with respect to p gives

(9.2.21) $$\int_0^\infty \frac{x^s e^{px} \, dx}{(e^{px} + 1)^2} = \frac{s(1 - 2^{1-s})}{p^{1+s}} \Gamma(s)\zeta(s)$$

and $p = 1/2$ produces

(9.2.22) $$\int_0^\infty \frac{x^s \, dx}{e^{x/2} + e^{-x/2} + 2} = 2^{s+1}(1 - 2^{1-s})\Gamma(s + 1)\zeta(s).$$

The change of variables $u = x/2$ and $\nu = s + 1$ gives the result.

The limiting case $\nu \to 2$

(9.2.23) $$\int_0^\infty \frac{x \, dx}{\cosh x + 1} = 2\ln 2,$$

which is also part of entry **3.531.6**, appears from the limiting behavior

(9.2.24) $$\zeta(s) = \frac{1}{s - 1} + F(s)$$

where $F(s)$ is an entire function.

EXAMPLE 9.2.5. Let $t = 2\pi a$ in Theorem 9.1.1 and take $\nu = 2m + 1$ with $m \in \mathbb{N}$ to obtain

(9.2.25) $$\int_0^\infty \frac{x^{2m} \, dx}{\cosh x - \cos 2\pi a} = \frac{2(2m)!}{\sin 2\pi a} \sum_{k=1}^\infty \frac{\sin 2\pi ka}{k^{2m+1}}.$$

This is entry **3.531.5** in [35]. The hypotheses of the theorem restrict a to $0 < a < 1/2$, but the symmetry about $a = 1/2$ implies that (9.2.25) also holds for $1/2 < a < 1$.

In the special case $a = \frac{1}{2}$, replacing $\sin 2\pi ka / \sin 2\pi a$ by its limiting value produces

(9.2.26) $$\int_0^\infty \frac{x^{2m} \, dx}{\cosh x + 1} = 2(1 - 2^{1-2m})(2m)!\zeta(2m),$$

in agreement with (9.2.19). For positive integer m, the relation

(9.2.27) $$\zeta(2m) = \frac{2^{2m-1}\pi^{2m}|B_{2m}|}{(2m)!}$$

expresses the integral in (9.2.26) in terms of the Bernoulli numbers B_{2m} as

(9.2.28) $$\int_0^\infty \frac{x^{2m} \, dx}{\cosh x + 1} = 2(2^{2m-1} - 1)\pi^{2m}|B_{2m}|.$$

CHAPTER 10

Evaluation by series

10.1. Introduction

The table of integrals [**35**] contains a large variety of definite integrals that can be evaluated by expanding the integrand. The idea is remarkably simple: to evaluate

$$(10.1.1) \qquad I = \int_a^b f(x)\,dx,$$

one chooses a set of functions $\{f_n : n \in \mathbb{N}\}$ for which it is possible to expand

$$(10.1.2) \qquad f(x) = \sum_{n=1}^{\infty} a_n f_n$$

uniformly on $[a, b]$. Then, with

$$(10.1.3) \qquad b_n = \int_a^b f_n(x)\,dx,$$

it follows that

$$(10.1.4) \qquad \int_a^b f(x)\,dx = \sum_{n=1}^{\infty} a_n b_n.$$

In order to obtain a simpler form of the integral I, it is required to identify the series in (10.1.4).

10.2. A hypergeometric example

The first example is entry **3.311.4** in [**35**]

$$(10.2.1) \qquad \int_0^{\infty} \frac{e^{-qx}\,dx}{1 - ae^{-px}} = \sum_{k=0}^{\infty} \frac{a^k}{q + kp}.$$

Expanding the integrand as a geometric series produces

$$(10.2.2) \qquad \frac{1}{1 - ae^{-px}} = \sum_{k=0}^{\infty} a^k e^{-kpx},$$

and integrating over $[0, \infty)$ gives

$$(10.2.3) \qquad I = \sum_{k=0}^{\infty} a^k \int_0^{\infty} e^{-(q+kp)x}\,dx = \sum_{k=0}^{\infty} \frac{a^k}{q + kp}.$$

The resulting series may be identified as a hypergeometric sum. Recall that the **hypergeometric function** is defined by

$$(10.2.4) \qquad {}_pF_q\left(a_1, \cdots a_p; b_1, \cdots, b_q; x\right) := \sum_{k=0}^{\infty} \frac{(a_1)_k \cdots (a_p)_k}{(b_1)_k \cdots (b_q)_k} \frac{x^k}{k!},$$

where the Pochhammer symbol $(a)_k$ is

$$(10.2.5) \qquad (a)_k = \begin{cases} a(a+1)(a+2)\cdots(a+k-1), & \text{if } k > 0 \\ 1 & \text{if } k = 0. \end{cases}$$

The reader will find in [**41**] a selection of entries in [**35**] that are evaluated in terms of these functions.

To identify the series in (10.2.3), write it as

$$(10.2.6) \qquad \sum_{k=0}^{\infty} \frac{a^k}{q+kp} = \frac{1}{p}\sum_{k=0}^{\infty} \frac{a^k}{k+c},$$

where $c = q/p$. Now use $k! = (1)_k$ and

$$(10.2.7) \qquad k+c = \frac{c\,(c+1)_k}{(c)_k}$$

to write

$$(10.2.8) \qquad \sum_{k=0}^{\infty} \frac{a^k}{q+kp} = \frac{1}{q}\sum_{k=0}^{\infty} \frac{(c)_k\,(1)_k}{(1+c)_k} \frac{a^k}{k!}.$$

It follows that

$$(10.2.9) \qquad \int_0^{\infty} \frac{e^{-qx}\,dx}{1-ae^{-px}} = \frac{1}{q}\,{}_2F_1\left(\frac{q}{p}, 1; 1+\frac{q}{p}; a\right).$$

10.3. An integral involving the binomial theorem

Entry **3.194.8**

$$(10.3.1) \qquad \int_0^1 \frac{x^{n-1}\,dx}{(1+x)^m} = 2^{-n}\sum_{k=0}^{\infty} \binom{m-n-1}{k} \frac{(-2)^{-k}}{n+k}$$

is now evaluated using the binomial theorem

$$(10.3.2) \qquad (1-t)^{-a} = \sum_{k=0}^{\infty} \frac{(a)_k}{k!}t^k.$$

Indeed, the change of variables $t = x/(1+x)$ produces

$$(10.3.3) \qquad \int_0^1 \frac{x^{n-1}\,dx}{(1+x)^m} = \int_0^{1/2} \frac{t^{n-1}\,dt}{(1-t)^{n-m+1}}.$$

The integrand is expanded by the binomial theorem (10.3.2) in the form

$$(10.3.4) \qquad (1-t)^{m-n-1} = \sum_{k=0}^{\infty} \binom{m-n-1}{k}(-t)^k$$

and replacing the series in (10.3.3) produces the stated result.

10.4. A product of logarithms

This section considers the evaluation of entries in [35] where the integrand is the product of two logarithmic functions. The entries are evaluated by expanding the integrand in series. Alternative proofs are sometimes offered.

EXAMPLE 10.4.1. The value of entry **4.221.1**

$$(10.4.1) \qquad \int_0^1 \ln x \, \ln(1-x) \, dx = 2 - \frac{\pi^2}{6}$$

can be obtained from the expansion

$$(10.4.2) \qquad \ln(1-x) = -\sum_{k=1}^{\infty} \frac{x^k}{k}.$$

It follows that

$$(10.4.3) \qquad \int_0^1 \ln x \, \ln(1-x) \, dx = -\sum_{k=1}^{\infty} \frac{1}{k} \int_0^1 x^k \ln x \, dx$$

and the integral can be evaluated by integration by parts to produce

$$(10.4.4) \qquad \int_0^1 x^k \ln x \, dx = -\frac{1}{(k+1)^2}.$$

Therefore

$$(10.4.5) \qquad \int_0^1 \ln x \, \ln(1-x) \, dx = \sum_{k=1}^{\infty} \frac{1}{k(k+1)^2}$$

and the partial fraction decomposition

$$(10.4.6) \qquad \frac{1}{k(k+1)^2} = \frac{1}{k} - \frac{1}{k+1} - \frac{1}{(k+1)^2}$$

gives the result.

EXAMPLE 10.4.2. The evaluation of entry **4.221.2**

$$(10.4.7) \qquad \int_0^1 \ln x \, \ln(1+x) \, dx = 2 - \frac{\pi^2}{12} - 2\ln 2$$

can be obtained by using the expansion

$$(10.4.8) \qquad \ln(1+x) = \sum_{k=1}^{\infty} \frac{(-1)^{k-1}}{k} x^k$$

and replacing the series in the integral to obtain

$$(10.4.9) \qquad \int_0^1 \ln x \, \ln(1+x) \, dx = \sum_{k=1}^{\infty} \frac{(-1)^{k-1}}{k} \int_0^1 x^k \ln x \, dx.$$

Integration by parts produces

$$(10.4.10) \qquad \int_0^1 x^k \ln x\, dx = -\frac{1}{(k+1)^2},$$

and this leads to

$$(10.4.11) \qquad \int_0^1 \ln x \ln(1+x)\, dx = \sum_{k=1}^{\infty} \frac{(-1)^k}{k(k+1)^2}.$$

Expanding

$$(10.4.12) \qquad \frac{1}{k(k+1)^2} = \frac{1}{k} - \frac{1}{k+1} - \frac{1}{(k+1)^2}$$

and using the values

$$(10.4.13) \qquad \sum_{k=1}^{\infty} \frac{(-1)^k}{k} = -\ln 2 \ \text{ and } \ \sum_{k=1}^{\infty} \frac{(-1)^k}{k^2} = -\frac{\pi^2}{12}$$

produces the result.

EXAMPLE 10.4.3. The evaluation of entry **4.221.3**

$$(10.4.14) \qquad \int_0^1 \ln\left(\frac{1-ax}{1-a}\right) \frac{dx}{\ln x} = -\sum_{k=1}^{\infty} \frac{a^k}{k} \ln(1+k)$$

is obtained via the change of variables $x = e^{-t}$ to produce

$$(10.4.15) \qquad \int_0^1 \ln\left(\frac{1-ae^{-t}}{1-a}\right) \frac{dx}{\ln x} = -\int_0^{\infty} \ln\left(\frac{1-ae^{-t}}{1-a}\right) \frac{e^{-t}}{t}\, dt.$$

The expansions

$$(10.4.16) \qquad \ln(1-ae^{-t}) = -\sum_{k=1}^{\infty} \frac{a^k}{k} e^{-kt} \ \text{ and } \ \ln(1-a) = -\sum_{k=1}^{\infty} \frac{a^k}{k}$$

produce

$$(10.4.17) \qquad \int_0^1 \ln\left(\frac{1-ae^{-t}}{1-a}\right) \frac{dx}{\ln x} = \sum_{k=1}^{\infty} \frac{a^k}{k} \int_0^{\infty} \frac{e^{-kt}-1}{t} e^{-t} dt.$$

The integral

$$(10.4.18) \qquad g(k) = \int_0^{\infty} \frac{e^{-kt}-1}{t} e^{-t}\, dt,$$

appearing above, satisfies $g(0) = 0$ and $g'(k) = -1/(k+1)$. Thus $g(k) = -\ln(1+k)$ as required.

The series

$$(10.4.19) \qquad h(a) = -\sum_{k=1}^{\infty} \frac{a^k}{k} \ln(1+k)$$

in the formula (10.4.14) is related to the **polylogarithm function**

$$(10.4.20) \qquad \mathrm{Li}_b(x) := \sum_{k=1}^{\infty} \frac{x^k}{k^b}.$$

Indeed, $h(a)$ satisfies
(10.4.21)

$$h'(a) = -\sum_{k=1}^{\infty} a^{k-1} \ln(1+k) = -\frac{1}{a^2} \sum_{k=2}^{\infty} a^k \ln k = -\frac{1}{a^2} \frac{\partial}{\partial b} \mathrm{Li}_{-b}(a) \Big|_{b=0}.$$

10.5. Some integrals involving the exponential function

This section presents some examples involving the exponential function.

EXAMPLE 10.5.1. The evaluation of entry **3.342**

$$(10.5.1) \qquad \int_0^1 \exp(-px \ln x)\, dx = \int_0^1 x^{-px}\, dx = \frac{1}{p} \sum_{k=1}^{\infty} \left(\frac{p}{k}\right)^k$$

can be established by expanding the integrand in series. Indeed,

$$(10.5.2) \qquad \int_0^1 \exp(-px \ln x)\, dx \;=\; \sum_{k=0}^{\infty} \frac{(-1)^k p^k}{k!} \int_0^1 x^k \ln^k x\, dx.$$

The change of variables $x = e^{-t}$ gives

$$\int_0^1 x^k \ln^k x\, dx \;=\; (-1)^k \int_0^\infty t^k e^{-(k+1)t}\, dt$$

$$= \frac{(-1)^k}{(k+1)^{k+1}} \int_0^\infty s^k e^{-s}\, ds$$

$$= \frac{(-1)^k k!}{(k+1)^{k+1}}.$$

Replacing this expression in (10.5.2) gives the result.

EXAMPLE 10.5.2. A similar procedure provides the evaluation of entry **3.466.3**

$$(10.5.3) \qquad \int_0^1 \frac{e^{x^2} - 1}{x^2}\, dx = \sum_{k=1}^{\infty} \frac{1}{k!\,(2k-1)}.$$

Expand the exponential in the integrand and integrate term by term. The resulting series can be identified as

$$(10.5.4) \qquad \sum_{k=1}^{\infty} \frac{1}{k!\,(2k-1)} = 1 - e + \sqrt{\pi}\,\mathrm{erfi}(1),$$

where

$$(10.5.5) \qquad \mathrm{erfi}(x) = \frac{2}{\sqrt{\pi}} \int_0^x e^{t^2}\, dt.$$

The generalization

$$(10.5.6) \qquad d_n := \int_0^1 \frac{1}{x^{2n+2}} \left(e^{x^2} - \sum_{k=0}^n \frac{x^{2k}}{k!} \right) dx$$

is evaluated as

$$(10.5.7) \qquad d_n = \sum_{k=n+1}^\infty \frac{1}{k! \, (2k - 2n - 1)}.$$

The first few values are

$$
\begin{aligned}
d_0 &= 1 - e + \sqrt{\pi}\,\mathrm{erfi}(1) \\
d_1 &= \frac{4}{3} - e + \frac{2}{3}\sqrt{\pi}\,\mathrm{erfi}(1) \\
d_2 &= \frac{31}{30} - \frac{3e}{5} + \frac{4}{15}\sqrt{\pi}\,\mathrm{erfi}(1) \\
d_3 &= \frac{71}{105} - \frac{11e}{35} + \frac{8}{105}\sqrt{\pi}\,\mathrm{erfi}(1) \\
d_4 &= \frac{379}{840} - \frac{19e}{105} + \frac{16}{945}\sqrt{\pi}\,\mathrm{erfi}(1).
\end{aligned}
$$

The reader will check that the coefficient of $\sqrt{\pi}\,\mathrm{erfi}(1)$ in d_n is $2^n/(2n+1)!!$. The remaining coefficients will be explored in a future article.

10.6. Some combinations of powers and algebraic functions

This section considers entries of the form

$$(10.6.1) \qquad \int_0^\infty x^n A\left(e^{-x}\right) dx$$

where A is an algebraic function; that is, it satisfies $P(x, A(x)) = 0$, for some polynomial P.

THEOREM 10.6.1. *Let* $a > 0$. *Then*

$$(10.6.2) \qquad \int_0^\infty x e^{-x} \sqrt{1 - e^{-ax}} \, dx = 1 - \sum_{k=0}^\infty \frac{(2k)!}{2^{2k+1}\,(k+1)!\,k!\,(b+ak)^2}$$

with $b = 1 + a$.

PROOF. The binomial theorem shows that

$$
\begin{aligned}
\sqrt{1 - e^{-ax}} &= \sum_{k=0}^\infty \binom{\frac{1}{2}}{k} (-1)^k e^{-akx} \\
&= -\sum_{k=0}^\infty \frac{(2k)!}{(2k-1)\,2^{2k}k!^2} e^{-akx}.
\end{aligned}
$$

Integration yields

$$\int_0^\infty xe^{-x}\sqrt{1-e^{-ax}}\,dx = -\sum_{k=0}^\infty \frac{(2k)!}{(2k-1)\,2^{2k}k!^2} \int_0^\infty xe^{-(1+ak)x}\,dx$$

$$= -\sum_{k=0}^\infty \frac{(2k)!}{(2k-1)\,2^{2k}k!^2}\frac{1}{(1+ak)^2}.$$

Now shift the index of summation to obtain the stated form. □

In the examples below, the notation

$$(10.6.3) \qquad I(a) = \int_0^\infty xe^{-x}\sqrt{1-e^{-ax}}\,dx = 1 - S(a)$$

where

$$(10.6.4) \qquad S(a) := \sum_{k=0}^\infty \frac{(2k)!}{2^{2k+1}\,(k+1)!\,k!\,(b+ak)^2}$$

is employed.

EXAMPLE 10.6.2. Entry **3.451.1** states that

$$(10.6.5) \qquad \int_0^\infty xe^{-x}\sqrt{1-e^{-x}}\,dx = \frac{4}{3}\left(\frac{4}{3}-\ln 2\right).$$

This entry corresponds to $I(1) = 1 - S(1)$, where

$$(10.6.6) \qquad S(1) = \sum_{k=0}^\infty \frac{(2k)!}{2^{2k+1}(k+2)^2(k+1)k!^2}.$$

To evaluate this sum by elementary means, start with

$$(10.6.7) \qquad f(x) = \sum_{k=0}^\infty \frac{(2k)!}{k!^2}x^k = \frac{1}{\sqrt{1-4x}},$$

where the last evaluation comes from the binomial theorem. Define

$$(10.6.8) \qquad g(x) = \int_0^x f(t)\,dt = \sum_{k=0}^\infty \frac{(2k)!}{k!^2}\frac{x^{k+1}}{k+1}.$$

An elementary calculation shows that

$$(10.6.9) \qquad g(x) = \frac{1}{2}(1-\sqrt{1-4x}).$$

Now define
$$(10.6.10)$$
$$h(x) = \int_0^x g(t)\,dt = \sum_{k=0}^\infty \frac{(2k)!}{k!^2}\frac{x^{k+2}}{(k+1)(k+2)} = \frac{x}{2} - \frac{1}{12} + \frac{1}{12}(1-4x)^{3/2}$$

and

$$(10.6.11) \qquad w(x) = \int_0^x \frac{h(t)}{t}\,dt = \sum_{k=0}^\infty \frac{(2k)!}{k!^2}\frac{x^{k+2}}{(k+1)(k+2)^2}.$$

The relation $S(1) = 8w(1/4)$ comes by comparing this last series to the one defining $S(1)$ in (10.6.6). Now observe that

$$(10.6.12) \qquad w(x) = \int_0^x \frac{h(t)}{t}\, dt = \frac{x}{2} + \frac{1}{12} J(x),$$

where

$$(10.6.13) \qquad J(x) = \int_0^x \frac{(1 - 4t)^{3/2} - 1}{t}\, dt.$$

The change of variables $u = \sqrt{1 - 4t}$ gives

$$(10.6.14) \qquad J(x) = -2 \int_{\sqrt{1-4x}}^1 \frac{u(1 + u + u^2)}{1 + u}\, du,$$

and the further change of variables $v = 1 + u$ gives

$$(10.6.15) \qquad J(x) = -2 \int_\sigma^2 (v^2 - 2v + 2 - 1/v)\, dv,$$

where $\sigma = 1 + \sqrt{1 - 4x}$. This last integral can be evaluated in elementary terms to produce

$$w(x) = \frac{1}{18} \left(-4 + 4\sqrt{1 - 4x} + x(9 - 4\sqrt{1 - 4x}) + 3 \ln 2 - 3 \ln(1 + \sqrt{1 - 4x}) \right).$$

In particular, $w\left(\frac{1}{4}\right) = \frac{1}{18} \left(3 \ln 2 - \frac{7}{4} \right)$ and then

$$(10.6.16) \qquad I(1) = 1 - S(1) = 1 - 8w\left(\frac{1}{4}\right) = \frac{4}{9}(4 - 3 \ln 2)$$

as required.

EXAMPLE 10.6.3. The second entry in this family is **3.451.2**, which states

$$(10.6.17) \qquad \int_0^\infty x e^{-x} \sqrt{1 - e^{-2x}}\, dx = \frac{\pi}{4} \left(\frac{1}{2} + \ln 2 \right).$$

The same technique used in the previous example is now used to evaluate $I(2) = 1 - S(2)$, where

$$(10.6.18) \qquad S(2) = \sum_{k=0}^\infty \frac{(2k)!}{2^{2k+1}\, (k + 1)!\, k!\, (2k + 3)^2}.$$

Start with

$$(10.6.19) \qquad f(x) = \sum_{k=0}^\infty \frac{(2k)!}{k!^2} x^k = \frac{1}{\sqrt{1 - 4x}},$$

and then evaluate

$$(10.6.20) \qquad g(x) := \int_0^x f(t)dt = \sum_{k=0}^\infty \frac{(2k)!}{k!\, (k + 1)!} x^{k+1},$$

and as before

$$(10.6.21) \qquad g(x) = \frac{1 - \sqrt{1 - 4x}}{2}.$$

The next step is to form

$$(10.6.22) \qquad h(x) = \int_0^x g(t^2)\, dt = \sum_{k=0}^{\infty} \frac{(2k)!}{(k+1)!\, k!} \frac{x^{2k+3}}{2k+3}.$$

Now

$$
\begin{aligned}
h(x) &= \frac{1}{2} \int_0^x \left(1 - \sqrt{1 - 4t^2}\right) dt \\
&= \frac{x}{2} - \frac{1}{2} \int_0^x \sqrt{1 - 4t^2}\, dt.
\end{aligned}
$$

Elementary changes of variables yield

$$(10.6.23) \qquad h(x) = \frac{x}{2} - \frac{1}{8} \sin^{-1}(2x) - \frac{x}{4} \sqrt{1 - 4x^2}.$$

Define

$$(10.6.24) \qquad w(x) = \int_0^x \frac{h(t)}{t}\, dt = \sum_{k=0}^{\infty} \frac{(2k)!}{(k+1)!\, k!} \frac{x^{2k+3}}{(2k+3)^2},$$

so that $S_2 = 4w(1/2)$. Now,

$$
\begin{aligned}
w(x) &= \int_0^x \left(\frac{1}{2} - \frac{1}{8} \frac{\sin^{-1}(2t)}{t} - \frac{1}{4} \sqrt{1 - 4t^2} \right) dt \\
&= \frac{x}{2} - \frac{1}{8} x \sqrt{1 - 4x^2} - \frac{1}{16} \sin^{-1}(2x) - \frac{1}{8} \int_0^x \frac{\sin^{-1}(2t)}{t}\, dt.
\end{aligned}
$$

The change of variables $\varphi = \sin^{-1}(2t)$ yields

$$(10.6.25) \qquad \int_0^x \frac{\sin^{-1}(2t)}{t}\, dt = \int_0^{\sin^{-1}(2x)} \varphi \cot \varphi\, d\varphi.$$

It follows that

$$(10.6.26) \qquad w(\tfrac{1}{2}) = \frac{1}{4} - \frac{\pi}{12} - \frac{1}{8} \int_0^{\pi/2} \frac{\varphi\, d\varphi}{\tan \varphi}.$$

The evaluation

$$(10.6.27) \qquad \int_0^{\pi/2} \frac{\varphi\, d\varphi}{\tan \varphi} = \frac{\pi}{2} \ln 2$$

appears as entry **3.747.7** and it has been presented in [**10**]. Therefore

$$(10.6.28) \qquad I(2) = \frac{\pi}{8}(1 + 2\ln 2),$$

as claimed.

The integral in Theorem 10.6.1 is evaluated in an alternative form. The answer involves the digamma function

$$(10.6.29) \qquad \psi(x) = \frac{\Gamma'(x)}{\Gamma(x)} = \frac{d}{dx} \log \Gamma(x).$$

The reader will find in [47] a variety of entries from [35] evaluated in terms of this function.

THEOREM 10.6.4. *Let $a > 0$. Then*

$$(10.6.30) \qquad \int_0^\infty x e^{-x}\sqrt{1 - e^{-ax}}\, dx = \frac{\sqrt{\pi}\,\Gamma\left(\frac{1}{a}\right)}{2a^2\Gamma\left(\frac{3}{2} + \frac{1}{a}\right)} \left[\psi\left(\tfrac{3}{2} + \tfrac{1}{a}\right) - \psi\left(\tfrac{1}{a}\right)\right].$$

PROOF. The change of variables $t = e^{-x}$ gives

$$(10.6.31) \qquad \int_0^\infty x e^{-x}\sqrt{1 - e^{-ax}}\, dx = -\int_0^1 \ln t \sqrt{1 - t^a}\, dt.$$

This last form of the integral is evaluated by differentiating the identity

$$(10.6.32) \qquad \int_0^1 (1 - t^a)^{1/2} t^b\, dt = \frac{\sqrt{\pi}\,\Gamma\left(\frac{1+b}{a}\right)}{2a\Gamma\left(\frac{3}{2} + \frac{1+b}{a}\right)}$$

at $b = 0$. $\qquad\qquad\qquad\qquad\qquad\qquad\qquad\qquad\qquad\qquad\qquad\qquad\square$

The special values required for the evaluations of the entries discussed in this section are

$$(10.6.33) \qquad \Gamma(n) = (n-1)! \text{ and } \Gamma\left(n + \tfrac{1}{2}\right) = \frac{\sqrt{\pi}\,(2n)!}{2^{2n}\, n!},$$

which appear as entries **8.339.1** and **8.339.2**, respectively, and also

$$(10.6.34)\ \ \psi(n) = -\gamma + \sum_{k=1}^{n-1}\frac{1}{k} \text{ and } \psi\left(n + \tfrac{1}{2}\right) = -\gamma + 2\left(\sum_{k=1}^{n}\frac{1}{2k-1} - \ln 2\right),$$

which are found as entries **8.365.4** and **8.366.3**, respectively.

10.7. Some examples related to geometric series

The paper [48] contains a variety of entries in [35] that are obtained by manipulating the geometric series

$$(10.7.1) \qquad\qquad \frac{1}{1-x} = \sum_{k=0}^\infty x^k$$

and the alternating version

$$(10.7.2) \qquad\qquad \frac{1}{1+x} = \sum_{k=0}^\infty (-1)^k x^k.$$

A couple of examples are presented here.

Integrating term by term yields

$$(10.7.3) \qquad \int_0^1 \frac{x^m}{1+x}\ln x\, dx = \sum_{k=0}^\infty (-1)^k \int_0^1 x^{k+m}\ln x\, dx$$

Integration by parts gives

$$(10.7.4) \qquad \int_0^1 x^{k+m}\ln x\, dx = -\frac{1}{(k+m+1)^2},$$

therefore

$$
\begin{aligned}
\int_0^1 \frac{x^m}{1+x} \ln x \, dx &= \sum_{k=0}^{\infty} \frac{(-1)^{k+1}}{(k+m+1)^2} \\
&= (-1)^m \sum_{k=m+1}^{\infty} \frac{(-1)^k}{k^2} \\
&= (-1)^{m+1} \left[\frac{\pi^2}{12} + \sum_{k=1}^{m} \frac{(-1)^k}{k^2} \right],
\end{aligned}
$$

using

(10.7.5)
$$
\sum_{k=1}^{\infty} \frac{(-1)^k}{k^2} = -\frac{\pi^2}{12}.
$$

This establishes

(10.7.6)
$$
\int_0^1 \frac{x^m}{1+x} \ln x \, dx = (-1)^{m+1} \left[\frac{\pi^2}{12} + \sum_{k=1}^{m} \frac{(-1)^k}{k^2} \right].
$$

EXAMPLE 10.7.1. Entry **4.251.5** states that

(10.7.7)
$$
\int_0^1 \frac{x^{2n}}{1+x} \ln x \, dx = -\frac{\pi^2}{12} - \sum_{k=1}^{2n} \frac{(-1)^k}{k^2}.
$$

This is the case $m = 2n$, an even integer of (10.7.6).

EXAMPLE 10.7.2. Entry **4.251.6** states that

(10.7.8)
$$
\int_0^1 \frac{x^{2n-1}}{1+x} \ln x \, dx = \frac{\pi^2}{12} + \sum_{k=1}^{2n-1} \frac{(-1)^k}{k^2}.
$$

This is the case $m = 2n - 1$, an odd integer of (10.7.6).

EXAMPLE 10.7.3. The integral

(10.7.9)
$$
I(\alpha) = \int_0^1 \left(\ln \frac{1}{x} \right)^{\alpha} \frac{dx}{1+x^2}
$$

is evaluated by expanding $1/(1+x^2)$ as a geometric series to obtain

(10.7.10)
$$
I(\alpha) = \sum_{j=0}^{\infty} (-1)^j \int_0^1 x^{2j} (-\ln x)^{\alpha} \, dx.
$$

The changes of variables $u = -\ln x$ and $v = (2j+1)u$ give

$$
\begin{aligned}
I(\alpha) &= \sum_{j=0}^{\infty} (-1)^j \int_0^{\infty} u^{\alpha} e^{-(2j+1)u} \, du \\
&= \sum_{j=0}^{\infty} \frac{(-1)^j}{(2j+1)^{\alpha+1}} \int_0^{\infty} v^{\alpha} e^{-v} \, dv \\
&= \Gamma(\alpha+1) \sum_{j=0}^{\infty} \frac{(-1)^j}{(2j+1)^{\alpha+1}}.
\end{aligned}
$$

Entry **4.269.1** is the special case $\alpha = \frac{1}{2}$ that produces

$$
(10.7.11) \qquad \int_0^1 \sqrt{\ln \frac{1}{x}} \, \frac{dx}{1+x^2} = \frac{\sqrt{\pi}}{2} \sum_{k=0}^{\infty} \frac{(-1)^j}{\sqrt{(2j+1)^3}}
$$

and entry **4.269.2**

$$
(10.7.12) \qquad \int_0^1 \frac{dx}{\sqrt{\ln \frac{1}{x}} \, (1+x^2)} = \sqrt{\pi} \sum_{k=0}^{\infty} \frac{(-1)^j}{\sqrt{2j+1}}
$$

corresponds to $\alpha = -\frac{1}{2}$.

CHAPTER 11

The exponential integral

11.1. Introduction

The *exponential integral* function is defined by

$$\text{(11.1.1)} \qquad \text{Ei}(x) = \int_{-\infty}^{x} \frac{e^t}{t}\, dt$$

for $x < 0$. In the case $x > 0$ we use the Cauchy principal value

$$\text{(11.1.2)} \qquad \text{Ei}(x) = -\lim_{\epsilon \to 0^+} \left[\int_{-x}^{-\epsilon} \frac{e^{-t}}{t}\, dt + \int_{\epsilon}^{\infty} \frac{e^{-t}}{t}\, dt \right].$$

This appears as entry **3.351.6** in [**35**].

Another function defined by an integral is the *logarithmic integral*

$$\text{(11.1.3)} \qquad \text{li}(u) := \int_{0}^{u} \frac{dx}{\ln x}.$$

This is entry **4.211.2**. The change of variables $t = \ln x$ shows that

$$\text{(11.1.4)} \qquad \text{li}(u) = \text{Ei}(\ln u).$$

Observe that the integral defining li diverges as $u \to \infty$. Indeed, entry **4.211.1** states that

$$\text{(11.1.5)} \qquad \int_{e}^{\infty} \frac{dx}{\ln x} = +\infty.$$

This is evident from the change of variables $t = \ln x$, which yields

$$\text{(11.1.6)} \qquad \int_{e}^{\infty} \frac{dx}{\ln x} = \int_{1}^{\infty} \frac{e^t\, dt}{t} \geq \int_{1}^{\infty} \frac{dt}{t} = \infty.$$

11.2. Some simple changes of variables

The change of variables $t = -as$ yields

$$\text{(11.2.1)} \qquad \int_{-x/a}^{\infty} \frac{e^{-as}}{s}\, ds = -\text{Ei}(x).$$

Replacing x by ax, this gives

$$\text{(11.2.2)} \qquad \int_{-ax}^{\infty} \frac{e^{-t}}{t}\, dt = -\text{Ei}(ax).$$

The special choice $x = -a$ in (11.2.1) yields entry **3.351.5**

(11.2.3) $$\int_1^\infty \frac{e^{-as}}{s} ds = -\mathrm{Ei}(-a).$$

The expression

(11.2.4) $$\mathrm{Ei}(-a) = -\int_1^\infty \frac{e^{-as}}{s} ds$$

is an analytic function of a for $\mathrm{Re}\, a > 0$. This provides an analytic extension of $\mathrm{Ei}(z)$ to the left half-plane $\mathrm{Re}\, z < 0$. Several entries of [**35**] are derived from here.

EXAMPLE 11.2.1. For any β such that $u + \beta > 0$

(11.2.5) $$\mathrm{Ei}(-au - a\beta) = \mathrm{Ei}(-a(u + \beta)) = -\int_{u+\beta}^\infty \frac{e^{-ax}}{x} dx$$

and then the shift $x \mapsto x + \beta$ produces

(11.2.6) $$\mathrm{Ei}(-au - a\beta) = -e^{-a\beta} \int_u^\infty \frac{e^{-ax}}{x + \beta} dx,$$

which can be written as

(11.2.7) $$\int_u^\infty \frac{e^{-ax}}{x + \beta} dx = -e^{a\beta}\mathrm{Ei}(-au - a\beta).$$

This appears as entry **3.352.2**. This representation is valid for $\beta \in \mathbb{C}$ outside the half-line $(-\infty, u]$.

EXAMPLE 11.2.2. The special case $u = 0$ and $\beta \notin (-\infty, 0]$ gives

(11.2.8) $$\int_0^\infty \frac{e^{-ax}}{x + \beta} dx = -e^{a\beta}\mathrm{Ei}(-a\beta).$$

This is entry **3.352.4** in [**35**].

EXAMPLE 11.2.3. The difference of (13.2.16) and (11.2.8) produces

(11.2.9) $$\int_0^u \frac{e^{-ax}}{x + \beta} dx = e^{au}\left[\mathrm{Ei}(-au - a\beta) - \mathrm{Ei}(-a\beta)\right].$$

This is entry **3.352.1**.

EXAMPLE 11.2.4. Entry **3.352.3** states that

(11.2.10) $$\int_u^v \frac{e^{-ax}}{x + \beta} dx = e^{a\beta}\left[\mathrm{Ei}(-a(v + \beta)) - \mathrm{Ei}(-a(u + \beta))\right].$$

This comes directly from (13.2.16)

(11.2.11) $$\int_u^v \frac{e^{-ax}\, dx}{x + \beta} = \int_u^\infty \frac{e^{-ax}\, dx}{x + \beta} - \int_v^\infty \frac{e^{-ax}\, dx}{x + \beta}$$
$$= -e^{a\beta}\mathrm{Ei}(-au - a\beta) + e^{a\beta}\mathrm{Ei}(-av - a\beta).$$

This is the result.

EXAMPLE 11.2.5. In the expression (13.2.16), when $u > 0$, the parameter β may be taken in the range $\beta < u$, so that $x - \beta > 0$ for all $x \geq u$. This produces entry **3.352.5**

$$(11.2.12) \qquad \int_u^\infty \frac{e^{-ax}\,dx}{x - \beta} = -e^{-a\beta}\text{Ei}(-a(u - \beta)).$$

EXAMPLE 11.2.6. In the case $u = 0$ and $\beta < 0$, the entry in Example 11.2.5 can be written as

$$(11.2.13) \qquad \int_0^\infty \frac{e^{-ax}\,dx}{\beta - x} = e^{-a\beta}\text{Ei}(a\beta).$$

This is entry **3.352.6** in [**35**].

11.3. Entries obtained by differentiation

This section presents proofs of some entries in [**35**] obtained by manipulations of derivatives of the exponential integral function.

EXAMPLE 11.3.1. Entry **3.353.3** is

$$(11.3.1) \qquad \int_0^\infty \frac{e^{-ax}\,dx}{(x + \beta)^2} = \frac{1}{\beta} + ae^{-a\beta}\text{Ei}(-a\beta).$$

To establish this, differentiate (13.2.16) and use

$$(11.3.2) \qquad \frac{d}{dt}\text{Ei}(u) = \frac{e^u}{u}\frac{du}{dt}$$

to obtain

$$(11.3.3) \qquad \int_u^\infty \frac{e^{-ax}\,dx}{(x + \beta)^2} = \frac{e^{-au}}{u + \beta} + ae^{a\beta}\text{Ei}(-au - a\beta).$$

The choice $u = 0$ with $\text{Re}\,\beta > 0$ and $\text{Re}\,a > 0$ gives the result.

EXAMPLE 11.3.2. Entry **3.353.1** states that
$$(11.3.4)$$
$$\int_u^\infty \frac{e^{-ax}\,dx}{(x + \beta)^n} = e^{-au}\sum_{k=1}^{n-1}\frac{(k - 1)!(-a)^{n-k-1}}{(n - 1)!(u + \beta)^k} - \frac{(-a)^{n-1}}{(n - 1)!}e^{a\beta}\text{Ei}(-au - a\beta).$$

This can be easily established by induction. The initial step $n = 2$ is (11.3.3). Simply differentiate (11.3.4) with respect to β to move from n to $n + 1$. The details are left to the reader.

EXAMPLE 11.3.3. The special case $u = 0$ of (11.3.4) gives

$$(11.3.5) \qquad \int_0^\infty \frac{e^{-ax}\,dx}{(x + \beta)^n} = \sum_{k=1}^{n-1}\frac{(k - 1)!(-a)^{n-k-1}}{(n - 1)!\beta^k} - \frac{(-a)^{n-1}}{(n - 1)!}e^{a\beta}\text{Ei}(-a\beta).$$

This is entry **3.353.2** in [**35**].

EXAMPLE 11.3.4. Entry **3.351.4** states that

$$(11.3.6) \quad \int_u^\infty \frac{e^{-ax}\,dx}{x^{n+1}} = e^{-au} \sum_{k=1}^n \frac{(k-1)!(-a)^{n-k}}{n!u^k} + (-1)^{n+1}\frac{a^n}{n!}\mathrm{Ei}(-au).$$

This results follows directly from (11.3.4) by taking $\beta = 0$ and $u > 0$ and then replacing n by $n+1$. Changing the index of summation $k \mapsto n-k$, this may be written as it appears in [**35**]

$$(11.3.7) \quad \int_u^\infty \frac{e^{-ax}\,dx}{x^{n+1}} = \frac{e^{-au}}{u^n} \sum_{k=1}^n \frac{(-1)^k a^k u^k}{n(n-1)\cdots(n-k)} + (-1)^{n+1}\frac{a^n}{n!}\mathrm{Ei}(-au).$$

EXAMPLE 11.3.5. Entry **3.353.5** states that

$$(11.3.8) \quad \int_0^\infty \frac{x^n e^{-ax}}{x+\beta}\,dx = (-1)^{n-1}\beta^n e^{a\beta}\mathrm{Ei}(-a\beta) + \sum_{k=1}^n (k-1)!(-\beta)^{n-k}\mu^{-k}.$$

In the special case $n = 1$, this reduces to

$$(11.3.9) \quad \int_0^\infty \frac{x e^{-ax}}{x+\beta}\,dx = \beta e^{a\beta}\mathrm{Ei}(-a\beta) + \frac{1}{a},$$

which follows by differentiating (11.2.8) with respect to a. The general formula (11.3.8) is obtained directly by further differentiation.

NOTE 11.3.6. The entry **3.353.4**

$$(11.3.10) \quad \int_0^1 \frac{x e^x\,dx}{(x+1)^2} = \frac{e}{2} - 1,$$

which does not involve the exponential integral function, can be evaluated by simple integration by parts. This entry has been included in Section 10 of [**8**].

11.4. Entries with quadratic denominators

This section considers the entries in [**35**] where the integrand is an exponential term divided by a quadratic polynomial.

EXAMPLE 11.4.1. Entry **3.354.3** is

$$(11.4.1) \quad \int_0^\infty \frac{e^{-ax}\,dx}{\beta^2 - x^2} = \frac{1}{2\beta}\left[e^{-a\beta}\mathrm{Ei}(a\beta) - e^{a\beta}\mathrm{Ei}(-a\beta) \right].$$

To evaluate this integral, assume $\beta \notin \mathbb{R}$ and use the partial fraction decomposition

$$(11.4.2) \quad \frac{1}{\beta^2 - x^2} = \frac{1}{2\beta}\left(\frac{1}{\beta - x} - \frac{1}{\beta + x} \right)$$

to obtain

$$(11.4.3) \quad \int_0^\infty \frac{e^{-ax}\,dx}{\beta^2 - x^2} = \frac{1}{2\beta}\left(\int_0^\infty \frac{e^{-ax}\,dx}{\beta - x} + \int_0^\infty \frac{e^{-ax}\,dx}{\beta + x} \right)$$

and now the result comes from (11.2.8) and (11.2.13). For $\beta \in \mathbb{R}$ the results are valid as a Cauchy principal value integral.

EXAMPLE 11.4.2. Differentiating (11.4.1) with respect to a produces

$$(11.4.4) \qquad \int_0^\infty \frac{xe^{-ax}\,dx}{\beta^2 - x^2} = \frac{1}{2}\left[e^{-a\beta}\operatorname{Ei}(a\beta) - e^{a\beta}\operatorname{Ei}(-a\beta)\right].$$

This appears as entry **3.354.4** in [**35**].

EXAMPLE 11.4.3. Entry **3.354.1**

$$(11.4.5) \qquad \int_0^\infty \frac{e^{-ax}\,dx}{\beta^2 + x^2} = \frac{1}{\beta}\left[\operatorname{ci}(a\beta)\sin a\beta - \operatorname{si}(a\beta)\cos a\beta\right]$$

involves the cosine and sine integrals defined by

$$(11.4.6) \qquad \operatorname{ci}(u) = -\int_u^\infty \frac{\cos t}{t}\,dt \text{ and } \operatorname{si}(u) = -\int_u^\infty \frac{\sin t}{t}\,dt.$$

Start by replacing β by $i\beta$ in (11.4.1) to obtain

$$(11.4.7) \qquad \int_0^\infty \frac{e^{-ax}\,dx}{\beta^2 + x^2} = \frac{1}{2i\beta}\left[e^{ia\beta}\operatorname{Ei}(-ia\beta) - e^{-ia\beta}\operatorname{Ei}(ia\beta)\right].$$

The classical identity of Euler

$$(11.4.8) \qquad e^{\pm i\beta} = \cos a\beta \pm i\sin a\beta$$

gives the relation

$$(11.4.9) \qquad \operatorname{Ei}(\pm ia\beta) = \operatorname{ci}(a\beta) \pm i\operatorname{si}(a\beta).$$

Replacing these expressions in (11.4.7) gives the result.

EXAMPLE 11.4.4. Differentiation of the entry in Example 11.4.3 gives

$$(11.4.10) \qquad \int_0^\infty \frac{xe^{-ax}\,dx}{\beta^2 + x^2} = -\operatorname{ci}(a\beta)\sin a\beta - \operatorname{si}(a\beta)\cos a\beta.$$

This is entry **3.354.2** in [**35**].

The entries in Sections **3.355** and **3.356** are obtained by differentiation of the entries in Section **3.354** given above.

EXAMPLE 11.4.5. Entry **3.355.1** is

$$(11.4.11) \quad \int_0^\infty \frac{e^{-ax}\,dx}{(\beta^2 + x^2)^2} = \frac{1}{2\beta^2}\{\operatorname{ci}(a\beta)\sin(a\beta) - \operatorname{si}(a\beta)\cos(a\beta) - a\beta\left[\operatorname{ci}(a\beta)\cos(a\beta) + \operatorname{si}(a\beta)\sin(a\beta)\right]\}.$$

This is obtained by differentiation of Entry **3.354.1** given in (11.4.5).

EXAMPLE 11.4.6. Entry **3.355.2** is

$$(11.4.12) \quad \int_0^\infty \frac{xe^{-ax}\,dx}{(\beta^2 + x^2)^2} = \frac{1}{2\beta^2}\left[1 - a\beta\left(\operatorname{ci}(a\beta)\sin(a\beta) - \operatorname{si}(a\beta)\cos(a\beta)\right)\right].$$

This entry appeared with a typo in [**35**]. This entry is obtained by direct differentiation of (11.4.11).

EXAMPLE 11.4.7. Differentiation of entries **3.354.3** and **3.354.4** produces
(11.4.13)
$$\int_0^\infty \frac{e^{-ax}\,dx}{(\beta^2 - x^2)^2} = \frac{1}{4\beta^3}\left[(a\beta - 1)e^{a\beta}\mathrm{Ei}(-a\beta) + (1 + a\beta)e^{-a\beta}\mathrm{Ei}(a\beta)\right]$$

and

(11.4.14) $$\int_0^\infty \frac{xe^{-ax}\,dx}{(\beta^2 - x^2)^2} = \frac{1}{4\beta^2}\left[-2 + a\beta\left(e^{-a\beta}\mathrm{Ei}(a\beta) - e^{a\beta}\mathrm{Ei}(-a\beta)\right)\right].$$

These are entries **3.355.3** and **3.355.4**, respectively.

EXAMPLE 11.4.8. Differentiating (11.4.5) $2n$ times with respect to a gives

(11.4.15)
$$\int_0^\infty \frac{x^{2n}e^{-ax}\,dx}{\beta^2 + x^2} = (-1)^{n-1}\beta^{2n}\left[\mathrm{ci}(a\beta)\cos(a\beta) + \mathrm{si}(a\beta)\sin(a\beta)\right] +$$
$$+ \frac{1}{\beta^{2n}}\sum_{k=1}^n (2n - 2k + 1)!(-a^2\beta^2)^{k-1}.$$

This appears as Entry **3.356.2**. The identity

(11.4.16) $$\int_0^\infty \frac{x^{2n}e^{-ax}\,dx}{\beta^2 - x^2} = \frac{1}{2}\beta^{2n-1}\left[e^{-a\beta}\mathrm{Ei}(a\beta) - e^{a\beta}\mathrm{Ei}(-a\beta)\right]$$
$$- \frac{1}{\beta^{2n-1}}\sum_{k=1}^n (2n - 2k)!(a^2\beta^2)^{k-1}$$

is obtained by differentiating (11.4.1). This appears as Entry **3.356.4**.

EXAMPLE 11.4.9. The entries **3.356.1**

(11.4.17) $$\int_0^\infty \frac{x^{2n+1}e^{-ax}\,dx}{\beta^2 + x^2} = (-1)^{n-1}\beta^{2n}\left[\mathrm{ci}(a\beta)\cos a\beta + \mathrm{si}(a\beta)\sin a\beta\right]$$
$$+ \frac{1}{a^{2n}}\sum_{k=1}^n (2n - 2k + 1)!(-a^2\beta^2)^{k-1}$$

and **3.356.3**

(11.4.18) $$\int_0^\infty \frac{x^{2n+1}e^{-ax}\,dx}{\beta^2 - x^2} = \frac{1}{2}\beta^{2n}\left[e^{a\beta}\mathrm{Ei}(-a\beta) + e^{-a\beta}\mathrm{Ei}(a\beta)\right]$$
$$- \frac{1}{a^{2n}}\sum_{k=1}^n (2n - 2k + 1)!(a^2\beta^2)^{k-1}$$

are obtained by differentiating the entries in Example 11.4.8.

11.5. Some higher degree denominators

This section evaluates a series of entries in [**35**] where the integrand is an exponential times a rational function with a denominator of degree larger than 2.

EXAMPLE 11.5.1. Entry **3.358.1** is

(11.5.1)
$$\int_0^\infty \frac{e^{-ax}\,dx}{\beta^4 - x^4} =$$
$$\frac{1}{4\beta^3}\left\{e^{-a\beta}\mathrm{Ei}(a\beta) - e^{a\beta}\mathrm{Ei}(-a\beta) + 2\,\mathrm{ci}(a\beta)\sin(a\beta) - 2\,\mathrm{si}(a\beta)\cos(a\beta)\right\}.$$

Start with the partial fraction decomposition

(11.5.2)
$$\frac{1}{\beta^4 - x^4} = \frac{1}{2\beta^2}\left(\frac{1}{\beta^2 - x^2} + \frac{1}{\beta^2 + x^2}\right),$$

which shows that the integral in question is a combination of (11.4.1) and (11.4.5). The result follows from here.

EXAMPLE 11.5.2. Entry **3.358.2**

(11.5.3)
$$\int_0^\infty \frac{xe^{-ax}\,dx}{\beta^4 - x^4} =$$
$$\frac{1}{4\beta^2}\left\{e^{a\beta}\mathrm{Ei}(-a\beta) + e^{-a\beta}\mathrm{Ei}(a\beta) - 2\,\mathrm{ci}(a\beta)\cos(a\beta) - 2\,\mathrm{si}(a\beta)\sin(a\beta)\right\}$$

is obtained by differentiation of (11.5.1). The entries **3.358.3**

(11.5.4)
$$\int_0^\infty \frac{x^2 e^{-ax}\,dx}{\beta^4 - x^4} =$$
$$\frac{1}{4\beta}\left\{e^{-a\beta}\mathrm{Ei}(a\beta) - e^{a\beta}\mathrm{Ei}(-a\beta) - 2\,\mathrm{ci}(a\beta)\sin(a\beta) + 2\,\mathrm{si}(a\beta)\cos(a\beta)\right\}$$

and **3.358.4**

(11.5.5)
$$\int_0^\infty \frac{x^3 e^{-ax}\,dx}{\beta^4 - x^4} =$$
$$\frac{1}{4}\left\{e^{a\beta}\mathrm{Ei}(-a\beta) + e^{-a\beta}\mathrm{Ei}(a\beta) + 2\,\mathrm{ci}(a\beta)\cos(a\beta) + 2\,\mathrm{si}(a\beta)\sin(a\beta)\right\}$$

come from further differentiation.

The entries in Section **3.357** can be established by algebraic manipulations of the examples given above.

EXAMPLE 11.5.3. Entry **3.357.1** states that

(11.5.6)
$$\int_0^\infty \frac{e^{-ax}\,dx}{\beta^3 + \beta^2 x + \beta x^2 + x^3} = \frac{1}{2\beta^2}\left\{\mathrm{ci}(a\beta)(\sin a\beta + \cos(a\beta)) + \right.$$
$$\left. \mathrm{si}(a\beta)(\sin a\beta - \cos(a\beta)) - e^{a\beta}\mathrm{Ei}(-a\beta)\right\}.$$

This formula is obtained from (11.5.1) and (11.5.3) and the algebraic identity

(11.5.7)
$$\frac{1}{\beta^3 + \beta^2 x + \beta x^2 + x^3} = \frac{\beta - x}{\beta^4 - x^4}.$$

EXAMPLE 11.5.4. Differentiation of (11.5.6) gives

(11.5.8)

$$\int_0^\infty \frac{xe^{-ax}\,dx}{\beta^3 + \beta^2 x + \beta x^2 + x^3} = \frac{1}{2\beta}\{\mathrm{ci}(a\beta)(\sin a\beta - \cos(a\beta))$$
$$-\mathrm{si}(a\beta)(\sin a\beta + \cos(a\beta)) - e^{a\beta}\mathrm{Ei}(-a\beta)\}.$$

This is entry **3.357.2** in [**35**].

EXAMPLE 11.5.5. Differentiating (11.5.8) produces entry **3.357.3**

(11.5.9)

$$\int_0^\infty \frac{x^2 e^{-ax}\,dx}{\beta^3 + \beta^2 x + \beta x^2 + x^3} = \frac{1}{2}\{-\mathrm{ci}(a\beta)(\sin a\beta + \cos(a\beta))$$
$$-\mathrm{si}(a\beta)(\sin a\beta - \cos(a\beta)) - e^{a\beta}\mathrm{Ei}(-a\beta)\}.$$

The identity

(11.5.10) $$\frac{1}{\beta^3 - \beta^2 x + \beta x^2 - x^3} = \frac{\beta + x}{\beta^4 - x^4}$$

and the method used to establish the last three entries produce proofs of the next three.

EXAMPLE 11.5.6. Entry **3.357.4** is

$$\int_0^\infty \frac{e^{-ax}\,dx}{\beta^3 - \beta^2 x + \beta x^2 - x^3} = \frac{1}{2\beta^2}\{\mathrm{ci}(a\beta)(\sin a\beta - \cos(a\beta))$$
$$-\mathrm{si}(a\beta)(\sin a\beta + \cos(a\beta)) + e^{-a\beta}\mathrm{Ei}(a\beta)\}$$

and **3.357.5** is

$$\int_0^\infty \frac{xe^{-ax}\,dx}{\beta^3 - \beta^2 x + \beta x^2 - x^3} = \frac{1}{2\beta}\{-\mathrm{ci}(a\beta)(\sin a\beta + \cos(a\beta))$$
$$-\mathrm{si}(a\beta)(\sin a\beta - \cos(a\beta)) + e^{-a\beta}\mathrm{Ei}(a\beta)\}$$

and, finally, **3.357.6** is

$$\int_0^\infty \frac{x^2 e^{-ax}\,dx}{\beta^3 - \beta^2 x + \beta x^2 - x^3} = \frac{1}{2}\{\mathrm{ci}(a\beta)(\cos a\beta - \sin(a\beta))$$
$$+\mathrm{si}(a\beta)(\cos a\beta + \sin(a\beta)) + e^{-a\beta}\mathrm{Ei}(a\beta)\}.$$

11.6. Entries involving absolute values

This section presents the evaluation of some entries in [**35**] where the integrand contains variations of the function $\ln|x|$.

EXAMPLE 11.6.1. Entry **4.337.3** states that

(11.6.1) $$\int_0^\infty e^{-\mu x}\ln|a - x|\,dx = \frac{1}{\mu}\left[\ln a - e^{-a\mu}\mathrm{Ei}(a\mu)\right].$$

To establish this entry, observe that the singularity at $x = a$ is integrable and that

$$(11.6.2) \qquad \frac{d}{dx} \ln |a - x| = \frac{1}{a - x}.$$

Integration by parts produces

$$
\begin{aligned}
\int_0^\infty e^{-\mu x} \ln |a - x| \, dx &= -\frac{1}{\mu} \int_0^\infty \ln |x - a| de^{-\mu x} \\
&= -\frac{1}{\mu} \left(-\log a - e^{-\mu a} \int_0^\infty \frac{e^{-\mu x}}{x - a} \, dx \right) \\
&= \frac{1}{\mu} \left(\ln a + e^{-\mu t} \int_{-\mu a}^\infty \frac{e^{-u}}{u} \, du \right) \\
&= \frac{1}{\mu} \left(\ln a - e^{-\mu a} \mathrm{Ei}(\mu a) \right).
\end{aligned}
$$

This is the result.

EXAMPLE 11.6.2. Entry **4.337.4** states that

$$(11.6.3) \qquad \int_0^\infty e^{-\mu x} \ln \left| \frac{\beta}{\beta - x} \right| \, dx = \frac{1}{\mu} e^{-\beta \mu} \mathrm{Ei}(\beta \mu).$$

This evaluation is obtained directly from (11.6.1) and the identity
(11.6.4)

$$\int_0^\infty e^{-\mu x} \ln \left| \frac{\beta}{\beta - x} \right| \, dx = \ln |\beta| \int_0^\infty e^{-\mu x} \, dx - \int_0^\infty e^{-\mu x} \ln |\beta - x| \, dx.$$

11.7. Some integrals involving the logarithm function

The exponential integral function Ei allows the evaluation of a variety of entries in [**35**] containing a logarithmic term. For instance, **4.212.1**

$$(11.7.1) \qquad \int_0^1 \frac{dx}{a + \ln x} = e^{-a} \mathrm{Ei}(a)$$

follows from the change of variables $t = a + \ln x$. Similarly, **4.212.2**

$$(11.7.2) \qquad \int_0^1 \frac{dx}{a - \ln x} = -e^a \mathrm{Ei}(-a)$$

is evaluated using $t = a - \ln x$.

We now consider the family

$$(11.7.3) \qquad f_n(a) := \int_0^1 \frac{dx}{(a + \ln x)^n}.$$

The change of variables $t = a + \ln x$ gives

$$(11.7.4) \qquad f_n(a) = e^{-a} \int_{-\infty}^a t^{-n} e^t \, dt.$$

Integrate by parts to produce

$$(11.7.5) \qquad \int_{-\infty}^{a} \frac{e^t\, dt}{t^n} = \frac{e^a a^{1-n}}{1-n} - \frac{1}{1-n} \int_{-\infty}^{a} \frac{e^t\, dt}{t^{n-1}}.$$

This yields a recurrence for the integrals $f_n(a)$

$$(11.7.6) \qquad f_n(a) = -\frac{a^{1-n}}{n-1} + \frac{1}{n-1} f_{n-1}(a).$$

The initial value is given in **4.212.1**. From here we deduce and prove by induction formula **4.212.8**

$$(11.7.7) \qquad \int_0^1 \frac{dx}{(a+\ln x)^n} = \frac{e^{-a}}{(n-1)!}\mathrm{Ei}(a) - \frac{1}{(n-1)!}\sum_{k=1}^{n-1}\frac{(n-k-1)!}{a^{n-k}}.$$

Using (11.7.4), we obtain **3.351.4**

$$(11.7.8) \qquad \int_a^{\infty} \frac{e^{-px}\, dx}{x^{n+1}} = \frac{(-1)^{n+1}p^n}{n!}\mathrm{Ei}(-ap) + \frac{e^{-ap}}{a^n n!}\sum_{k=0}^{n-1}(-1)^k p^k a^k (n-k-1)!$$

The integral **4.212.3**

$$(11.7.9) \qquad \int_0^1 \frac{dx}{(a+\ln x)^2} = -\frac{1}{a} + e^{-a}\mathrm{Ei}(a)$$

is the special case $n=2$ of (11.7.7). The integral **4.212.5**

$$(11.7.10) \qquad \int_0^1 \frac{\ln x\, dx}{(a+\ln x)^2} = 1 + (1-a)e^{-a}\mathrm{Ei}(a)$$

can be obtained from

$$(11.7.11) \qquad \frac{\ln x}{(a+\ln x)^2} = \frac{1}{a+\ln x} - \frac{a}{(a+\ln x)^2}.$$

Similar arguments produce **4.212.9**

$$(11.7.12)$$
$$\int_0^1 \frac{dx}{(a+\ln x)^n} = \frac{(-1)^n e^a \mathrm{Ei}(-a)}{(n-1)!} + \frac{(-1)^{n-1}}{(n-1)!}\sum_{k=1}^{n-1}(n-k-1)!(-a)^{k-n}.$$

The formula **4.212.4**

$$(11.7.13) \qquad \int_0^1 \frac{dx}{(a-\ln x)^2} = \frac{1}{a} + e^a \mathrm{Ei}(-a)$$

is the special case $n=2$. Writing

$$(11.7.14) \qquad \ln x = a - (a - \ln x),$$

we obtain the evaluation of **4.212.6**

$$(11.7.15) \qquad \int_0^1 \frac{\ln x\, dx}{(a-\ln x)^2} = 1 + (1+a)e^a \mathrm{Ei}(-a).$$

11.8. The exponential scale

Several of the entries in [**35**] contain integrals that can be reduced to the definition of the exponential integral. This section contains some of them.

EXAMPLE 11.8.1. Entry **4.331.2** states that

$$(11.8.1) \qquad \int_1^\infty e^{-\mu x} \ln x \, dx = -\frac{1}{\mu} \mathrm{Ei}(-\mu), \text{ for } \mathrm{Re}\,\mu > 0.$$

To evaluate this entry, assume $\mu > 0$ and integrate by parts to obtain

$$(11.8.2) \qquad \int_1^\infty e^{-\mu x} \ln x \, dx = \frac{1}{\mu} \int_1^\infty \frac{e^{-\mu x}}{x} \, dx.$$

The change of variables $s = -\mu x$ now gives the result for $\mu \in \mathbb{R}$. The case $\mu \in \mathbb{C}$ follows by analytic continuation.

EXAMPLE 11.8.2. Entry **4.337.1**

(11.8.3)
$$\int_0^\infty e^{-\mu x} \ln(\beta + x) \, dx = \frac{1}{\mu} \left[\ln \beta - e^{\mu\beta} \mathrm{Ei}(-\beta\mu) \right], \text{ for } |\arg \beta| < \pi, \, \mathrm{Re}\,\mu > 0$$

can be transformed to **4.331.2** by simple changes of variables. Start with $\beta > 0$ and make the change of variables $x = \beta t$ to obtain

$$(11.8.4) \qquad \int_0^\infty e^{-\mu x} \ln(\beta + x) \, dx = \frac{\ln \beta}{\mu} + \beta \int_0^\infty e^{-\mu\beta t} \ln(1 + t) \, dt.$$

The change of variables $s = t + 1$ and Entry **4.331.2** give the result.

EXAMPLE 11.8.3. Entry **4.337.2** is

$$(11.8.5) \qquad \int_0^\infty e^{-\mu x} \ln(1 + \beta x) \, dx = -\frac{1}{\mu} e^{\mu/\beta} \mathrm{Ei}(-\mu/\beta).$$

The change of variables $t = \beta x$ reduces this integral to **4.337.1** with $\mu \mapsto \mu/\beta$ and $\beta \mapsto 1$.

The change of variables $t = -ae^{nu}$ produces

$$(11.8.6) \qquad \mathrm{Ei}(x) = -n \int_c^\infty \exp\left(-ae^{nu}\right) du,$$

where $c = \frac{1}{n} \ln(-x/a)$. The choice $x = -a$ produces

$$(11.8.7) \qquad \mathrm{Ei}(-a) = -n \int_0^\infty \exp\left(-ae^{nu}\right).$$

This appears as **3.327** in [**35**].

Some further examples of entries in [**35**] containing the exponential integral function will be described in a future publication.

CHAPTER 12

More logarithmic integrals

12.1. Introduction

The compendium [**35**] contains a large collection of evaluation of integrals of the form

$$(12.1.1) \qquad \int_a^b R_1(x) \ln R_2(x)\, dx$$

where R_1 and R_2 are rational functions. The first paper in this series [**49**] considered the family

$$(12.1.2) \qquad f_n(a) = \int_0^\infty \frac{\ln^{n-1} x\, dx}{(x-1)(x+a)}, \text{ for } n \geq 2 \text{ and } a > 0.$$

The function $f_n(a)$ is given explicitly by

$$
\begin{aligned}
(12.1.3)\; f_n(a) \;=\;& \frac{(-1)^n (n-1)!}{1+a}\left[1 + (-1)^n\right] \zeta(n) \\
+\;& \frac{1}{n(1+a)} \sum_{j=0}^{\lfloor n/2 \rfloor} \binom{n}{2j} (2^{2j} - 2)(-1)^{j-1} B_{2j} \pi^{2j} (\log a)^{n-2j}.
\end{aligned}
$$

Here $\zeta(s)$ is the Riemann zeta function and B_{2j} is the Bernoulli number. In particular, (12.1.3) shows that $(1+a)f_n(a)$ is a polynomial in $\log a$.

Other papers in this series [**10, 48, 50**] and also [**46**] considered examples of integrals of this type. The results in [**10**] can be used to provide explicit expressions for an integral of the type considered here, when the poles of the rational function R_2 in (12.1.1) have real or purely imaginary parts. The present paper is a continuation of this work.

12.2. Some examples involving rational functions

This section considers integrals of the form

$$(12.2.1) \qquad \int_a^b R_1(x) \ln R_2(x)\, dx$$

where R_1 and R_2 are rational functions.

EXAMPLE 12.2.1. Entry **4.234.4** is

$$(12.2.2) \qquad \int_0^\infty \frac{1-x^2}{(1+x^2)^2} \ln x\, dx = -\frac{\pi}{2}.$$

To evaluate this entry, observe that

(12.2.3)
$$\frac{d}{dx}\frac{x}{1+x^2} = \frac{1-x^2}{(1+x^2)^2},$$

and integrating by parts gives

(12.2.4)
$$\int_0^\infty \frac{1-x^2}{(1+x^2)^2} \ln x \, dx = -\int_0^\infty \frac{dx}{1+x^2} = -\frac{\pi}{2}.$$

EXAMPLE 12.2.2. Entry **4.234.5** states that

(12.2.5)
$$\int_0^1 \frac{x^2 \ln x \, dx}{(1-x^2)(1+x^4)} = -\frac{\pi^2}{16(2+\sqrt{2})}.$$

To prove this, use the method of partial fractions to obtain
(12.2.6)
$$\int_0^1 \frac{x^2 \ln x \, dx}{(1-x^2)(1+x^4)} = \frac{1}{4}\int_0^1 \frac{\ln x \, dx}{1-x} + \frac{1}{4}\int_0^1 \frac{\ln x \, dx}{1+x} + \frac{1}{2}\int_0^1 \frac{(x^2-1)\ln x \, dx}{1+x^4}.$$

The first integral is $-\pi^2/6$ according to Entry **4.231.2** and the second one is $-\pi^2/12$ from Entry **4.231.1**. These entries were established in [**4**]. This gives

(12.2.7)
$$\int_0^1 \frac{x^2 \ln x \, dx}{(1-x^2)(1+x^4)} = -\frac{\pi^2}{16} + \frac{1}{2}\int_0^1 \frac{(x^2-1)\ln x \, dx}{1+x^4}.$$

To evaluate the last integral, observe that

(12.2.8)
$$\frac{x^2-1}{1+x^4} = \sum_{n=0}^\infty (-1)^{n-1}x^{4n} + \sum_{n=0}^\infty (-1)^n x^{4n+2}.$$

Now recall the *digamma function* $\psi(z) = \Gamma'(z)/\Gamma(z)$ and the expansion of its derivative

(12.2.9)
$$\psi'(x) = \sum_{n=0}^\infty \frac{1}{(x+n)^2}.$$

Details about this function may be found in [**17**] and [**65**]. This gives
(12.2.10)
$$\int_0^1 \frac{(x^2-1)\ln x \, dx}{1+x^4} = \frac{1}{64}\left[\psi'\left(\frac{1}{8}\right) - \psi'\left(\frac{3}{8}\right) - \psi'\left(\frac{5}{8}\right) + \psi'\left(\frac{7}{8}\right)\right].$$

The classical relation

(12.2.11)
$$\Gamma(x)\Gamma(1-x) = \frac{\pi}{\sin \pi x}$$

can be shifted to produce

(12.2.12)
$$\Gamma\left(\tfrac{1}{2}+x\right)\Gamma\left(\tfrac{1}{2}-x\right) = \frac{\pi}{\cos \pi x}.$$

Logarithmic differentiation shows that the digamma function satisfies

(12.2.13)
$$\psi\left(\frac{1}{2}+x\right) - \psi\left(\frac{1}{2}-x\right) = \pi \tan \pi x.$$

This appears as Entry **8.365.9** in [**35**]. Differentiation produces

(12.2.14) $$\psi'\left(\frac{1}{2}+x\right)+\psi'\left(\frac{1}{2}-x\right)=\pi^2\sec^2\pi x.$$

Now use (12.2.14) and group $1/8$ with $7/8$ and $3/8$ with $5/8$ to produce

(12.2.15) $$\int_0^1\frac{(x^2-1)\ln x\,dx}{1+x^4}=\frac{1}{64}\left(\frac{4\pi^2}{2-\sqrt{2}}-\frac{4\pi^2}{2+\sqrt{2}}\right)=\frac{\pi^2}{8\sqrt{2}}.$$

NOTE 12.2.3. The reader should evaluate the family of integrals

(12.2.16) $$I_n=\int_0^1\frac{x^{2n}\ln x}{(1-x^2)(1+x^4)^n}\,dx,\quad n\in\mathbb{N},$$

by the method described here. The computation of the first few special values indicates an interesting arithmetic structure of the answer.

12.3. An entry involving the Poisson kernel for the disk

This section discusses a single entry in [**35**], where the integrand involves the Poisson kernel for the disk. Further examples of this type will be presented in a future publication.

EXAMPLE 12.3.1. The next evaluation is Entry **4.233.5**

(12.3.1) $$\int_0^\infty\frac{\ln x\,dx}{x^2+2xa\cos t+a^2}=\frac{t}{\sin t}\frac{\ln a}{a}.$$

The integrand is related to the *Poisson kernel* for $D=\{z\in\mathbb{C}:|z|<1\}$.

THEOREM 12.3.2. *Define*

(12.3.2) $$\mathcal{P}_r(\theta)=\operatorname{Re}\frac{1+re^{i\theta}}{1-re^{i\theta}}.$$

Then $\mathcal{P}_r(\theta)$ is given by

(12.3.3) $$\mathcal{P}_r(\theta)=\sum_{n=-\infty}^\infty r^{|n|}e^{in\theta}=\frac{1-r^2}{1-2r\cos\theta+r^2}.$$

Moreover, given f defined on the boundary of D, the expression

(12.3.4) $$u(re^{i\theta})=\frac{1}{2\pi}\int_{-\pi}^\pi\mathcal{P}_r(\theta-t)f(e^{it})\,dt$$

for $0\le r<1$ is a harmonic function on D and it has a radial limit which agrees with f almost everywhere on the boundary of D.

The form of the Poisson kernel can be used to establish the next result.

LEMMA 12.3.3. *For $a,x\in\mathbb{R}$ with $|x|<|a|$,*

(12.3.5) $$\sum_{k=0}^\infty\frac{(-1)^k\sin((k+1)t)x^k}{a^k}=\frac{a^2\sin t}{x^2+2ax\cos t+a^2}.$$

NOTE 12.3.4. The Chebyshev polynomial of the second kind $U_n(t)$ is defined by the identity

$$(12.3.6) \qquad \frac{\sin((n+1)\theta)}{\sin\theta} = U_n(\cos\theta).$$

The result of Lemma 12.3.3 can be written as

$$(12.3.7) \qquad \sum_{k=0}^{\infty} U_k(t)x^k = \frac{1}{x^2 - 2x\cos t + 1}.$$

Lemma 12.3.3 produces

$$(12.3.8) \qquad \int_0^R \frac{x^s\,dx}{x^2 + 2ax\cos t + a^2} = \frac{1}{a^2\sin t} \sum_{k=0}^{\infty} \frac{(-1)^k \sin((k+1)t)R^{k+s+1}}{a^k\,(k+s+1)}.$$

Now write $\sin((k+1)t)$ in terms of exponentials to obtain an expression for the previous integral as

$$\int_0^R \frac{x^s\,dx}{x^2 + 2ax\cos t + a^2} = \frac{R^{s+1}}{2ia^2\sin t}\left(e^{it}\Phi\left(-\frac{R}{ae^{it}}, 1, s+1\right) - e^{-it}\Phi\left(-\frac{R}{ae^{-it}}, 1, s+1\right)\right)$$

where

$$(12.3.9) \qquad \Phi(z,s,a) = \sum_{k=0}^{\infty} \frac{z^k}{(a+k)^s}$$

is the *Lerch Phi function*.

Now differentiate with respect to s and let $s \to 0$ to produce

$$\int_0^R \frac{\ln x\,dx}{x^2 + 2ax\cos t + a^2} = \frac{i\ln R}{2a\sin t}\left(\log(1 + e^{-it}R/a) - \log(1 + e^{it}R/a)\right)$$
$$+ \frac{i}{2a\sin t}\left(\mathrm{Li}_2(-e^{-it}R/a) - \mathrm{Li}_2(-e^{-it}R/a)\right),$$

where

$$(12.3.10) \qquad \mathrm{Li}_2(z) = \sum_{k=1}^{\infty} \frac{z^k}{k^2}$$

is the *dilogarithm* function. Then use the identity
(12.3.11)
$$i\left(\mathrm{Li}_2(-e^{-it}R/a) - \mathrm{Li}_2(-e^{-it}R/a)\right) = -\int_0^t \ln\left(\frac{a^2 + 2Ra\cos z + R^2}{a^2}\right)dz$$

to obtain

(12.3.12)
$$\int_0^R \frac{\ln x \, dx}{x^2 + 2ax\cos t + a^2} = \frac{i\ln R}{2a\sin t}\left(\log(1 + e^{-it}R/a) - \log(1 + e^{it}R/a)\right)$$
$$-\frac{1}{2a\sin t}\int_0^t \ln\left(\frac{a^2 + 2Ra\cos z + R^2}{a^2}\right) dz.$$

The next step is to differentiate (12.3.12) with respect to t and let $R \to \infty$. The left-hand side produces

(12.3.13)
$$T_1(a,t) = \int_0^\infty \frac{2ax\ln x\sin t \, dx}{(x^2 + 2ax\cos t + a^2)^2}.$$

Direct differentiation of the right-hand side yields

(12.3.14)
$$T_2(a,t) = \lim_{R\to\infty} V_1(R;a,t) + V_2(R;a,t)$$

where
(12.3.15)
$$V_1(R;a,t) = \frac{R\ln R(R + a\cos t)}{a\sin t(a^2 + 2aR\cos t + R^2)} - \frac{1}{2a\sin t}\ln\left(\frac{a^2 + 2aR\cos t + R^2}{a^2}\right)$$

and

$$V_2(R;a,t) = \frac{i\cos t\ln R}{2a\sin^2 t}\left(\log(1 + e^{it}R/a) - \log(1 + e^{-it}R/a)\right)$$
$$+\frac{\cos t}{2a\sin^2 t}\int_0^t \ln\left(\frac{a^2 + 2Ra\cos z + R^2}{a^2}\right) dz.$$

PROPOSITION 12.3.5. *The function $T_2(a,t)$ is given*

(12.3.16)
$$T_2(a,t) = -\frac{\ln a}{2a\sin t}\left(t\cot t - 1\right).$$

PROOF. Start with the computation of the limiting behavior of $V_1(R;a,t)$. The claim that

(12.3.17)
$$\lim_{R\to\infty} V_1(R;a,t) = \frac{\ln a}{a\sin(t)}$$

is verified first.

First note that since

(12.3.18)
$$\lim_{R\to\infty} \frac{R\ln R}{a^2 + 2aR\cos(t) + R^2} = 0,$$

then

$$\lim_{R\to\infty} V_1(R;a,t)$$
$$= \frac{1}{a\sin t}\lim_{R\to\infty}\left(\frac{R^2\ln R}{a^2 + 2aR\cos t + R^2} - \frac{1}{2}\ln(a^2 + 2aR\cos t + R^2) + \ln a\right).$$

The claim is equivalent to

$$(12.3.19) \qquad \lim_{R \to \infty} \left(\frac{R^2 \ln R}{a^2 + 2aR \cos t + R^2} - \frac{1}{2} \ln(a^2 + 2aR \cos t + R^2) \right) = 0.$$

The identities

$$(12.3.20) \qquad \frac{R^2 \ln R}{a^2 + 2aR \cos t + R^2} = \frac{\ln R}{a^2/R^2 + 2a \cos t/R + 1}$$

and

$$(12.3.21) \quad \frac{1}{2} \ln(a^2 + 2aR \cos t + R^2) = \ln R + \frac{1}{2} \ln(a^2/R^2 + 2a \cos t/R + 1)$$

can be used to see that the left-hand side of (12.3.19) is equivalent to

$$\lim_{R \to \infty} \left(\ln R \left(\frac{1}{a^2/R^2 + 2a \cos t/R + 1} - 1 \right) - \frac{1}{2} \ln(a^2/R^2 + 2a \cos t/R + 1) \right) = 0.$$

It is clear that the second term vanishes as $R \to \infty$. For the first term, observe that

$$(12.3.22) \qquad \frac{1}{a^2/R^2 + 2a \cos(t)/R + 1} - 1 = -\frac{2a \cos t}{R} + O\left(\frac{1}{R^2} \right)$$

and thus the first term also vanishes as $R \to \infty$. This concludes the proof.

The next step is to verify that

$$(12.3.23)(R; a, t) = \frac{i \cot t \ln R}{2a \sin^2 t} \left(\log(1 + e^{it} R/a) - \log(1 + e^{-it} R/a) \right)$$
$$+ \frac{\cos t}{2a \sin^2 t} \int_0^t \ln \left(\frac{a^2 + 2aR \cos z + R^2}{a^2} \right) dz$$

satisfies

$$(12.3.24) \qquad \lim_{R \to \infty} V_2(R; a, t) = -\frac{t \cos t}{a \sin^2 t} \ln a.$$

The proof begins with the identity

$$(12.3.25) \qquad \log(1 + b/x) = \log(b/x) + \sum_{n=1}^{\infty} (-1)^{n-1} \frac{x^n}{n b^n}$$

to obtain
(12.3.26)
$$\log(1 + e^{it} R/a) - \log(1 + e^{-it} R/a) = \log(e^{it}) - \log(e^{-it}) + O(a/R), \text{ as } R \to \infty.$$

The bounds $0 < t < \pi$ imply $\log(e^{it}) - \log(e^{-it}) = 2it$. This gives

$$\lim_{R \to \infty} V_2(R; a, t) = \lim_{R \to \infty} \left(\frac{\cos t}{2a \sin^2 t} \int_0^t \ln \left(\frac{a^2 + 2aR \cos z + R^2}{a^2} \right) dz - \frac{t \cos z \ln R}{a \sin^2 t} \right)$$

$$= \lim_{R \to \infty} \frac{\cos t}{2a \sin^2 t} \left(\int_0^t \ln \left(\frac{a^2 + 2aR \cos z + R^2}{a^2} \right) dz - 2t \ln R \right)$$

$$= \lim_{R \to \infty} \frac{\cos t}{2a \sin^2 t} \left(\int_0^t \ln \left(\frac{a^2 + 2aR \cos z + R^2}{a^2} \right) - \ln(R^2) dz \right)$$

$$= \lim_{R \to \infty} \frac{\cos t}{2a \sin^2 t} \left(\int_0^t [\ln \left(a^2 + 2aR \cos z + R^2 \right) - \ln(R^2)] dz - 2t \ln a \right).$$

The identity

$$\int_0^t [\ln \left(a^2 + 2aR \cos z + R^2 \right) - \ln(R^2)] dz = \int_0^t \ln \left(\frac{a^2}{R^2} + \frac{2a \cos z}{R} + 1 \right) dz$$

gives the result. The proof of the proposition is finished.   □

The evaluation of Entry **4.233.5** is now obtained from the identity $T_1(a, t) = T_2(a, t)$. Observe that this implies

$$(12.3.27) \qquad \int_0^\infty \frac{2ax \ln x \sin t \, dx}{(x^2 + 2ax \cos t + a^2)^2} = -\frac{\ln a}{a \sin t} (t \cot t - 1).$$

Integrating with respect to t gives (12.3.1). Entry **4.231.8** in [**35**], established in [**10**],

$$(12.3.28) \qquad \int_0^\infty \frac{\ln x \, dx}{x^2 + a^2} = \frac{\pi \ln a}{2a}$$

can be used to show that the implicit constant of integration actually vanishes. The evaluation is complete.

12.4. Some rational integrands with a pole at $x = 1$

This section contains proofs of the four entries appearing in Section 4.235. These are integrals of the form

$$(12.4.1) \qquad f(a, b, c) := \int_0^\infty \frac{x^b - x^c}{1 - x^a} \ln x \, dx$$

where $a, b, c \in \mathbb{N}$. These integrals are evaluated using Entry **4.254.2**

$$(12.4.2) \qquad \int_0^\infty \frac{x^{p-1} \ln x}{1 - x^q} dx = -\frac{\pi^2}{q^2 \sin^2 \frac{\pi p}{q}}.$$

To obtain this formula, start from **3.231.6**

$$(12.4.3) \qquad \int_0^\infty \frac{x^{p-1} - x^{q-1}}{1 - x} dx = \pi \left(\cot \pi p - \cot \pi q \right),$$

established in [**47**], and make the change of variables $t = x^q$ to produce

$$\int_0^\infty \frac{x^{p-1} - 1}{1 - x^q}\, dx = -\frac{1}{q} \int_0^\infty \frac{t^{1/q - 1} - t^{p/q - 1}}{1 - t}\, dt$$

$$= -\frac{\pi}{q} \left(\cot \frac{\pi}{q} - \cot \frac{\pi p}{q} \right).$$

Differentiating with respect to p gives (12.4.2).

LEMMA 12.4.1. *Let* a, b, $c \in \mathbb{R}$. *Then*

$$(12.4.4) \qquad \int_0^\infty \frac{x^{b-1} - x^{c-1}}{1 - x^a} \ln x\, dx = -\frac{\pi^2}{a^2} \frac{\sin(c_1 - b_1) \sin(c_1 + b_1)}{\sin^2 b_1 \sin^2 c_1}$$

where $b_1 = \pi b / a$ *and* $c_1 = \pi c / a$.

PROOF. Simply write

$$\int_0^\infty \frac{x^{b-1} - x^{c-1}}{1 - x^a} \ln x\, dx = \int_0^\infty \frac{x^{b-1}}{1 - x^a} \ln x\, dx - \int_0^\infty \frac{x^{c-1}}{1 - x^a} \ln x\, dx$$

and use (12.4.2). □

The four entries in Section 4.235 are established next.

EXAMPLE 12.4.2. Entry **4.235.1** states that

$$(12.4.5) \qquad \int_0^\infty \frac{(1 - x)x^{n-2}}{1 - x^{2n}} \ln x\, dx = -\frac{\pi^2}{4n^2} \tan^2 \frac{\pi}{2n}.$$

Lemma 12.4.1 is used with $a = 2n$, $b = n - 1$, and $c = n$. This gives

$$(12.4.6) \qquad b_1 = \frac{\pi}{2} - \frac{\pi}{2n} \text{ and } c_1 = \frac{\pi}{2}$$

and

$$\int_0^\infty \frac{(1 - x)x^{n-2}}{1 - x^{2n}} \ln x\, dx = -\frac{\pi^2}{4n^2} \frac{\sin\left(\frac{\pi}{2} - \frac{\pi}{2n}\right) \sin\left(\frac{\pi}{2} + \frac{\pi}{2n}\right)}{\sin^2\left(\frac{\pi}{2} - \frac{\pi}{2n}\right)} = -\frac{\pi^2}{4n^2} \tan^2 \frac{\pi}{2n}.$$

EXAMPLE 12.4.3. Entry **4.235.2** is

$$(12.4.7) \qquad \int_0^\infty \frac{(1 - x^2)x^{m-1}}{1 - x^{2n}} \ln x\, dx = -\frac{\pi^2}{4n^2} \frac{\sin\left(\frac{m+1}{n}\pi\right) \sin\left(\frac{\pi}{n}\right)}{\sin^2\left(\frac{\pi m}{2n}\right) \sin^2\left(\frac{(m+2)}{2n}\pi\right)}.$$

Lemma 12.4.1 is now used with $a = 2n$, $b = m$, and $c = m + 2$. This gives

$$(12.4.8) \qquad c_1 - b_1 = \frac{\pi}{n} \text{ and } c_1 + b_1 = \frac{\pi}{n}(m + 1)$$

to produce the result.

EXAMPLE 12.4.4. Entry **4.235.3** states that

$$(12.4.9) \qquad \int_0^\infty \frac{(1 - x^2)x^{n-3}}{1 - x^{2n}} \ln x\, dx = -\frac{\pi^2}{4n^2} \tan^2 \frac{\pi}{n}.$$

The values $a = 2n$, $b = n - 2$, and $c = n$ give

$$(12.4.10) \qquad b_1 = \frac{\pi}{2} - \frac{\pi}{n} \text{ and } c_1 = \frac{\pi}{2}.$$

This verifies the claim.

EXAMPLE 12.4.5. Entry **4.235**.4 appears as

$$(12.4.11) \qquad \int_0^1 \frac{x^{m-1} + x^{n-m-1}}{1 - x^n} \ln x \, dx = -\frac{\pi^2}{n^2 \sin^2 \frac{\pi m}{n}}.$$

The change of variables $t = 1/x$ shows that the integral over $[1, \infty)$ is equal to that over $[0, 1]$; therefore, this entry should be written as

$$(12.4.12) \qquad \int_0^\infty \frac{x^{m-1} + x^{n-m-1}}{1 - x^n} \ln x \, dx = -\frac{2\pi^2}{n^2 \sin^2 \frac{\pi m}{n}}$$

to be consistent with the other entries in this section. The proof comes from Lemma 12.4.1 with $a = n$, $b = m$, and $c = n - m$.

12.5. Some singular integrals

The table [**35**] contains a variety of singular integrals of the form being discussed here. The examples considered in this section are evaluated employing the formula

$$(12.5.1) \qquad \int_0^\infty \frac{t^{\mu-1} \, dt}{1 - t} = \pi \cot \pi\mu.$$

To verify this evaluation, transform the integral over $[1, \infty)$ to $[0, 1]$ by the change of variables $x \mapsto 1/x$. This gives

$$(12.5.2) \qquad \int_0^\infty \frac{t^{\mu-1} \, dt}{1 - t} = \int_0^1 \frac{t^{\mu-1} - t^{-\mu}}{1 - t} \, dt.$$

This is Entry **3.231**.1. It was established in [**47**].

Differentiating with respect to μ, the formula (12.5.1) gives

$$(12.5.3) \qquad \int_0^\infty \frac{t^{\mu-1} \ln t}{1 - t} \, dt = -\frac{\pi^2}{\sin^2 \pi\mu},$$

and the change of variables $t = x^a$ gives

$$(12.5.4) \qquad \omega(a, b) := \int_0^\infty \frac{x^{b-1} \ln x}{1 - x^a} \, dx = -\frac{\pi^2}{a^2 \sin^2 \left(\frac{\pi b}{a}\right)}.$$

EXAMPLE 12.5.1. Entry **4.251**.2 states that

$$(12.5.5) \qquad \int_0^\infty \frac{x^{\mu-1} \ln x}{a - x} = \pi a^{\mu-1} \left(\ln a \, \cot(\pi\mu) - \frac{\pi}{\sin^2 \pi\mu} \right).$$

The change of variables $x = at$ yields

$$(12.5.6) \qquad \int_0^\infty \frac{x^{\mu-1} \ln x}{a - x} = a^{\mu-1} \int_0^\infty \frac{t^{\mu-1} \ln t}{1 - t} \, dt + a^{\mu-1} \ln a \int_0^\infty \frac{t^{\mu-1} \, dt}{1 - t}.$$

The result now follows from (12.5.1) and (12.5.3). It is probably clearer to write this entry as

$$(12.5.7) \qquad \int_0^\infty \frac{x^{\mu-1} \ln x}{a - x} = \pi a^{\mu-1} \left(\frac{\ln a}{\tan \pi\mu} - \frac{\pi}{\sin^2 \pi\mu} \right).$$

to avoid possible confusion.

EXAMPLE 12.5.2. Entry **4.252.3** is

$$(12.5.8) \qquad \int_0^\infty \frac{x^{p-1} \ln x}{1 - x^2} \, dx = -\frac{\pi^2}{4} \operatorname{cosec}^2 \frac{\pi p}{2}.$$

This is $\omega(2, p)$ and the result follows from (12.5.4).

EXAMPLE 12.5.3. Entry **4.255.3** states that

$$(12.5.9) \qquad \int_0^\infty \frac{1 - x^p}{1 - x^2} \ln x \, dx = \frac{\pi^2}{4} \tan^2 \left(\frac{\pi p}{2} \right).$$

This is $\omega(1, 2) - \omega(p + 1, 2)$ and the result comes from (12.5.4).

EXAMPLE 12.5.4. Entry **4.252.1** is written as

$$\int_0^\infty \frac{x^{\mu-1} \ln x \, dx}{(x + a)(x + b)} = \frac{\pi}{(b - a) \sin \pi \mu} \left[a^{\mu-1} \ln a - b^{\mu-1} \ln b - \pi \frac{a^{\mu-1} - b^{\mu-1}}{\tan \pi \mu} \right].$$

This value follows from the partial fraction decomposition

$$(12.5.10) \qquad \frac{1}{(x + a)(x + b)} = \frac{1}{b - a} \frac{1}{x + a} - \frac{1}{b - a} \frac{1}{x + b}$$

and Entry **4.251.1**

$$(12.5.11) \qquad \int_0^\infty \frac{x^{\mu-1} \ln x}{x + c} \, dx = \frac{\pi c^{\mu-1}}{\sin \pi \mu} \left(\ln c - \pi \cot \pi \mu \right),$$

established in [**51**]. Differentiating (12.5.11) with respect to c yields

$$(12.5.12) \qquad \int_0^\infty \frac{x^{\mu-1} \ln x}{(x + c)^2} \, dx = -\frac{(\mu - 1) c^{\mu-2} \pi}{\sin \pi \mu} \left(\ln c - \pi \cot \pi \mu + \frac{1}{\mu - 1} \right).$$

This is Entry **4.252.4**.

EXAMPLE 12.5.5. Entry **4.257.1**

$$(12.5.13) \qquad \int_0^\infty \frac{x^\mu \ln (x/a) \, dx}{(x + a)(x + b)} = \frac{\pi [b^\mu \ln (b/a) + \pi (a^\mu - b^\mu) \cot \pi \mu]}{(b - a) \sin \pi \mu}$$

follows from (12.5.11) and the beta integral

$$(12.5.14) \qquad \int_0^\infty \frac{x^{\mu-1} \, dx}{x + a} = \frac{\pi a^{\mu-1}}{\sin \pi \mu}.$$

This appears as Entry **3.194.3** and it was established in [**51**].

EXAMPLE 12.5.6. The change of variables $t = x^q$ gives

$$(12.5.15) \qquad \int_0^\infty \frac{x^{p-1} \, dx}{1 - x^q} = \frac{1}{q} \int_0^\infty \frac{t^{p/q-1} \, dx}{1 - t} = \frac{\pi}{q} \cot \left(\frac{\pi p}{q} \right)$$

from (12.5.3). This is Entry **3.241.3**. The special case $q = 1$ gives

$$(12.5.16) \qquad \int_0^\infty \frac{x^{p-1} \, dx}{1 - x} = \pi \cot \pi p.$$

Differentiating with respect to p produces

(12.5.17)
$$\int_0^\infty \frac{x^{p-1} \ln x}{1-x} \, dx = -\frac{\pi^2}{\sin^2 \pi p}.$$

The partial fraction decomposition

(12.5.18)
$$\frac{1}{(x+a)(x-1)} = \frac{1}{a+1}\frac{1}{x-1} - \frac{1}{a+1}\frac{1}{x+a}$$

then produces Entry **4.252.2**

(12.5.19)
$$\int_0^\infty \frac{x^{\mu-1} \ln x}{(x+a)(x-1)} \, dx = \frac{\pi}{(a+1)\sin^2 \pi\mu} \left[\pi - a^{\mu-1} \left(\ln a \, \sin \pi\mu - \pi \cos \pi\mu \right) \right].$$

EXAMPLE 12.5.7. The change of variables $t = x^q$ produces

(12.5.20)
$$\int_0^\infty \frac{\ln x \, dx}{x^p(x^q - 1)} = -\frac{1}{q^2} \int_0^\infty \frac{t^{(1-p)/q-1} \ln t \, dt}{1-t}.$$

Then, (12.5.3) gives

(12.5.21)
$$\int_0^\infty \frac{\ln x \, dx}{x^p(x^q - 1)} = \frac{\pi^2}{q^2} \frac{1}{\sin^2 \left(\frac{p-1}{q} \pi \right)}.$$

This is Entry **4.254.3**.

EXAMPLE 12.5.8. Entry **4.255.2** is

(12.5.22)
$$\int_0^1 \frac{(1+x^2)x^{p-2}}{1-x^{2p}} \ln x \, dx = - \left(\frac{\pi}{2p} \right)^2 \sec^2 \frac{\pi}{2p}.$$

The evaluation of this entry starts with Entry **3.231.5**

(12.5.23)
$$\int_0^1 \frac{x^{\mu-1} - x^{\nu-1}}{1-x} \, dx = -\psi(\mu) + \psi(\nu),$$

which was established in [**47**]. The special case $\mu = 1$

(12.5.24)
$$\int_0^1 \frac{1 - x^{\nu-1}}{1-x} \, dx = -\psi(1) + \psi(\nu)$$

is differentiated with respect to ν to produce

(12.5.25)
$$\int_0^1 \frac{x^{\nu-1} \ln x}{1-x} \, dx = -\psi'(\nu).$$

The change of variables $x = t^b$ gives

(12.5.26)
$$\int_0^1 \frac{t^{c-1} \ln t}{1-t^b} \, dt = -\frac{1}{b^2} \psi' \left(\frac{c}{b} \right).$$

Therefore

$$\int_0^1 \frac{(1-x^2)x^{p-2}}{1-x^{2p}} \ln x \, dx = \int_0^1 \frac{x^{p-2}}{1-x^{2p}} \ln x \, dx + \int_0^1 \frac{x^p}{1-x^{2p}} \ln x \, dx$$
$$= -\frac{1}{4p^2} \left[\psi' \left(\frac{1}{2} - \frac{1}{2p} \right) + \psi' \left(\frac{1}{2} + \frac{1}{2p} \right) \right].$$

The result now follows from the reflection formula for the polygamma function ψ' given in (12.2.14).

12.6. Combinations of logarithms and algebraic functions

This section presents the evaluation of some entries in [35] of the form

$$(12.6.1) \qquad \int_a^b E_1(x) \ln E_2(x)\, dx$$

where E_1 or E_2 is an algebraic function. Some of these have appeared in previous papers in this series. For example, Entry **4.241.11**

$$(12.6.2) \qquad \int_0^1 \frac{\ln x\, dx}{\sqrt{x(1-x^2)}} = -\frac{\sqrt{2\pi}}{8} \Gamma^2\left(\frac{1}{4}\right)$$

and Entry **4.241.5**

$$(12.6.3) \qquad \int_0^1 \ln x \,\sqrt{(1-x^2)^{2n-1}}\, dx = -\frac{(2n-1)!!}{4(2n)!!} \pi \left[\psi(n+1) + \gamma + \ln 4\right]$$

were evaluated in [47]. Here $\psi(x)$ is the digamma function and γ is Euler's constant.

NOTE 12.6.1. Define the family of integrals

$$(12.6.4) \qquad f_n(a) := \int_0^1 \frac{x^a \ln^n x\, dx}{\sqrt{1-x^2}}.$$

Special cases include Entry **4.241.7**

$$(12.6.5) \qquad \int_0^1 \frac{\ln x\, dx}{\sqrt{1-x^2}} = -\frac{\pi}{2} \ln 2$$

which was evaluated in [47], and Entry **4.261.9**

$$(12.6.6) \qquad \int_0^1 \frac{\ln^2 x\, dx}{\sqrt{1-x^2}} = \frac{\pi}{2}\left(\ln^2 2 + \frac{\pi^2}{12}\right).$$

A trigonometric form of the family is obtained by the change of variables $x = \sin t$

$$(12.6.7) \qquad f_n(a) = \int_0^{\pi/2} \sin^a t \,\ln^n \sin t\, dt.$$

THEOREM 12.6.2. *The integral $f_n(a)$ is given by*

$$(12.6.8) \qquad f_n(a) = \lim_{s \to a} \left(\frac{d}{ds}\right)^n h(s),$$

where

$$(12.6.9) \qquad h(s) = \int_0^{\pi/2} \sin^s t\, dt = \frac{1}{2} B\left(\frac{s+1}{2}, \frac{1}{2}\right) = \frac{\sqrt{\pi}\, \Gamma\left(\frac{s}{2} + \frac{1}{2}\right)}{2\, \Gamma\left(\frac{s}{2} + 1\right)}.$$

*This appears as Entry **3.621.5**. Therefore, the evaluation of $f_n(a)$ requires the values of $\Gamma^{(k)}(x)$ for $0 \le k \le n$ at $x = (a+1)/2$ and $x = a/2 + 1$.*

EXAMPLE 12.6.3. For example,

$$f_1(0) = \int_0^1 \frac{\ln x\, dx}{\sqrt{1-x^2}} = \lim_{s \to 0} \frac{d}{ds} \left[\frac{\sqrt{\pi}\, \Gamma\left(\frac{s}{2} + \frac{1}{2}\right)}{2\, \Gamma\left(\frac{s}{2} + 1\right)} \right]$$

$$= \frac{\sqrt{\pi}}{4} \frac{\Gamma'(1/2)\Gamma(1) - \Gamma'(1)\Gamma(1/2)}{\Gamma^2(1)}.$$

The values

(12.6.10) $\Gamma'\left(\frac{1}{2}\right) = -\sqrt{\pi}\,(\gamma + 2\ln 2),\ \Gamma'(1) = -\gamma,\ \Gamma\left(\frac{1}{2}\right) = \sqrt{\pi}$ and $\Gamma(1) = 1$

give

(12.6.11) $$f_1(0) = -\frac{\pi}{2} \ln 2.$$

PROPOSITION 12.6.4. *The derivatives of the gamma function satisfy the recurrence*

(12.6.12) $$\Gamma^{(n+1)}(x) = \sum_{k=0}^n \binom{n}{k} \Gamma^{(k)}(x) \psi^{(n-k)}(x).$$

EXAMPLE 12.6.5. A direct application of formula (12.6.8) evaluates Entry **4.261.9**

(12.6.13) $$f_2(0) = \int_0^1 \frac{\ln^2 x\, dx}{\sqrt{1-x^2}}.$$

Indeed, using $\Gamma(1) = 1$ gives
(12.6.14)
$$f_2(0) = \frac{\sqrt{\pi}}{2} \left[-\tfrac{1}{2}\Gamma'\left(\tfrac{1}{2}\right)\Gamma'(1) + \tfrac{1}{2}\Gamma\left(\tfrac{1}{2}\right)\Gamma'(1)^2 + \tfrac{1}{4}\Gamma''\left(\tfrac{1}{2}\right) - \tfrac{1}{4}\Gamma\left(\tfrac{1}{2}\right)\Gamma''(1) \right].$$

The values

(12.6.15) $\Gamma''(1) = \gamma^2 + \dfrac{\pi^2}{6}$ and $\Gamma''\left(\tfrac{1}{2}\right) = \dfrac{1}{2}\pi^{5/2} + \sqrt{\pi}(\gamma + 2\ln 2)^2$

give the identity (12.6.6).

It remains to explain the values given in (12.6.10) and (12.6.15). The recurrence (12.6.12) reduces the computation of the derivatives of $\Gamma(x)$ to those of $\psi(x)$. The special values given above come from the next result.

LEMMA 12.6.6. *The digamma function satisfies*

$$\psi^{(n)}(1) = (-1)^{n+1} n!\, \zeta(n+1)$$
$$\psi^{(n)}\left(\tfrac{1}{2}\right) = (-1)^{n+1} n!\, (2^{n+1} - 1)\zeta(n+1).$$

PROOF. This comes directly from (12.2.9).   □

EXAMPLE 12.6.7. The values given in Lemma 12.6.6 yield

$$f_3(0) = \int_0^1 \frac{\ln^3 x\, dx}{\sqrt{1-x^2}} = -\frac{\pi}{8}\left(\pi^2 \ln 2 + 4\ln^3 2 + 6\zeta(3)\right)$$

$$f_4(0) = \int_0^1 \frac{\ln^4 x\, dx}{\sqrt{1-x^2}} = \frac{\pi}{480}\left(19\pi^4 + 120\pi^2 \ln^2 2 + 240\ln^4 2 + 1440\ln 2\,\zeta(3)\right)$$

and

$$f_1\left(\tfrac{1}{2}\right) = \int_0^1 \frac{\sqrt{x}\,\ln x\,dx}{\sqrt{1-x^2}} = \frac{(\pi-4)}{\sqrt{2\pi}}\Gamma^2\left(\frac{3}{4}\right)$$

$$f_2\left(\tfrac{1}{2}\right) = \int_0^1 \frac{\sqrt{x}\,\ln^2 x\,dx}{\sqrt{1-x^2}} = \frac{1}{2\sqrt{2\pi}}\Gamma^2\left(\frac{3}{4}\right)(32-16G+\pi(\pi-8)),$$

where G is **Catalan's constant**

(12.6.16) $$G = \sum_{n=0}^{\infty} \frac{(-1)^n}{(2n+1)^2}.$$

EXAMPLE 12.6.8. Entry **4.261.15** states that

(12.6.17) $$\int_0^1 \frac{x^{2n}\ln^2 x}{\sqrt{1-x^2}}\,dx =$$

$$\frac{(2n-1)!!}{2(2n)!!}\pi\left\{\frac{\pi^2}{12} + \sum_{k=1}^{2n}\frac{(-1)^k}{k^2} + \left[\sum_{k=1}^{2n}\frac{(-1)^k}{k} + \ln 2\right]^2\right\}.$$

This is obtained by differentiating $h(s)$ twice with respect to s to produce

$$\int_0^1 \frac{x^s \ln^2 x\,dx}{\sqrt{1-x^2}} =$$

$$\frac{\sqrt{\pi}}{8}\frac{\Gamma\left(\frac{s+1}{2}\right)}{\Gamma\left(\frac{s}{2}+1\right)}\left[\left(\psi\left(\frac{s}{2}+1\right) - \psi\left(\frac{s+1}{2}\right)\right)^2 + \psi'\left(\frac{s+1}{2}\right) - \psi'\left(\frac{s}{2}+1\right)\right].$$

Therefore

$$\int_0^1 \frac{x^{2n}\ln^2 x\,dx}{\sqrt{1-x^2}} = \frac{\sqrt{\pi}}{8}\frac{\Gamma\left(n+\frac{1}{2}\right)}{\Gamma(n+1)}\left[\left(\psi(n+1) - \psi\left(n+\frac{1}{2}\right)\right)^2\right.$$

$$\left. + \psi'\left(n+\frac{1}{2}\right) - \psi'(n+1)\right].$$

The special values

(12.6.18) $$\Gamma\left(n+\tfrac{1}{2}\right) = \frac{(2n-1)!!}{2^n}\sqrt{\pi} \text{ and } \Gamma(n+1) = n!$$

give

$$\int_0^1 \frac{x^{2n}\ln^2 x\,dx}{\sqrt{1-x^2}}$$

$$= \frac{\pi}{8}\frac{(2n-1)!!}{(2n)!!}\left[(\psi(n+1) - \psi(n+\tfrac{1}{2}))^2 + \psi'(n+\tfrac{1}{2}) - \psi'(n+1)\right].$$

Now use the special values

(12.6.19) $\psi(n+1) = -\gamma + \sum_{k=1}^{n}\frac{1}{k}$ and $\psi(n+\tfrac{1}{2}) = -\gamma - 2\ln 2 + 2\sum_{k=1}^{n}\frac{1}{2k-1}$

as well as

(12.6.20) $\psi'(n+1) = \dfrac{\pi^2}{6} - \displaystyle\sum_{k=1}^{n} \dfrac{1}{k^2}$ and $\psi'(n+\tfrac{1}{2}) = \dfrac{\pi^2}{2} - 4\displaystyle\sum_{k=1}^{n} \dfrac{1}{(2k-1)^2}$

to obtain

(12.6.21) $$\psi(n+1) - \psi(n+\tfrac{1}{2}) = 2\sum_{k=1}^{2n} \frac{(-1)^k}{k} + 2\ln 2$$

and

(12.6.22) $$\psi'(n+\tfrac{1}{2}) - \psi'(n+1) = \frac{\pi^2}{3} + 4\sum_{k=1}^{2n} \frac{(-1)^k}{k^2}.$$

This gives the result.

EXAMPLE 12.6.9. A similar analysis gives Entry **4.261.16**

$$\int_0^1 \frac{x^{2n+1}\ln^2 x}{\sqrt{1-x^2}}\,dx =$$

$$-\frac{(2n)!!}{(2n+1)!!}\left\{ \frac{\pi^2}{12} + \sum_{k=1}^{2n+1}\frac{(-1)^k}{k^2} - \left[\sum_{k=1}^{2n+1}\frac{(-1)^k}{k} + \ln 2\right]^2 \right\}.$$

EXAMPLE 12.6.10. Entry **4.241.6** states that

(12.6.23) $$\int_0^{1/\sqrt{2}} \frac{\ln x\,dx}{\sqrt{1-x^2}} = -\frac{\pi}{4}\ln 2 - \frac{G}{2}.$$

The change of variables $x = \sin t$ gives

(12.6.24) $$\int_0^{1/\sqrt{2}} \frac{\ln x\,dx}{\sqrt{1-x^2}} = \int_0^{\pi/4} \ln \sin t\,dt.$$

This integral is entry **4.224.2** and it has been evaluated in [**10**].

12.7. An example producing a trigonometric answer

The next example contains, in the logarithmic part, a quotient of linear functions. The evaluation of this entry requires a different approach.

EXAMPLE 12.7.1. Entry **4.297.8** states that

(12.7.1) $$\int_0^1 \ln\frac{1+ax}{1-ax}\frac{dx}{x\sqrt{1-x^2}} = \pi\sin^{-1}a.$$

This evaluation starts with the expansion

(12.7.2) $$\frac{1}{x}\ln\frac{1+ax}{1-ax} = \sum_{n=0}^{\infty}\frac{2a^{2n+1}}{2n+1}x^{2n}$$

to obtain

$$(12.7.3) \qquad \int_0^1 \ln \frac{1+ax}{1-ax} \frac{dx}{x\sqrt{1-x^2}} = \sum_{n=0}^{\infty} \frac{2a^{2n+1}}{2n+1} \int_0^1 \frac{x^{2n} \, dx}{\sqrt{1-x^2}}.$$

The change of variables $x = \sin\theta$ gives

$$(12.7.4) \qquad \int_0^1 \frac{x^{2n} \, dx}{\sqrt{1-x^2}} = \int_0^{\pi/2} \sin^{2n}\theta \, d\theta = \frac{\pi}{2^{2n+1}} \binom{2n}{n}.$$

The last evaluation is the famous Wallis formula. It appears as Entry **3.621.3** and it was established in [**7**] and [**52**]. Therefore

$$(12.7.5) \qquad \int_0^1 \ln \frac{1+ax}{1-ax} \frac{dx}{x\sqrt{1-x^2}} = \sum_{n=0}^{\infty} \frac{\pi}{2^{2n}} \frac{a^{2n+1}}{2n+1} \binom{2n}{n}.$$

The series is now identified from the classical expansion

$$\sin^{-1} x \;=\; \sum_{n=0}^{\infty} \frac{\left(\frac{1}{2}\right)_n}{(2n+1)\,n!} x^{2n+1}$$

$$= \sum_{n=0}^{\infty} \frac{1}{2^{2n}(2n+1)} \binom{2n}{n} x^{2n+1}$$

obtained by expanding the integrand in

$$(12.7.6) \qquad \sin^{-1} x = \int_0^x \frac{dt}{\sqrt{1-t^2}}$$

as a binomial series and integrating term by term.

Further examples in [**35**] of the class considered here will be presented in a future publication.

CHAPTER 13

Confluent hypergeometric and Whittaker functions

13.1. Introduction

The confluent hypergeometric function, denoted by $_1F_1(a;c;z)$, is defined by

$$(13.1.1) \qquad _1F_1(a;c;z) = \sum_{n=0}^{\infty} \frac{(a)_n z^n}{(c)_n n!}$$

with $(a)_n$ being the rising factorial

$$(a)_n := a(a+1)\cdots(a+n-1) = \frac{\Gamma(a+n)}{\Gamma(a)},$$

for $a \in \mathbb{C}$. It arises when two of the regular singular points of the differential equation for the Gauss hypergeometric function $_2F_1(a,b;c;z)$, given by

$$(13.1.2) \qquad z(1-z)y'' + (c-(a+b+1)z)y' - aby = 0,$$

are allowed to merge into one singular point. More specifically, if we replace z by z/b in $_2F_1(a,b;c;z)$, then the corresponding differential equation has singular points at 0, b, and ∞. Now let $b \to \infty$ so as to have infinity as a confluence of two singularities. This results in the function $_1F_1(a;c;z)$ so that

$$(13.1.3) \qquad _1F_1(a;c;z) = \lim_{b\to\infty} {_2F_1}\left(a,b;c;\frac{z}{b}\right),$$

and the corresponding differential equation

$$(13.1.4) \qquad zy'' + (c-z)y' - ay = 0,$$

known as the confluent hypergeometric equation. The following two transformation formulas for $_1F_1$, due to Kummer, are very useful:

$$_1F_1(a;c;z) = e^z {_1F_1}(c-a;c;-z) \qquad (b \neq 0, -1, -2, \cdots),$$

$$(13.1.5)$$

$$_1F_1(a;2a;2z) = e^z {_0F_1}\left(-;a+\frac{1}{2};\frac{z^2}{4}\right) \qquad (2a \text{ is not an odd integer} < 0).$$

The confluent hypergeometric function has many different notations other than $_1F_1(a;c;z)$, for example, $\Phi(a;c;z)$ [**35**, p. 1023] or $M(a;c;z)$ [**65**]. Closely associated to $_1F_1(a;c;z)$ are the Whittaker functions $M_{k,\mu}(z)$ and $W_{k,\mu}(z)$ defined by [35, p. 1024]

$$(13.1.6) M_{k,\mu}(z) = z^{\mu+\frac{1}{2}} e^{-z/2} {}_1F_1\left(\mu - k + \frac{1}{2}; 2\mu + 1; z\right),$$

$$(13.1.7) W_{k,\mu}(z) = \frac{\Gamma(-2\mu)}{\Gamma(\frac{1}{2} - \mu - k)} M_{k,\mu}(z) + \frac{\Gamma(2\mu)}{\Gamma(\frac{1}{2} + \mu - k)} M_{k,-\mu}(z).$$

In this paper, the formulas in Sections 7.612 and 7.621 of [35] are established. These involve the confluent hypergeometric function and the Whittaker functions. The remaining entries involving these functions will be considered in the future.

The asymptotic formulas for these functions have been well studied in the literature. Some are collected here for the benefit of the reader. The first one is an asymptotic expansion for $_1F_1$

$$(13.1.8) \quad {}_1F_1(a;c;z) \sim \frac{\Gamma(c)e^z z^{a-c}}{\Gamma(a)} \sum_{n=0}^{\infty} \frac{(c-a)_n(1-a)_n}{n!} z^{-n} \qquad \text{as } z \to \infty,$$

for $|\arg(z)| < \frac{\pi}{2}$. This appears in [65, p. 174, Equation (7.9)]. The more general asymptotic expansion

$$(13.1.9) {}_1F_1(a;c;z) \quad \sim \quad e^z z^{a-c} \frac{\Gamma(c)}{\Gamma(a)} \sum_{n=0}^{\infty} \frac{(c-a)_n(1-a)_n}{n!} z^{-n}$$

$$+ \frac{\Gamma(c)\, e^{\pm i\pi a}}{\Gamma(c-a)} z^{-a} \sum_{n=0}^{\infty} \frac{(a)_n(1+a-c)_n}{n!} (-z)^{-n},$$

where the upper sign is taken of $-\frac{1}{2}\pi < \arg z < \frac{3}{2}\pi$ and the lower sign in the case $-\frac{3}{2}\pi < \arg z < \frac{1}{2}\pi$. The first part dominates the second when $\text{Re}(z) > 0$ corresponding with (13.1.8). The second part is dominant for $\text{Re}(z) < 0$. This appears in [65, p. 189, Exercise (7.7)].

The Ψ-function is defined in [35, p. 1023] by
(13.1.10)
$$\Psi(a;c;z) = \frac{\Gamma(1-c)}{\Gamma(a-c+1)} {}_1F_1(a;c;z) + \frac{\Gamma(c-1)}{\Gamma(a)} z^{1-c} {}_1F_1(a-c+1;2-c;z)$$

and its asymptotic behavior is given by
(13.1.11)
$$\Psi(a;c;z) \sim z^{-a} \sum_{n=0}^{\infty} \frac{(a)_n(a-c+1)_n}{n!}(-z)^{-n}, \quad \text{as } z \to \infty \text{ for } |\arg z| < \frac{3}{2}\pi,$$

(see [65, p. 175, formula (7.13)]).

The first formula established here is Entry 7.612.1. This is a standard result for the Mellin transform of $_1F_1(a;c;-t)$ [13, p. 192]. A proof is presented here to make the results self-contained. The argument begins with an entry in [35].

Entry **7.612.1** states that, for $0 < \mathrm{Re}\,(b) < \mathrm{Re}\,(a)$,

$$(13.1.12) \qquad \int_0^\infty t^{b-1}\,{}_1F_1(a;c;-t)\,dt = \frac{\Gamma(b)\Gamma(c)\Gamma(a-b)}{\Gamma(a)\Gamma(c-b)}.$$

Proof. We need $\mathrm{Re}\,(b) > 0$ for the convergence of the integral near $t = 0$. Since the argument of ${}_1F_1(a;c;-t)$ is in the left half-plane, the second expression in the asymptotic expansion in (13.1.9) becomes dominant. Hence we require $\mathrm{Re}\,(b) < \mathrm{Re}\,(a)$ for the convergence of the integral near ∞.

Apply Kummer's first transformation in (13.1.5) on the left-hand side of (13.1.12) so that

$$\int_0^\infty t^{b-1}\,{}_1F_1(a;c;-t)\,dt = \int_0^\infty t^{b-1}e^{-t}\,{}_1F_1(c-a;c;t)\,dt$$

$$= \int_0^\infty t^{b-1}e^{-t}\sum_{n=0}^\infty \frac{(c-a)_n t^n}{(c)_n n!}\,dt$$

$$= \sum_{n=0}^\infty \frac{(c-a)_n}{(c)_n n!}\int_0^\infty t^{b+n-1}e^{-t}\,dt$$

$$= \sum_{n=0}^\infty \frac{(c-a)_n \Gamma(b+n)}{(c)_n n!}$$

$$= \Gamma(b)\,{}_2F_1(c-a,b;c;1)$$

$$= \frac{\Gamma(b)\Gamma(c)\Gamma(a-b)}{\Gamma(a)\Gamma(c-b)},$$

where the last equality follows from Gauss' formula for evaluating ${}_2F_1$ at 1 [**13**, p. 66, Theorem 2.2.2].

The next evaluation is Entry **7.612.2**. It states that, for $0 < \mathrm{Re}\,(b) < \mathrm{Re}\,(a)$ and $\mathrm{Re}\,(c) < \mathrm{Re}\,(b+1)$,

$$(13.1.13) \qquad \int_0^\infty t^{b-1}\Psi(a;c;t)\,dt = \frac{\Gamma(b)\Gamma(a-b)\Gamma(b-c+1)}{\Gamma(a)\Gamma(a-c+1)},$$

where the function $\Psi(a;c;t)$ is defined in (13.1.10).

Proof. Use (13.1.10) in the integrand and write the given integral as the sum of two integrals. The condition $\mathrm{Re}\,(b) > 0$ is required for the convergence of the first integral near $t = 0$, whereas the second integral requires $\mathrm{Re}\,(b-c+1) > 0$. The behavior of the integral in (13.1.13) near ∞ requires $\mathrm{Re}\,(b-a) < 0$, as can be seen by using (13.1.11).

The following integral representation for $\Psi(a;c;z)$, valid for $\mathrm{Re}\,(a) > 0$ and $\mathrm{Re}\,(z) > 0$, is used in the argument

$$(13.1.14) \qquad \Psi(a;c;z) = \frac{1}{\Gamma(a)}\int_0^\infty e^{-zt}t^{a-1}(1+t)^{c-a-1}\,dt.$$

This is Entry **9.211.4** in [**35**, p. 1023]. A proof appears in [**65**, p. 174–175].

Using (13.1.14) on the left-hand side of (13.1.13), it follows that
(13.1.15)
$$\int_0^\infty t^{b-1}\Psi(a;c;t)\,dt = \frac{1}{\Gamma(a)}\int_0^\infty t^{b-1}\,dt\int_0^\infty e^{-tx}x^{a-1}(1+x)^{c-a-1}\,dx.$$

Interchanging the order of integration, one obtains

$$\begin{aligned}
\int_0^\infty t^{b-1}\Psi(a;c;t)\,dt &= \frac{1}{\Gamma(a)}\int_0^\infty x^{a-1}(1+x)^{c-a-1}\,dx\int_0^\infty e^{-tx}t^{b-1}\,dt\\
&= \frac{\Gamma(b)}{\Gamma(a)}\int_0^\infty x^{a-b-1}(1+x)^{c-a-1}\,dx\\
&= \frac{\Gamma(b)B(a-b,b-c+1)}{\Gamma(a)}\\
&= \frac{\Gamma(b)\Gamma(a-b)\Gamma(b-c+1)}{\Gamma(a)\Gamma(a-c+1)},
\end{aligned}$$

where the last two steps follow from the classical representation for Euler's beta function $B(x,y)$

(13.1.16) $$B(a,b) = \int_0^\infty \frac{x^{a-1}}{(1+x)^{a+b}}\,dx \quad \text{for Re}\,(a)>0,\,\text{Re}\,(b)>0$$

and its expression in terms of the gamma function

(13.1.17) $$B(a,b) = \frac{\Gamma(a)\Gamma(b)}{\Gamma(a+b)}.$$

This completes the proof.

13.2. A sample of formulas

This section collects a selection of formulas from [35] involving the confluent hypergeometric function. The first example is **7.621.4**.

EXAMPLE 13.2.1. Entry **7.621.4** states

$$\int_0^\infty e^{-st}t^{b-1}\,{}_1F_1(a;c;kt)\,dt$$

(13.2.1)
$$= \begin{cases} \Gamma(b)s^{-b}\,{}_2F_1\left(\begin{matrix}a & b\\ & c\end{matrix}\middle|\frac{k}{s}\right) & \text{if } |s|>|k|\\[2ex] \Gamma(b)(s-k)^{-b}\,{}_2F_1\left(\begin{matrix}c-a & b\\ & c\end{matrix}\middle|\frac{k}{k-s}\right) & \text{if } |s-k|>|k|, \end{cases}$$

where Re $(b)>0$ and Re $(s)>\max\{0,\text{Re}\,(k)\}$.

Proof. Assume first Re $(k)>0$. Using (13.1.8), it follows that, as $t\to\infty$, the integrand behaves like $e^{(k-s)t}$, and in order to ensure convergence, the condition Re $(k-s)<0$ is needed. This explains the condition Re $(s)>$ Re (k). As $t\to 0$, the integrand behaves like t^{b-1}. The condition Re $(b)>0$ is required for the convergence of the integral.

The discussion above guarantees the validity of interchange of summation and integration in the next steps:

$$(13.2.2) \int_0^\infty e^{-st} t^{b-1} {}_1F_1(a; c; kt)\, dt \; = \; \sum_{n=0}^{\infty} \frac{(a)_n k^n}{(c)_n n!} \int_0^\infty e^{-st} t^{b+n-1}\, dt$$

$$= \; \Gamma(b) \sum_{n=0}^{\infty} \frac{(a)_n}{(c)_n} \frac{k^n}{n!} \frac{\Gamma(b+n)}{\Gamma(b)} \frac{1}{s^{b+n}}$$

$$= \; \frac{\Gamma(b)}{s^b} \sum_{n=0}^{\infty} \frac{(b)_n (a)_n}{(c)_n n!} \left(\frac{k}{s}\right)^n$$

$$= \; \frac{\Gamma(b)}{s^b} {}_2F_1 \left(\begin{matrix} a & b \\ & c \end{matrix} \,\bigg|\, \frac{k}{s} \right)$$

provided $|k/s| < 1$, i.e., if $|s| > |k|$. This proves the first part.

In the case $\operatorname{Re}(k) < 0$ one has Kummer's relation

$$(13.2.3) \qquad {}_1F_1(a; c; w) = e^w {}_1F_1(c - a; c; -w)$$

(see [**13**, p. 191, Equation (4.1.11)]). Therefore, as $t \to \infty$, the integrand behaves like e^{-s} and convergence requires the condition $\operatorname{Re}(s) > 0$. Then

$$\int_0^\infty e^{-st} t^{b-1} {}_1F_1(a; c; kt)\, dt \; = \; \int_0^\infty e^{-(s-k)t} t^{b-1} {}_1F_1(c - a; c; -kt)\, dt$$

$$= \; \Gamma(b) \sum_{n=0}^{\infty} \frac{(c - a)_n}{(c)_n} \frac{\Gamma(b+n)}{\Gamma(b)} \frac{(-k)^n}{(s - k)^{b+n} n!}$$

$$= \; \frac{\Gamma(b)}{(s - k)^b} {}_2F_1 \left(\begin{matrix} c - a & b \\ & c \end{matrix} \,\bigg|\, \frac{k}{k - s} \right).$$

Now apply Pfaff's transformation ([**13**, p.68, Equation (2.2.6)])

$$(13.2.4) \qquad {}_2F_1 \left(\begin{matrix} a & b \\ & c \end{matrix} \,\bigg|\, z \right) = (1 - z)^{-b} {}_2F_1 \left(\begin{matrix} c - a & b \\ & c \end{matrix} \,\bigg|\, \frac{z}{z - 1} \right)$$

to the hypergeometric series above to obtain the result. This establishes the first formula when $\operatorname{Re}(k) < 0$. The second formula, for the range $|s - k| > |k|$, is established along similar lines.

A direct application of the more general asymptotic expansion (13.1.9) then reduces the case $\operatorname{Re} k = 0$ to the previous two cases (according to the sign of $\operatorname{Im} k$).

EXAMPLE 13.2.2. Entry **7.621.5** states that

$$(13.2.5) \qquad \int_0^\infty t^{c-1} {}_1F_1(a; c; t) e^{-st}\, dt = \Gamma(c) s^{-c} (1 - s^{-1})^{-a}$$

for $\operatorname{Re}(c) > 0$ and $\operatorname{Re}(s) > 1$.

Proof. This is actually the special case $k = 1$ and $b = c$ in the first part of **7.621.4**. The condition $|s| > 1$ implies $\mathrm{Re}\,(s) > 1$. Then

$$
\int_0^\infty t^{c-1}\,_1F_1(a;c;t)e^{-st}\,dt \;=\; \Gamma(c)s^{-c}\,_2F_1\left(\begin{matrix} a & c \\ & c \end{matrix}\,\middle|\,\frac{1}{s}\right)
$$
$$
= \Gamma(c)s^{-c}(1 - 1/s)^{-a}
$$

using

$$
_2F_1\left(\begin{matrix} a & c \\ & c \end{matrix}\,\middle|\,u\right) \;=\; \sum_{n=0}^{\infty} \frac{(a)_n(c)_n}{(c)_n n!}\,u^n
$$
$$
= (1 - u)^{-a}
$$

by the binomial theorem for $|u| < 1$.

EXAMPLE 13.2.3. Entry **7.621.6** states that, for $\mathrm{Re}\,(c) < \mathrm{Re}\,(b) + 1$,

$$
(13.2.6) \quad \int_0^\infty t^{b-1}\Psi(a;c;t)e^{-st}\,dt
$$

$$
= \frac{\Gamma(b)\Gamma(b - c + 1)}{\Gamma(a + b - c + 1)}
\begin{cases}
{}_2F_1\left(\begin{matrix} b & b-c+1 \\ & a+b-c+1 \end{matrix}\,\middle|\,1 - s\right) & \mathrm{Re}\,(b) > 0,\, |1 - s| < 1, \\[2ex]
s^{-b}\,{}_2F_1\left(\begin{matrix} a & b \\ & a+b-c+1 \end{matrix}\,\middle|\,1 - \frac{1}{s}\right) & \mathrm{Re}\,(s) > \frac{1}{2}.
\end{cases}
$$

Proof. The integral in question is now evaluated in two cases, according to the conditions given in (13.2.3). The assumptions on the parameters will appear as conditions in the proof.

(i) *First part.* Using the expression for $\Psi(a;c;t)$ in (13.1.10) gives

$$
(13.2.7)
$$
$$
\int_0^\infty t^{b-1}\Psi(a;c;t)e^{-st}\,dt \;=\; \frac{\Gamma(1 - c)}{\Gamma(a - c + 1)} \int_0^\infty t^{b-1}\,_1F_1(a;c;t)e^{-st}\,dt
$$
$$
+ \frac{\Gamma(c - 1)}{\Gamma(a)} \int_0^\infty t^{(b-c+1)-1}\,_1
$$
$$
\times F_1(a - c + 1; 2 - c; t)e^{-st}\,dt.
$$

The first integral requires $\mathrm{Re}\,(b) > 0$ for convergence near $t = 0$ and the second integral requires $\mathrm{Re}\,(c) < \mathrm{Re}\,(b) + 1$ in order to apply the first formula in **7.621.4**, with $k = 1$. This also requires the condition $|s| > 1$. However, the behavior of the integrand on the left-hand side at infinity renders the integral convergent when $\mathrm{Re}\,(s) > 0$, and, as will be seen below, the result holds for $|1 - s| < 1$ by analytic continuation.

A direct application of **7.621.4** gives the value

(13.2.8) $\displaystyle\int_0^\infty t^{b-1}\Psi(a;c;t)e^{-st}\,dt = \frac{\Gamma(1-c)\Gamma(b)s^{-b}}{\Gamma(a-c+1)}\,{}_2F_1\left(\begin{array}{cc|c} a & b & 1 \\ & c & s \end{array}\right)$

$\displaystyle + \frac{\Gamma(c-1)\Gamma(b-c+1)}{\Gamma(a)}s^{c-b-1}\,{}_2F_1\left(\begin{array}{cc|c} a-c+1 & b-c+1 & 1 \\ & 2-c & s \end{array}\right).$

The answer is simplified using the identity (see [**65**, p.113, (5.12)],

(13.2.9) $\displaystyle {}_2F_1\left(\begin{array}{cc|c} a & b & z \\ & c & \end{array}\right) = \frac{\Gamma(c)\Gamma(b-a)}{\Gamma(b)\Gamma(c-a)}(1-z)^{-a}\,{}_2F_1\left(\begin{array}{cc|c} a & c-b & 1 \\ a-b+1 & & 1-z \end{array}\right)$

$\displaystyle + \frac{\Gamma(c)\Gamma(a-b)}{\Gamma(a)\Gamma(c-b)}(1-z)^{-b}\,{}_2F_1\left(\begin{array}{cc|c} b & c-a & 1 \\ b-a+1 & & 1-z \end{array}\right)$

to produce
(13.2.10)

$\displaystyle\int_0^\infty t^{b-1}\Psi(a;c;t)e^{-st}\,dt = \frac{\Gamma(b)\Gamma(b-c+1)}{\Gamma(a+b-c+1)}\,{}_2F_1\left(\begin{array}{cc|c} b & b-c+1 & 1-s \\ a+b-c+1 & & \end{array}\right),$

as claimed. Note that the presence of the hypergeometric function on the left of (13.2.9) requires $|z| < 1$, that is, $|1-s| < 1$ in this example. The above identity also requires $|\arg(1-z)| < \pi$, that is, $|\arg(s)| < \pi$, which is satisfied when $|1-s| < 1$.

(ii) *Second part.* The formula (13.1.10) gives

(13.2.11)

$\displaystyle\int_0^\infty t^{b-1}\Psi(a;c;t)e^{-st}\,dt = \frac{\Gamma(1-c)}{\Gamma(a-c+1)}\int_0^\infty t^{b-1}\,{}_1F_1(a;c;t)e^{-st}\,dt$

$\displaystyle + \frac{\Gamma(c-1)}{\Gamma(a)}\int_0^\infty t^{(b-c+1)-1}\,{}_1$

$\displaystyle \times F_1(a-c+1;2-c;t)e^{-st}\,dt,$

and then the second part of **7.621.4** gives

(13.2.12)

$\displaystyle\int_0^\infty t^{b-1}\Psi(a;c;t)e^{-st}\,dt$

$\displaystyle = \frac{\Gamma(1-c)\Gamma(b)}{\Gamma(a-c+1)(s-1)^b}\,{}_2F_1\left(\begin{array}{cc|c} c-a & b & 1 \\ & c & 1-s \end{array}\right)$

$\displaystyle + \frac{\Gamma(c-1)\Gamma(b-c+1)}{\Gamma(a)(s-1)^{b-c+1}}\,{}_2F_1\left(\begin{array}{cc|c} 1-a & b-c+1 & 1 \\ & 2-c & 1-s \end{array}\right),$

which is valid for $|1 - s| > 1$. To reduce this expression to the form stated in (13.2.6), one uses the identity

(13.2.13)
$$
{}_2F_1\left(\begin{matrix} a & b \\ & c \end{matrix}\middle| z\right) = \frac{\Gamma(c)\Gamma(c-a-b)}{\Gamma(c-a)\Gamma(c-b)} z^{-a} {}_2F_1\left(\begin{matrix} a & a-c+1 \\ a+b-c+1 \end{matrix}\middle| 1-\frac{1}{z}\right)
$$
$$
+ \frac{\Gamma(c)\Gamma(a+b-c)}{\Gamma(a)\Gamma(b)} (1-z)^{c-a-b} z^{a-c}
$$
$$
\times {}_2F_1\left(\begin{matrix} c-a & 1-a \\ c-a-b+1 \end{matrix}\middle| 1-\frac{1}{z}\right)
$$

valid for $|\arg(1-z)| < \pi$ and $|\arg z| < \pi$. This appears in [**65**, p. 113, (5.13)]. Now take $z = 1 - 1/s$ and replace c by $a+b-c+1$ to obtain

(13.2.14)
$$
{}_2F_1\left(\begin{matrix} a & b \\ a+b-c+1 \end{matrix}\middle| 1-\frac{1}{s}\right) =
$$
$$
\frac{\Gamma(a+b-c+1)\Gamma(1-c)}{\Gamma(b-c+1)\Gamma(a-c+1)}\left(1-\frac{1}{s}\right)^{-a} {}_2F_1\left(\begin{matrix} a & c-b \\ & c \end{matrix}\middle| \frac{1}{1-s}\right) +
$$
$$
+\frac{\Gamma(a+b-c+1)\Gamma(c-1)}{\Gamma(a)\Gamma(b)}\left(1-\frac{1}{s}\right)^{c-b-1} s^{1-c}{}_2F_1\left(\begin{matrix} b-c+1 & 1-a \\ 2-c \end{matrix}\middle| \frac{1}{1-s}\right)
$$

for $|s| > |s-1|$, $|\arg(1/s)| < \pi$, and $|\arg(1-1/s)| < \pi$. Euler's relation

(13.2.15)
$$
{}_2F_1\left(\begin{matrix} a & b \\ & c \end{matrix}\middle| z\right) = (1-z)^{c-a-b}{}_2F_1\left(\begin{matrix} c-a & c-b \\ & c \end{matrix}\middle| z\right)
$$

on the first hypergeometric function on the right of (13.2.14) produces

(13.2.16) ${}_2F_1\left(\begin{matrix} a & c-b \\ & c \end{matrix}\middle| \frac{1}{1-s}\right) = \left(1-\frac{1}{1-s}\right)^{b-a}{}_2F_1\left(\begin{matrix} c-a & b \\ & c \end{matrix}\middle| \frac{1}{1-s}\right).$

Now use (13.2.14) and (13.2.16) to produce the desired result. It is easy to check that the conditions $|s| > |1-s| > 1$ and the principle of analytic continuation coupled with the fact that the integral on the left-hand side converges for $\operatorname{Re} s > 0$ implies that the result is true for $\operatorname{Re} s > 1/2$. Also, the conditions $|\arg(1/s)| < \pi$ and $|\arg(1-1/s)| < \pi$ are satisfied for these values of s.

EXAMPLE 13.2.4. Entry **7.621.1** states that

$$
\int_0^\infty e^{-st}t^\alpha M_{\mu,\nu}(t)\, dt = \frac{\Gamma(\alpha+\nu+3/2)}{(s+\frac{1}{2})^{\alpha+\nu+\frac{3}{2}}} {}_2F_1\left(\begin{matrix} \alpha+\nu+\frac{3}{2} & \nu-\mu+\frac{1}{2} \\ 2\nu+1 \end{matrix}\middle| \frac{2}{2s+1}\right)
$$

for $\operatorname{Re}\left(\alpha+\mu+\frac{3}{2}\right) > 0$ and $\operatorname{Re}(s) > \frac{1}{2}$.

Proof. Note that, from (13.1.6),

(13.2.17)
$$
M_{\mu,\nu}(t) = t^{\nu+1/2}e^{-t/2}{}_1F_1(\nu-\mu+\tfrac{1}{2}; 2\nu+1; t)
$$

and this gives

(13.2.18)
$$\int_0^\infty e^{-st} t^\alpha M_{\mu,\nu}(t)\, dt = \int_0^\infty e^{-(s+1/2)t} t^{(\alpha+\nu+3/2)-1} {}_1F_1(\nu-\mu+\tfrac{1}{2}; 2\nu+1; t)\, dt.$$

Now use the first part of **7.621.4** with $s \mapsto s+\tfrac{1}{2}$, $b \mapsto \alpha+\nu+\tfrac{3}{2}$ and $k = 1$, $a \mapsto \nu-\mu+\tfrac{1}{2}$, $c \mapsto 2\nu+1$. This gives the required result. The asymptotics of ${}_1F_1$ as $t \to \infty$ shows that the integrand behaves like $e^{(1/2-s)t}$. Therefore the condition $\mathrm{Re}\,(s) > \tfrac{1}{2}$ is imposed for convergence.

EXAMPLE 13.2.5. Entry **7.621.2** states that

$$\int_0^\infty e^{-st} t^{\mu-1/2} M_{\lambda,\mu}(qt)\, dt$$

$$= q^{\mu+1/2}\Gamma(2\mu+1)\left(s-\frac{q}{2}\right)^{\lambda-\mu-1/2}\left(s+\frac{q}{2}\right)^{-\lambda-\mu-1/2},$$

for $\mathrm{Re}\,(\mu) > -\tfrac{1}{2}$ and $\mathrm{Re}\,(s) > \tfrac{1}{2}|\mathrm{Re}\,(q)|$.

Proof. Assume first that $q > 0$. The change of variables $w = qt$ gives

(13.2.19) $$\int_0^\infty e^{-(s+q/2)t}(qt)^{\mu+1/2} t^{\mu-1/2} {}_1F_1(\mu-\lambda+\tfrac{1}{2}; 2\mu+1; qt)\, dt =$$

$$\frac{1}{q^{\mu+1/2}}\int_0^\infty e^{-(s/q+1/2)w} w^{(2\mu+1)-1} {}_1F_1(\mu-\lambda+\tfrac{1}{2}; 2\mu+1; w)\, dw.$$

The evaluation of this last integral uses Entry **7.621.5** with $c \mapsto 2\mu+1$, $a \mapsto \mu-\lambda+\tfrac{1}{2}$, $s \mapsto \tfrac{s}{q}+\tfrac{1}{2}$. This requires $\mathrm{Re}\,(c) > 1$, that is, $\mathrm{Re}\,(\mu) > -\tfrac{1}{2}$ and also $\mathrm{Re}\,(s) > 1$, which translates to $\mathrm{Re}\,(s) > q/2$. This gives the stated result.

This result is now extended to $q \in \mathbb{C}$ by analytic continuation. The proof uses the asymptotic expansion (13.1.9) for $z = qt$, where $-\tfrac{1}{2}\pi < \arg q < \tfrac{3}{2}\pi$. Note that, when $\mathrm{Re}\,q > 0$, the first term in the asymptotic expansion is dominant and convergence of the resulting integral requires the restriction $\mathrm{Re}\,s > \mathrm{Re}\,(q/2)$. Since the right-hand side is also analytic in the region $\mathrm{Re}\,s > \tfrac{1}{2}\mathrm{Re}\,q > 0$, analytic continuation established the formula. A similar argument can be made for $\mathrm{Re}\,q \le 0$. In the case $\mathrm{Re}\,q < 0$, the leading term in (13.1.9) is now the second one. The details are omitted.

EXAMPLE 13.2.6. Entry **7.621.3** states that, for $\mathrm{Re}\,\left(\alpha \pm \mu + \tfrac{3}{2}\right) > 0$, $\mathrm{Re}\,(s) > -\tfrac{q}{2}$ and $q > 0$,

(13.2.20)
$$\int_0^\infty e^{-st} t^\alpha W_{\lambda,\mu}(qt)\, dt = \frac{\Gamma(\alpha+\mu+\tfrac{3}{2})\Gamma(\alpha-\mu+\tfrac{3}{2})q^{\mu+\frac{1}{2}}}{\Gamma(\alpha-\lambda+2)}\left(s+\frac{q}{2}\right)^{-\alpha-\mu-\frac{3}{2}}$$

$$\times\ {}_2F_1\left(\begin{matrix}\alpha+\mu+\tfrac{3}{2} & \mu-\lambda+\tfrac{1}{2}\\ \alpha-\lambda+2\end{matrix}\ \middle|\ \frac{2s-q}{2s+q}\right).$$

Note that, from (13.1.7) and (13.1.10),

(13.2.21) $$W_{\lambda,\mu}(x) = x^{\mu+\frac{1}{2}} e^{-x/2} \Psi(\mu-\lambda+\tfrac{1}{2}; 2\mu+1; x).$$

The evaluation begins with the change of variables $x = qt$ to produce

$$\int_0^\infty e^{-st} t^\alpha W_{\lambda,\mu}(qt)\, dt = \frac{1}{q^{\alpha+1}} \int_0^\infty e^{-sx/q} x^\alpha W_{\lambda,\mu}(x)\, dx$$

$$= \frac{1}{q^{\alpha+1}} \int_0^\infty e^{-\left(\frac{s}{q}+\frac{1}{2}\right)x} x^{\left(\alpha+\mu+\frac{3}{2}\right)-1} \Psi(\mu - \lambda + \tfrac{1}{2}; 2\mu + 1; x)\, dx$$

$$= \frac{1}{q^{\alpha+1}} \frac{\Gamma(\alpha+\mu+\frac{3}{2})\Gamma(\alpha-\mu+\frac{3}{2})}{\Gamma(\alpha-\lambda+2)}$$

$$\times {}_2F_1\left(\begin{array}{c} \alpha+\mu+\frac{3}{2} \quad \alpha-\mu+\frac{3}{2} \\ \alpha-\lambda+2 \end{array}\middle| \frac{1}{2} - \frac{s}{q}\right),$$

using the first part of **7.621.6** in Example 13.2.3. The application of formula **7.621.6** requires the conditions

(13.2.22) $\mathrm{Re}\,(\alpha + \mu + \tfrac{3}{2}) > 0,\ \mathrm{Re}\,(\alpha - \mu + \tfrac{3}{2}) > 0$ and $\left|\frac{1}{2} - \frac{s}{q}\right| < 1.$

The last condition is more restrictive than the conditions given for the present entry. This can be relaxed to $\mathrm{Re}\left(\frac{s}{q}\right) > -\frac{1}{2}$, that is, $\mathrm{Re}\,(s) > -\frac{q}{2}$ using (13.1.11). This shows that the integrand behaves like $e^{-(s/q+1/2)x}$ at infinity. Convergence requires the stated restriction $\mathrm{Re}\,(s/q + 1/2) > 0$. The final form of the answer can now be produced by using Pfaff's transformation.

The special case $\alpha = \nu - 1$, $s = \frac{1}{2}$, and $q = 1$ produces

(13.2.23) $$\int_0^\infty e^{-x/2} x^{\nu-1} W_{\kappa,\mu}(x)\, dx = \frac{\Gamma\left(\nu + \frac{1}{2} - \mu\right)\Gamma\left(\nu + \frac{1}{2} + \mu\right)}{\Gamma(\nu - \kappa + 1)}.$$

This is Entry **7.621.11**. Observe that, given the specialized parameters, the hypergeometric term in **7.621.3** reduces to 1.

EXAMPLE 13.2.7. Entry **7.621.7** is evaluated next. This evaluation will show that the answer stated in [**35**] contains a typo. The entry, as stated in the table, is

(13.2.24) $$\int_0^\infty e^{-\frac{b}{2}x} x^{\nu-1} M_{\kappa,\mu}(bx)\, dx = \frac{\Gamma(1+2\mu)\Gamma(\kappa-\nu)\Gamma(\frac{1}{2}+\mu+\nu)}{\Gamma(\frac{1}{2}+\mu+\kappa)\Gamma(\frac{1}{2}+\mu-\nu)} b^\nu,$$

for $\mathrm{Re}\,(\nu + \mu + \tfrac{1}{2}) > 0$ and $\mathrm{Re}\,(\kappa - \nu) > 0$.

Proof. Assume first that $b > 0$. Then

$$\int_0^\infty e^{-\frac{b}{2}x} x^{\nu-1} M_{\kappa,\mu}(bx)\, dx = \frac{1}{b^\nu} \int_0^\infty e^{-t/2} t^{\nu-1} M_{\kappa,\mu}(t)\, dt$$

$$= \frac{1}{b^\nu} \int_0^\infty e^{-t/2} t^{\nu-1} t^{\mu+\frac{1}{2}} e^{-t/2} {}_1F_1(\mu - \kappa + \tfrac{1}{2}; 2\mu + 1; t)\, dt$$

$$= \frac{1}{b^\mu} \int_0^\infty e^{-t} t^{(\mu+\nu+\frac{1}{2})-1} {}_1F_1(\mu - \kappa + \tfrac{1}{2}; 2\mu + 1; t)\, dt.$$

The convergence at $t = 0$ requires $\mathrm{Re}\,(\mu + \nu + \frac{1}{2}) > 0$, and, near infinity, observe that

$$_1F_1(\mu - \kappa + \tfrac{1}{2}; 2\mu + 1; t) \sim \frac{\Gamma(2\mu + 1)}{\Gamma(\mu - \kappa + \frac{1}{2})} e^t t^{-\kappa - \mu - \frac{1}{2}}$$

(13.2.25)
$$\times \sum_{n=0}^{\infty} \frac{(\mu + \kappa + \frac{1}{2})_n (-\mu + \kappa + \frac{1}{2})_n}{n!} t^{-n}.$$

Since the exponential factors cancel, the leading order term at infinity is $t^{\nu - \kappa - 1}$; therefore, convergence requires $\mathrm{Re}\,(\kappa - \nu) > 0$.

The evaluation of the integral is obtained using the first part of Entry **7.621.4** in Example 13.2.1. This gives

(13.2.26) $\dfrac{1}{b^\mu} \displaystyle\int_0^\infty e^{-t} t^{(\mu + \nu + \frac{1}{2}) - 1} {}_1F_1(\mu - \kappa + \tfrac{1}{2}; 2\mu + 1; t)\, dt =$

$$\frac{\Gamma(\mu + \nu + \frac{1}{2})}{b^\nu} {}_2F_1\left(\begin{matrix} \mu - \kappa + \frac{1}{2} & \mu + \nu + \frac{1}{2} \\ 2\mu + 1 \end{matrix} \middle| 1\right).$$

The value of the hypergeometric function is obtained using Gauss' formula

(13.2.27) $$_2F_1\left(\begin{matrix} a & b \\ c \end{matrix} \middle| 1\right) = \frac{\Gamma(c)\Gamma(c - a - b)}{\Gamma(c - a)\Gamma(c - b)}$$

valid for $\mathrm{Re}\,(c - a - b) > 0$. In the case considered here, this condition becomes $\mathrm{Re}\,(\kappa - \nu) > 0$, which is satisfied. To complete the evaluation, the restriction $b > 0$ is now removed by analytic continuation.

The correct value of the entry is

(13.2.28) $\displaystyle\int_0^\infty e^{-\frac{b}{2}x} x^{\nu - 1} M_{\kappa,\mu}(bx)\, dx = \dfrac{\Gamma(1 + 2\mu)\Gamma(\kappa - \nu)\Gamma(\frac{1}{2} + \mu + \nu)}{\Gamma(\frac{1}{2} + \mu + \kappa)\Gamma(\frac{1}{2} + \mu - \nu)} b^{-\nu},$

i.e., the exponent of the parameter b has an error in (13.2.24).

EXAMPLE 13.2.8. Entry **7.621.8** states that, for $\mathrm{Re}\,(\mu + \frac{1}{2}) > 0$ and $\mathrm{Re}\,(s) > \frac{1}{2}$,
(13.2.29)

$$\int_0^\infty e^{-sx} M_{\kappa,\mu}(x)\, \frac{dx}{x} = \frac{2\Gamma(1 + 2\mu)}{\Gamma(\frac{1}{2} + \mu + \kappa)} e^{-\pi i \kappa} \left(\frac{s - \frac{1}{2}}{s + \frac{1}{2}}\right)^{\kappa/2} Q^\kappa_{\mu - \frac{1}{2}}(2s).$$

Proof. Start by using the definition of $M_{\kappa,\mu}(x)$ to obtain
(13.2.30)

$$\int_0^\infty e^{-sx} M_{\kappa,\mu}(x)\, \frac{dx}{x} = \int_0^\infty e^{-(s + \frac{1}{2})x} x^{\mu - \frac{1}{2}} {}_1F_1(\mu - \kappa + \tfrac{1}{2}; 2\mu + 1; x)\, dx.$$

The behavior of the integrand at 0 requires $\mathrm{Re}\,(\mu + \frac{1}{2}) > 0$ and at infinity $\mathrm{Re}\,(s) > \frac{1}{2}$.

To produce the evaluation of this entry, assume first $\left|s - \frac{1}{2}\right| > 1$, in order to use the second part of **7.621.4** in Example 13.2.1, with $s \mapsto s + \frac{1}{2}$, $b \mapsto$

$\mu + \frac{1}{2}$, $k = 1$, $a \mapsto \mu - \kappa + \frac{1}{2}$, and $c = 2\mu + 1$. This produces

(13.2.31)
$$\int_0^\infty e^{-sx} M_{\kappa,\mu}(x) \, \frac{dx}{x} = \Gamma(\mu + \tfrac{1}{2})(s - \tfrac{1}{2})^{-\mu - \frac{1}{2}} {}_2F_1 \left(\begin{matrix} \mu + \kappa + \frac{1}{2} & \mu + \frac{1}{2} \\ 2\mu + 1 \end{matrix} \middle| \frac{2}{1 - 2s} \right).$$

Now use the transformation [**13**, p. 127, equation (3.1.7)]

(13.2.32)
$$\,_2F_1 \left(\begin{matrix} a & b \\ 2a \end{matrix} \middle| x \right) = \left(1 - \frac{x}{2} \right)^{-b} {}_2F_1 \left(\begin{matrix} \frac{b}{2} & \frac{b+1}{2} \\ a + \frac{1}{2} \end{matrix} \middle| \left(\frac{x}{2 - x} \right)^2 \right)$$

with $a \mapsto \mu + \frac{1}{2}$, $b \mapsto \mu + \kappa + \frac{1}{2}$, and $x = 2/(1 - 2s)$ to obtain

(13.2.33)
$$\int_0^\infty e^{-sx} M_{\kappa,\mu}(x) \, \frac{dx}{x}$$
$$= \Gamma(\mu + \tfrac{1}{2})(s - \tfrac{1}{2})^\kappa s^{-\mu - \kappa - \frac{1}{2}} {}_2F_1 \left(\begin{matrix} \frac{1}{2}(\mu + \kappa + \frac{1}{2}) & \frac{1}{2}(\mu + \kappa + \frac{3}{2}) \\ \mu + 1 \end{matrix} \middle| \frac{1}{4s^2} \right).$$

Use the duplication formula for the gamma function

(13.2.34)
$$\Gamma(2u) = \frac{2^{2u-1}}{\sqrt{\pi}} \Gamma(u) \Gamma(u + \tfrac{1}{2})$$

(Entry **8.335.1** in [**35**]) to obtain

$$\int_0^\infty e^{-sx} M_{\kappa,\mu}(x) \, \frac{dx}{x} = \frac{2\Gamma(2\mu + 1)e^{-i\pi\kappa}}{\Gamma(\mu + \kappa + \frac{1}{2})} \left(\frac{s - \frac{1}{2}}{s + \frac{1}{2}} \right)^{\kappa/2}$$

(13.2.35)
$$\left\{ \frac{\sqrt{\pi} e^{i\pi\kappa} \Gamma(\mu + \kappa + \frac{1}{2})}{2^{\mu + \frac{1}{2}} \Gamma(\mu + 1)} (4s^2 - 1)^{\kappa/2} (2s)^{-(\mu - \frac{1}{2}) - \kappa - 1} \right.$$
$$\left. {}_2F_1 \left(\begin{matrix} \frac{1}{2}(\mu + \kappa + \frac{1}{2}) & \frac{1}{2}(\mu + \kappa + \frac{3}{2}) \\ \mu + 1 \end{matrix} \middle| \frac{1}{4s^2} \right) \right\}.$$

The convergence of the hypergeometric term requires $|s| > \frac{1}{2}$. But, as has been stated before, the convergence of the integral requires a more restrictive condition $\text{Re}(s) > \frac{1}{2}$.

The function $Q_\mu^\kappa(s)$ is called the *associated Legendre function of the second kind* and is defined in Entry **8.703** of [**35**, p. 959] as

$$Q_\nu^\mu(z) = \frac{e^{i\pi\mu} \Gamma(\nu + \mu + 1) \Gamma\left(\frac{1}{2}\right)}{2^{\nu+1} \Gamma(\nu + \frac{3}{2})} (z^2 - 1)^{\mu/2} z^{-\nu - \mu - 1}$$

(13.2.36)
$$\times \, {}_2F_1 \left(\begin{matrix} \frac{1}{2}(\mu + \nu + 2) & \frac{1}{2}(\mu + \nu + 1) \\ \nu + \frac{3}{2} \end{matrix} \middle| \frac{1}{z^2} \right).$$

EXAMPLE 13.2.9. Entry **7.621.9** states that

(13.2.37)
$$\int_0^\infty e^{-sx} W_{\kappa,\mu}(x) \frac{dx}{x} = \frac{\pi}{\cos\left(\frac{\pi\mu}{2}\right)} \left(\frac{s - \frac{1}{2}}{s + \frac{1}{2}} \right)^{\kappa/2} P_{\mu - \frac{1}{2}}^\kappa(2s)$$

for $\mathrm{Re}\left(\frac{1}{2} \pm \mu\right) > 0$, $\mathrm{Re}\,(s) > -\frac{1}{2}$. The function $P_\nu^\mu(z)$ is defined in Entry **8.702** of [**35**] by

$$(13.2.38) \qquad P_\nu^\mu(z) = \frac{1}{\Gamma(1-\mu)} \left(\frac{z+1}{z-1}\right)^{\mu/2} {}_2F_1\left(\begin{matrix} -\nu & \nu+1 \\ & 1-\mu \end{matrix} \,\middle|\, \frac{1-z}{2}\right).$$

Proof. Start with the expression

$$(13.2.39) \qquad W_{\kappa,\mu}(x) = e^{-x/2} x^{\mu+\frac{1}{2}} \Psi(\mu-\kappa+\tfrac{1}{2}; 2\mu+1; x)$$

to obtain
(13.2.40)
$$\int_0^\infty e^{-sx} W_{\kappa,\mu}(x) \frac{dx}{x} = \int_0^\infty e^{-(s+\frac{1}{2})x} x^{(\mu+\frac{1}{2})-1} \Psi(\mu-\kappa+\tfrac{1}{2}; 2\mu+1; x)\,dx.$$

The convergence of the integral at the origin requires $\mathrm{Re}\,(\mu) > -\frac{1}{2}$ and convergence at infinity requires $\mathrm{Re}\,(s) > -\frac{1}{2}$.

The first formula in Entry **7.621.6** shows that

$$(13.2.41) \quad \int_0^\infty e^{-(s+\frac{1}{2})x} x^{(\mu+\frac{1}{2})-1} \Psi(\mu-\kappa+\tfrac{1}{2}; 2\mu+1; x)\,dx =$$

$$\frac{\Gamma\left(\frac{1}{2}+\mu\right) \Gamma\left(\frac{1}{2}-\mu\right)}{\Gamma(1-\kappa)} {}_2F_1\left(\begin{matrix} \frac{1}{2}+\mu & \frac{1}{2}-\mu \\ & 1-\kappa \end{matrix} \,\middle|\, \frac{1}{2}-s\right).$$

The conditions on **7.621.6** require the restriction $\mathrm{Re}\,(\mu) < \frac{1}{2}$.

The expression (13.2.41) can be written in the form (13.2.37) by using the elementary identity

$$(13.2.42) \qquad \Gamma\left(\tfrac{1}{2}+\mu\right) \Gamma\left(\tfrac{1}{2}-\mu\right) = \frac{\pi}{\cos \pi\mu}, \quad \mu - \tfrac{1}{2} \notin \mathbb{Z},$$

which appears as Entry **8.334.2** in [**35**].

EXAMPLE 13.2.10. Entry **7.621.10** is
(13.2.43)
$$\int_0^\infty x^{\kappa+2\mu-1} e^{-3x/2} W_{\kappa,\mu}(x)\,dx = \frac{\Gamma(\kappa+\mu+\frac{1}{2}) \Gamma\left[\frac{1}{4}(2\kappa+6\mu+5)\right]}{(\kappa+3\mu+\frac{1}{2}) \Gamma\left[\frac{1}{4}(2\mu-2\kappa+3)\right]}$$

under the conditions $\mathrm{Re}\,(k+\mu) > -\frac{1}{2}$, $\mathrm{Re}\,(k+3\mu) > -\frac{1}{2}$.

As usual, the conditions on the parameters can be established by examining the convergence of the integral. The proof begins with
(13.2.44)
$$\int_0^\infty x^{\kappa+2\mu-1} e^{-3x/2} W_{\kappa,\mu}(x)\,dx = e^{-2x} x^{(3\mu+\kappa+\frac{1}{2})-1} \Psi\left(\mu-\kappa+\tfrac{1}{2}; 2\mu+1; x\right)\,dx.$$

The first part of Entry **7.621.6** gives the value of the integral as

$$(13.2.45) \qquad \frac{\Gamma(3\mu+\kappa+\frac{1}{2}) \Gamma(\mu+\kappa+\frac{1}{2})}{\Gamma(2\mu+1)} {}_2F_1\left(\begin{matrix} 3\mu+\kappa+\frac{1}{2} & \mu+\kappa+\frac{1}{2} \\ & 1+2\mu \end{matrix} \,\middle|\, -1\right).$$

The form of the answer given in [**35**] is obtained by using Kummer's theorem (see [**57**, p. 68])

$$(13.2.46) \qquad {}_2F_1\left(\begin{matrix} a & b \\ a-b+1 \end{matrix} \middle| -1\right) = \frac{\Gamma(a-b+1)\Gamma(1+\frac{a}{2})}{\Gamma(1+\frac{a}{2}-b)\Gamma(1+a)}.$$

EXAMPLE 13.2.11. The final entry established here is **7.621.12**. It states that, for $\operatorname{Re}\left(\nu + \frac{1}{2} \pm \mu\right) > 0$ and $\operatorname{Re}\left(\kappa + \nu\right) < 0$,

$$(13.2.47)$$
$$\int_0^\infty e^{x/2} x^{\nu-1} W_{\kappa,\mu}(x)\, dx = \frac{\Gamma(-\kappa-\mu)\,\Gamma\left(\frac{1}{2}+\mu+\nu\right)\Gamma\left(\frac{1}{2}-\mu+\nu\right)}{\Gamma\left(\frac{1}{2}-\mu-\kappa\right)\Gamma\left(\frac{1}{2}+\mu-\kappa\right)}.$$

The evaluation comes directly from Entry **7.612.2**. The convergence at $x = 0$ requires $\operatorname{Re}\left(\nu + \mu + \frac{1}{2}\right) > 0$ and at infinity $\psi(\mu - \kappa + \frac{1}{2}; 2\mu + 1; x) \sim x^{-\mu+\kappa-1/2}$ and this shows that the integrand is asymptotic to $x^{\nu+\kappa-1}$. Therefore $\operatorname{Re}\left(\kappa + \nu\right) < 0$ is needed for convergence.

CHAPTER 14

Evaluation of entries in Gradshteyn and Ryzhik employing the method of brackets

14.1. Introduction

The problem of providing a closed-form expression for a definite integral has been studied by a variety of methods. The corresponding problem for indefinite integrals has been solved, for a large class of integrands, by the methods developed by Risch [58, 59, 60]. The reader will find in [21] a modern description of these ideas and in [70] an interesting overwiew of techniques for integration.

The lack of a universal algorithm for the evaluation of definite integrals has created a collection of results saved in the form of Tables of Integrals. The volume created by I. S. Gradshteyn and I. M. Ryzhik [35], currently in its 8th edition, is widely used by the scientific community. Others include [14, 22, 26, 56]. The use of symbolic languages, such as Mathematica or Maple, for this task usually contains a *database search* as a preprocessing of the algorithms. The question of reliability of these tables is essential.

The method of brackets employed here was developed by one of the authors in [33, 34] in the context of evaluations of definite integrals obtained from the Schwinger parametrization of Feynman diagrams. The method is closely related to the so-called *negative dimensional integration method* developed by I. G. Halliday and R. M. Ricotta [36] and A. T. Suzuki et al. [61, 62, 63]. The reader will find a nice collection of examples in [11, 12]. The use of this method in the general framework of definite integrals has appeared in [31, 32]. In the present work, the flexibility of the method of brackets is illustrated with the evaluation of a selected list of examples from [35].

With just a few rules, the method can easily be automated. Code has been produced in [40] using Sage with calls to Mathematica. Testing this implementation against [35] has suggested adjustments to the original set of rules in the method. This modified set of rules is presented here.

The main rule of the method of brackets corresponds to one of Ramanujan's favorite methods to evaluate integrals of the form $\int_0^\infty dx\, x^{\nu-1} f(x)$. This is the so-called *Ramanujan Master Theorem*. It states that if $f(x)$ admits a series expansion of the form

(14.1.1)
$$f(x) = \sum_{n=0}^{\infty} \varphi(n) \frac{(-x)^n}{n!}$$

in a neighborhood of $x = 0$, with $f(0) = \varphi(0) \neq 0$, then

(14.1.2)
$$\int_0^{\infty} x^{\nu-1} f(x)\, dx = \Gamma(\nu)\varphi(-\nu).$$

The integral is the Mellin transform of $f(x)$ and the term $\varphi(-\nu)$ requires an extension of the function φ, initially defined only for $\nu \in \mathbb{N}$. Details on the natural unique extension of φ are given in [5]. Observe that, for $\nu > 0$, the condition $\varphi(0) \neq 0$ guarantees the convergence of the integral near $x = 0$. The proof of Ramanujan's Master Theorem and the precise conditions for its application appear in Hardy [39]. The reader will find in [5] many other examples.

14.2. The method of brackets

This is a method that evaluates definite integrals over the half line $[0, \infty)$. The application of the method consists of a small number of rules, deduced in heuristic form, some of which are placed on solid ground [5].

For $a \in \mathbb{R}$, the symbol

(14.2.1)
$$\langle a \rangle \mapsto \int_0^{\infty} x^{a-1}\, dx$$

is the *bracket* associated to the (divergent) integral on the right. The symbol

(14.2.2)
$$\phi_n := \frac{(-1)^n}{\Gamma(n+1)}$$

is called the *indicator* associated to the index n. The notation $\phi_{i_1 i_2 \cdots i_r}$, or simply $\phi_{12\cdots r}$, denotes the product $\phi_{i_1} \phi_{i_2} \cdots \phi_{i_r}$.

Rules for the production of bracket series

Rule P_1. Power series appearing in the integrand are converted into *bracket series* by the procedure

(14.2.3)
$$\sum_{n=0}^{\infty} a_n x^{\alpha n + \beta - 1} \mapsto \sum_{n \geq 0} a_n \langle \alpha n + \beta \rangle.$$

Rule P_2. For $\alpha \in \mathbb{C}$, the multinomial power $(a_1 + a_2 + \cdots + a_r)^{\alpha}$ is assigned the r-dimension bracket series

(14.2.4)
$$\sum_{n_1 \geq 0} \sum_{n_2 \geq 0} \cdots \sum_{n_r \geq 0} \phi_{n_1 n_2 \cdots n_r} a_1^{n_1} \cdots a_r^{n_r} \frac{\langle -\alpha + n_1 + \cdots + n_r \rangle}{\Gamma(-\alpha)}.$$

Rule P_3. Each representation of an integral by a bracket series has an associated *index of the representation* via

(14.2.5)   index = number of sums $-$ number of brackets.

It is important to observe that the index is attached to a specific representation of the integral and not just to the integral itself. The experience obtained by the authors using this method suggests that, among all representations of an integral as a bracket series, the one with the *minimal index* should be chosen.

Rules for the evaluation of a bracket series

Rule E_1. The one-dimensional bracket series is assigned the value

$$(14.2.6) \qquad \sum_{n\geq 0} \phi_n f(n)\langle an + b\rangle \mapsto \frac{1}{|a|} f(n^*)\Gamma(-n^*),$$

where n^* is obtained from the vanishing of the bracket; that is, n^* solves $an + b = 0$. This is precisely the Ramanujan Master Theorem.

The next rule provides a value for multi-dimensional bracket series of index 0, that is, the number of sums is equal to the number of brackets.

Rule E_2. Assuming the matrix $A = (a_{ij})$ is non-singular, then the assignment is

$$\sum_{n_1\geq 0} \cdots \sum_{n_r\geq 0} \phi_{n_1\cdots n_r} f(n_1, \cdots, n_r)$$

$$\langle a_{11}n_1 + \cdots + a_{1r}n_r + c_1\rangle \cdots \langle a_{r1}n_1 + \cdots + a_{rr}n_r + c_r\rangle$$

$$\mapsto \frac{1}{|\det(A)|} f(n_1^*, \cdots n_r^*)\Gamma(-n_1^*) \cdots \Gamma(-n_r^*)$$

where $\{n_i^*\}$ is the (unique) solution of the linear system obtained from the vanishing of the brackets. There is no assignment if A is singular.

Rule E_3. The value of a multi-dimensional bracket series of positive index is obtained by computing all the contributions of maximal rank by Rule E_2. These contributions to the integral appear as series in the free parameters. Series converging in a common region are added and divergent series are discarded. Any series producing a non-real contribution is also discarded. There is no assignment to a bracket series of negative index.

The next sections offer a variety of examples that illustrate these rules.

14.3. Examples of index 0

This section contains some integrals from [**35**] that lead to bracket series of index 0. The evaluation of these entries by the method of brackets illustrates the rules described in the previous section.

Example 3.1. Entry **3.310** states the elementary result

$$(14.3.1) \qquad I = \int_0^\infty e^{-x}\, dx = 1.$$

The method of brackets begins with the integral representation

$$(14.3.2) \qquad e^{-x} = \sum_{n_1=0}^\infty \frac{(-x)^{n_1}}{n_1!}$$

with its corresponding bracket series

(14.3.3)
$$\sum_{n_1=0}^{\infty} \frac{(-1)^{n_1}}{n_1!} \langle n_1 + 1 \rangle = \sum_{n_1=0}^{\infty} \phi_{n_1} \langle n_1 + 1 \rangle,$$

and the associated function $f(n_1) \equiv 1$. Therefore, this problem produces one sum and a single bracket giving a sum of index 0. The vanishing of the brackets gives $n_1^* = -1$. Rule E_1 gives the integral as

(14.3.4)
$$I = \Gamma(n_1^*) = \Gamma(1) = 1.$$

Example 3.2. The integrand now involves the Bessel function

(14.3.5)
$$J_\nu(x) := \left(\frac{x}{2}\right)^\nu \sum_{n=0}^{\infty} \frac{(-1)^n}{n!\,\Gamma(\nu + 1 + n)} \left(\frac{x}{2}\right)^{2n}.$$

The bracket series corresponding to the integral

(14.3.6)
$$I = \int_0^\infty J_\nu(bx)\,dx$$

is

(14.3.7)
$$S = \left(\frac{b}{2}\right)^\nu \sum_{n \geq 0} \phi_{n_1} \frac{1}{\Gamma(\nu + 1 + n_1)} \frac{b^{2n_1}}{4^{n_1}} \langle 2n_1 + \nu + 1 \rangle.$$

This bracket series also has index 0: one sum and one bracket. The vanishing of this bracket yields $n_1^* = -\frac{1}{2}(1 + \nu)$. Therefore, the integral is assigned the value

$$\begin{aligned}
I &= \frac{1}{2} \left(\frac{b}{2}\right)^\nu \frac{b^{2n_1^*}}{2^{2n_1^*}\,\Gamma(\nu + 1 + n_1^*)} \Gamma(-n_1^*) \\
&= \frac{1}{2} \left(\frac{b}{2}\right)^\nu \frac{b^{-1-\nu}}{2^{-1-\nu}\,\Gamma(\frac{\nu+1}{2})} \Gamma(\tfrac{\nu+1}{2}) \\
&= \frac{1}{b}.
\end{aligned}$$

This agrees with Entry **6.511.1** in [**35**].

Example 3.2. Entry **6.521.11** gives the identity

(14.3.8)
$$I = \int_0^\infty x^2 K_1(ax)\,dx = \frac{2}{a^3}$$

for $a > 0$. The integrand now involves the modified Bessel function of the second kind. Use of the integral representation

(14.3.9)
$$K_\nu(x) := \frac{2^\nu \Gamma\left(\nu + \frac{1}{2}\right)}{x^\nu \Gamma\left(\frac{1}{2}\right)} \int_0^\infty \frac{\cos(xt)\,dt}{(t^2 + 1)^{\nu + \frac{1}{2}}}$$

produces the double integral

(14.3.10)
$$I = \int_0^\infty x^2 K_1(ax)\,dx = \int_0^\infty \int_0^\infty \frac{x\,\cos(axt)}{a\,(t^2 + 1)^{\frac{3}{2}}}\,dt\,dx.$$

The $\cos(axt)$ factor is written as a series in n_1

$$(14.3.11) \qquad \cos(axt) = \sum_{n_1=0}^{\infty} \frac{\Gamma\left(\frac{1}{2}\right)}{\Gamma\left(\frac{1}{2}+n_1\right)} \left(-\frac{axt}{2}\right)^{2n_1}$$

and Rule P_2 assigns to the factor $(t^2+1)^{-\frac{3}{2}}$ the bracket series

$$(14.3.12) \qquad \sum_{n_2 \geq 0} \sum_{n_3 \geq 0} \phi_{n_2 \, n_3} t^{2n_2} 1^{n_3} \frac{\left\langle \frac{3}{2} + n_2 + n_3 \right\rangle}{\Gamma\left(\frac{3}{2}\right)}.$$

The final step in producing the bracket series is to replace $t^{2n_1+2n_2}$ with $\langle 1 + 2n_1 + 2n_2 \rangle$ and x^{1+2n_1} with $\langle 2 + 2n_1 \rangle$. The bracket series

$$(14.3.13)$$
$$\sum_{n_1 \geq 0} \sum_{n_2 \geq 0} \sum_{n_3 \geq 0} \phi_{1,2,3} \frac{2^{1-2n_1} a^{-1+2n_1}}{\Gamma\left(\frac{1}{2}+n_1\right)} \langle 2 + 2n_1 \rangle \langle 1 + 2n_1 + 2n_2 \rangle \left\langle \frac{3}{2} + n_2 + n_3 \right\rangle$$

is of index 0.

The linear system constructed from the vanishing of the brackets is

$$(14.3.14) \qquad \begin{pmatrix} 2 & 0 & 0 \\ 2 & 2 & 0 \\ 0 & 1 & 1 \end{pmatrix} \begin{pmatrix} n_1 \\ n_2 \\ n_3 \end{pmatrix} = \begin{pmatrix} -2 \\ -1 \\ -\frac{3}{2} \end{pmatrix}$$

with the matrix A having rank 3 and determinant 4. The solution of the system gives $n_1^* = -1$, $n_2^* = \frac{1}{2}$, and $n_3^* = -2$. The value of the integral by Rule E_2 is

$$
\begin{aligned}
I &= \frac{1}{4} \left(\frac{2^{1-2n_1^*} a^{-1+2n_1^*}}{\Gamma\left(\frac{1}{2}+n_1^*\right)} \right) \Gamma(-n_1^*)\Gamma(-n_2^*)\Gamma(-n_3^*) \\
&= \frac{1}{4} \left(\frac{2^3 a^{-3}}{\Gamma\left(-\frac{1}{2}\right)} \right) \Gamma(1)\Gamma\left(-\frac{1}{2}\right)\Gamma(2) \\
&= \frac{2}{a^3},
\end{aligned}
$$

verifying (14.3.8).

14.4. Examples of index 1

This section considers integrals that lead to representations of index 1.

Example 4.1 The first example provides an evaluation of the elementary entry **3.311.1** in [**35**]

$$(14.4.1) \qquad \int_0^{\infty} \frac{dx}{e^{px}+1} = \frac{\ln 2}{p}.$$

The method of brackets will reduce the problem to a triple series and two brackets, leading to a representation of index 1. Rule E_3 reduces the number of sums by two and the answer is expressed as a single series. The remaining series is elementary and is recognized as $\ln 2$.

The first step is to replace the integrand by its brackets series

$$\sum_{n_1 \geq 0} \sum_{n_2 \geq 0} \phi_{1,2}(e^{px})^{n_1} 1^{n_2} \frac{\langle 1 + n_1 + n_2 \rangle}{\Gamma(1)} = \sum_{n_1 \geq 0} \sum_{n_2 \geq 0} \phi_{1,2} e^{n_1 px} \langle 1 + n_1 + n_2 \rangle.$$

The power series representation of the exponential is now employed to produce

$$\sum_{n_1 \geq 0} \sum_{n_2 \geq 0} \phi_{1,2} \left(\sum_{n_3 \geq 0} \frac{(xn_1)^{n_3}}{\Gamma(n_3 + 1)} \right) \langle 1 + n_1 + n_2 \rangle =$$

$$\sum_{n_1, n_2, n_3 \geq 0} \phi_{1,2,3} (-n_1)^{n_3} p^{n_3} x^{n_3} \langle 1 + n_1 + n_2 \rangle.$$

This form of integrand produces the bracket series

(14.4.2) $$I = \sum_{n_1, n_2, n_3 \geq 0} \phi_{1,2,3} (-n_1)^{n_3} p^{n_3} \langle 1 + n_1 + n_2 \rangle \langle n_3 + 1 \rangle$$

for the integral.

There are now three sums and two brackets, giving a representation of index 1. The matrix equation associated to the vanishing of the brackets

(14.4.3) $$\begin{pmatrix} 1 & 1 & 0 \\ 0 & 0 & 1 \end{pmatrix} \begin{pmatrix} n_1 \\ n_2 \\ n_3 \end{pmatrix} = \begin{pmatrix} -1 \\ -1 \end{pmatrix}$$

has rank 2. It follows that the problem has 1 free parameter. Observe that the equation coming from the vanishing of the bracket $\langle n_3 + 1 \rangle$ determines $n_3^* = -1$. The system has reduced to the single equation $n_1 + n_2 = -1$. The choices of free indices are n_1 and n_2 and their contributions to the integral are described next.

Case 1: n_1 is free. The relation among the indices yields $n_2^* = -n_1 - 1$ and the corresponding determinant is -1. The contribution of this index to the integral is

(14.4.4) $$\sum_{n_1 \geq 0} \phi_1 \frac{1}{|-1|} (-n_1 p)^{n_3^*} \Gamma(-n_2^*) \Gamma(-n_3^*) = \sum_{n_1 \geq 0} \frac{(-1)^{n_1 + 1}}{n_1 p}.$$

The term $n_1 = 0$ yields the series divergent, so its contribution to the integral is discarded.

Case 2: n_2 is free. Then $n_1^* = -n_2 - 1$ with determinant -1. The contribution of this index to the integral is given by

(14.4.5) $$\sum_{n_2 \geq 0} \phi_2 \frac{1}{|-1|} (-n_1^* p)^{n_3^*} \Gamma(-n_1^*) \Gamma(-n_3^*) = \sum_{n_2 \geq 0} \frac{(-1)^{n_2}}{(n_2 + 1)p}.$$

Adding all the finite contributions of free indices gives the evaluation

$$(14.4.6) \qquad \int_0^\infty \frac{dx}{e^x + 1} = \sum_{n_2 \geq 0} \frac{(-1)^{n_2}}{(n_2 + 1)p}.$$

In order to present the integral in its simplest possible form, it is now required to identify this series. In this case this is elementary: the result is

$$(14.4.7) \qquad \sum_{n_2 \geq 0} \frac{(-1)^{n_2}}{(n_2 + 1)p} = \frac{\ln 2}{p}.$$

Thus

$$(14.4.8) \qquad \int_0^\infty \frac{dx}{e^{px} + 1} = \frac{\ln 2}{p},$$

as stated in [**35**].

Example 4.2. This example illustrates the fact that the method of brackets gives, as the value of a definite integral, a finite number of series. The question of reduction of these series to their simplest form is a separate issue. As of now, there is no algorithmic solution to this question.

Entry **3.452.1** states that

$$(14.4.9) \qquad I = \int_0^\infty \frac{x \, dx}{\sqrt{e^x - 1}} = 2\pi \ln 2.$$

The brackets series for the integral is obtained as before, with the result

$$(14.4.10) \qquad I = \sum_{n_1, n_2, n_3} \phi_{1,2,3} \frac{1}{\sqrt{\pi}} (-1)^{n_2 + n_3} \langle n_3 + 2 \rangle \langle n_1 + n_2 + \tfrac{1}{2} \rangle.$$

This is a representation of index 1 (three sums and two brackets).

The vanishing of the brackets shows that n_3 is fixed: $n_3^* = -2$ and the relation $n_1 + n_2 + \tfrac{1}{2} = 0$ must hold. Therefore, the integral is given in terms of a single series.

Case 1: if n_1 is free, then $n_2^* = -n_1 - \tfrac{1}{2}$ and the corresponding series is

$$(14.4.11) \qquad S_1 = \sum_{n_1 = 0}^\infty (-1)^{2n_1 + 5/2} \frac{\Gamma(n_1 + 1/2)}{\sqrt{\pi} n_1^2 \Gamma(n_1 + 1)}.$$

This series is discarded due to the presence of the singular term at $n_1 = 0$.

Case 2: if n_2 is free, then $n_1^* = -n_2 - \tfrac{1}{2}$ and the corresponding series is

$$(14.4.12) \qquad S_2 = \sum_{n_2 = 0}^\infty \frac{4\Gamma(n_2 + 1/2)}{(2n_2 + 1)^2 \sqrt{\pi} \Gamma(n_2 + 1)}.$$

The duplication formula for the gamma function

$$(14.4.13) \qquad \Gamma(m + \tfrac{1}{2}) = \frac{(2m)!}{2^{2m} m!} \sqrt{\pi}$$

reduces the series to

$$(14.4.14) \qquad S_2 = \sum_{n_2=0}^{\infty} \binom{2n_2}{n_2} \frac{2^{-2n_2}}{(2n_2+1)^2}.$$

The method of brackets now yields

$$(14.4.15) \qquad \int_0^{\infty} \frac{x\,dx}{\sqrt{e^x - 1}} = \sum_{n_2=0}^{\infty} \binom{2n_2}{n_2} \frac{2^{-2n_2}}{(2n_2+1)^2}.$$

To evaluate the series, start with

$$(14.4.16) \qquad \sum_{m=0}^{\infty} \binom{2m}{m} x^m = \frac{1}{\sqrt{1-4x}},$$

replace x by x^2, and integrate from 0 to $1/2$ to produce (after simplifications)

$$(14.4.17) \qquad \sum_{m=0}^{\infty} \frac{\binom{2m}{m}}{(2m+1)^2} 2^{-2m} = \int_0^1 \frac{\text{Arcsin } u}{u}\,du.$$

Finally, integrate by parts to obtain

$$(14.4.18) \qquad \sum_{m=0}^{\infty} \frac{\binom{2m}{m}}{(2m+1)^2} 2^{-2m} = -\int_0^{\pi/2} \ln \sin y\,dy.$$

Euler showed that this last integral evaluates to $-\frac{\pi}{2}\ln 2$. Details of this elementary evaluation can be found in Section 12.5 of [**17**]. Formula (14.4.9) has been verified.

Example 4.3. Entry **6.554.1** gives the evaluation

$$(14.4.19) \qquad \int_0^{\infty} x J_0(xy) \frac{dx}{(x^2+a^2)^{3/2}} = a^{-1} e^{-ay}$$

for $y > 0$ and $a \in \mathbb{C}$ with $\text{Re}\,a > 0$. Here J_0 is the Bessel function

$$(14.4.20) \qquad J_0(x) = \sum_{k=0}^{\infty} (-1)^k \frac{x^{2k}}{2^{2k}\,k!^2}.$$

The bracket representations of the terms in the integrand are

$$(14.4.21) \qquad (a^2+x^2)^{-3/2} = \sum_{n_1 \geq 0}\sum_{n_2 \geq 0} \phi_{12} a^{2n_1} x^{2n_2} \frac{\langle \frac{3}{2}+n_1+n_2 \rangle}{\Gamma\left(\frac{3}{2}\right)}$$

and

$$(14.4.22) \qquad J_0(xy) = \sum_{n_3 \geq 0} \phi_{n_3} \frac{1}{2^{2n_3}\,\Gamma(1+n_3)} (xy)^{2n_3}.$$

Therefore the integral is assigned the bracket series

$$(14.4.23) \quad \int_0^\infty x J_0(xy) \frac{dx}{(x^2 + a^2)^{3/2}} \mapsto$$

$$\sum_{n_1 \geq 0} \sum_{n_2 \geq 0} \sum_{n_3 \geq 0} \phi_{1,2,3} \frac{y^{2n_3} a^{2n_1}}{2^{2n_3} \Gamma(1 + n_3) \Gamma(\frac{3}{2})} \langle \tfrac{3}{2} + n_1 + n_2 \rangle \langle 2n_2 + 2n_3 + 2 \rangle.$$

This is a representation of index $+1$.

Case 1: n_1 free. The linear system from the vanishing of brackets is

$$(14.4.24) \quad \begin{pmatrix} 1 & 0 \\ 2 & 2 \end{pmatrix} \begin{pmatrix} n_2 \\ n_3 \end{pmatrix} = \begin{pmatrix} -n_1 - \frac{3}{2} \\ -2 \end{pmatrix}$$

with determinant 2 and solutions $n_2^* = -n_1 - \frac{3}{2}$ and $n_3^* = n_1 + \frac{1}{2}$. The contribution to the integral is given by

$$S_1 = \sum_{n_1 \geq 0} \frac{(-1)^{n_1} 2^{-2n_1 - 1} a^{2n_1} y^{2n_1 + 1} \Gamma(-n_1 - \frac{1}{2})}{\sqrt{\pi} \, \Gamma(n_1 + 1)}$$

$$= -y \sum_{n_1 = 0}^\infty \frac{(ay)^{2n_1}}{(2n_1 + 1)!}$$

$$= -\frac{\sinh ay}{a}.$$

Case 2: n_2 free. Proceeding as before, it is found that this case leads to a divergent series so its contribution is ignored.

Case 3: n_3 free. As in Case 1, the system has determinant -2 with solutions $n_1^* = n_3 - \frac{1}{2}$ and $n_2^* = -n_3 - \frac{1}{2}$. The contribution to the integral is

$$S_3 = \sum_{n_3 \geq 0} \frac{(-1)^{n_3} a^{2n_3 - 1} y^{2n_3} \Gamma(-n_3 + \frac{1}{2})}{\sqrt{\pi} \, 2^{2n_3} \Gamma(n_3 + 1)}$$

$$= \frac{\cosh ay}{a}.$$

Summing the finite contributions by Rule E_3 gives

$$(14.4.25) \quad \int_0^\infty x J_0(xy) \frac{dx}{(x^2 + a^2)^{3/2}} = S_1 + S_3 = \frac{e^{-ay}}{a},$$

as stated.

Example 4.4. Entry **6.512.1** provides the value for the integral

$$(14.4.26) \quad \int_0^\infty J_\mu(ax) J_\nu(bx) \, dx.$$

The answer in [**35**] is divided according to conditions on the parameters.

The integral is provided a brackets series by the usual method:

$$\int_0^\infty J_\mu(ax)J_\nu(bx)\,dx \quad = \int_0^\infty \left(\frac{\left(\frac{ax}{2}\right)^\mu}{\Gamma(\mu+1)} \sum_{n_1=0}^\infty \frac{\left(-\frac{(ax)^2}{4}\right)^{n_1}}{(\mu+1)_{n_1}\, n_1!} \right)$$

$$\times \left(\frac{\left(\frac{bx}{2}\right)^\nu}{\Gamma(\nu+1)} \sum_{n_2=0}^\infty \frac{\left(-\frac{(bx)^2}{4}\right)^{n_2}}{(\nu+1)_{n_2}\, n_2!} \right) dx$$

is given by the series

$$S \;=\; \frac{a^\mu b^\nu}{2^{\mu+\nu}\Gamma(\mu+1)\Gamma(\nu+1)}$$

$$\times \sum_{n_1,n_2} \phi_{12}\, \frac{a^{2n_1} b^{2n_2}}{2^{2n_1+2n_2}\,(\mu+1)_{n_1}(\nu+1)_{n_2}} \langle 2n_1 + 2n_2 + \mu + \nu + 1 \rangle.$$

This is a representation of index 1.

Case 1: n_2 free. Then $n_1^* = -\frac{1}{2}(2n_2 + \mu + \nu + 1)$ and the contribution to the integral is

$$S_1 = \frac{a^\mu b^\nu}{2^{\mu+\nu}\Gamma(\mu+1)\Gamma(\nu+1)} \sum_{n_2 \geq 0} \phi_2 \frac{b^{2n_2}}{(\nu+1)_{n_2} 2^{2n_2}} \left(\frac{1}{2} \left(\frac{a}{2}\right)^{2n_1^*} \frac{\Gamma(-n_1^*)}{(\mu+1)_{n_1^*}} \right)$$

$$= \frac{b^\nu a^{-\nu-1}}{\Gamma(\mu+1)\Gamma(\nu+1)} \sum_{n_2 \geq 0} \frac{(-1)^{n_2}}{n_2!} \left(\frac{b^2}{a^2}\right)^{n_2} \frac{\Gamma\left(\frac{\mu+\nu+1}{2}+n_2\right)}{(\nu+1)_{n_2}} \frac{\Gamma(\mu+1)}{\Gamma\left(\frac{\mu+1-\nu}{2}-n_2\right)}$$

$$= \frac{b^\nu a^{-\nu-1}\Gamma\left(\frac{\mu+\nu+1}{2}\right)}{\Gamma(\nu+1)} \sum_{n_2 \geq 0} \frac{\left(-\frac{b^2}{a^2}\right)^{n_2}}{n_2!} \frac{\left(\frac{\mu+\nu+1}{2}\right)_{n_2}}{(\nu+1)_{n_2}} \frac{1}{\left(\frac{\mu-\nu+1}{2}\right)_{-n_2} \Gamma\left(\frac{\mu+1-\nu}{2}\right)}.$$

Simplifying the Pochhammer with negative index using

$$(14.4.27) \qquad\qquad\qquad (a)_{-n} = \frac{(-1)^n}{(1-a)_n}$$

gives

$$S_1 \;=\; \frac{b^\nu a^{-\nu-1}\Gamma\left(\frac{\mu+\nu+1}{2}\right)}{\Gamma(\nu+1)\Gamma\left(\frac{\mu+1-\nu}{2}\right)} \sum_{n_2 \geq 0} \frac{\left(-\frac{b^2}{a^2}\right)^{n_2} \left(\frac{\mu+\nu+1}{2}\right)_{n_2}\left(1-\frac{\mu-\nu+1}{2}\right)_{n_2}}{n_2!(\nu+1)_{n_2}(-1)^{n_2}}$$

$$= \frac{b^\nu a^{-\nu-1}\Gamma\left(\frac{\mu+\nu+1}{2}\right)}{\Gamma(\nu+1)\Gamma\left(\frac{\mu+1-\nu}{2}\right)} \sum_{n_2 \geq 0} \frac{\left(\frac{b^2}{a^2}\right)^{n_2} \left(\frac{\mu+\nu+1}{2}\right)_{n_2}\left(\frac{-\mu+\nu+1}{2}\right)_{n_2}}{n_2!(\nu+1)_{n_2}}$$

$$= \frac{b^\nu a^{-\nu-1}\Gamma\left(\frac{\mu+\nu+1}{2}\right)}{\Gamma(\nu+1)\Gamma\left(\frac{\mu+1-\nu}{2}\right)}\, {}_2F_1\left(\begin{array}{c} \frac{\mu+\nu+1}{2}, \frac{\nu-\mu+1}{2} \\ \nu+1 \end{array} \bigg| \frac{b^2}{a^2} \right).$$

The series converges provided $|b| < |a|$.

Case 2: n_2 free. The calculation is done as in Case 1. The result is the formula in Case 1, with μ and ν interchanged and a and b interchanged.

Example 4.5. Entry **3.423.1** is

$$(14.4.28) \qquad \int_0^\infty \frac{x^{\nu-1}\, dx}{(e^x - 1)^2} = \Gamma(\nu)\left[\zeta(\nu - 1) - \zeta(\nu)\right].$$

The method of brackets provides a direct evaluation. The bracket series corresponding to the integral is

$$\int_0^\infty \frac{x^{\nu-1}\, dx}{(e^x - 1)^2} \mapsto \sum_{n_1 \geq 0}\sum_{n_2 \geq 0}\sum_{n_3 \geq 0} \phi_{1,2,3}(-1)^{n_2+n_3} n_1^{n_3}\langle n_3 + \nu \rangle \langle n_1 + n_2 + 2 \rangle.$$

This is a representation of index 1 and $n_3^* = -\nu$ is determined. Choosing n_1 as a free parameter gives the series

$$\sum_{n_1 \geq 0} \frac{(-1)^{\nu-2}\Gamma(n_1 + 2)\,\Gamma(\nu)}{n_1^\nu\,\Gamma(n_1 + 1)}.$$

The term $n_1 = 0$ makes the series diverge, so its contribution is ignored.

The choice of n_2 as a free parameter gives the series

$$(14.4.29) \qquad \sum_{n_2 \geq 0} \frac{(-1)^\nu \Gamma(n_2 + 2)\,\Gamma(\nu)}{(-n_2 - 2)^\nu\,\Gamma(n_2 + 1)} = \sum_{n_2 \geq 0} \frac{(n_2 + 1)\,\Gamma(\nu)}{(n_2 + 2)^\nu}.$$

The answer is obtained by writing $n_2 + 1 = (n_2 + 2) - 1$.

14.5. Examples of index 2

This section considers integrals that lead to representations of index 2.

Example 5.1. Entry **7.414.9** provides the value

$$(14.5.1) \qquad \int_0^\infty e^{-x} x^{a+b} L_m^a(x) L_n^b(x)\, dx = (-1)^{m+n}(a + b)! \binom{a + m}{n}\binom{b + n}{m},$$

with $\operatorname{Re} a + b > -1$. Here $L_n^\lambda(x)$ is the associated Laguerre polynomial

$$(14.5.2) \qquad L_n^\lambda(z) = \frac{(\lambda + 1)_n}{n!}\, {}_1F_1\left(\begin{array}{c}-n\\ \lambda + 1\end{array}\middle| z\right).$$

The resulting bracket series is of index $+2$:

$$\sum_{n_1}\sum_{n_2}\sum_{n_3} \phi_{1,2,3} \frac{(-1)^{n_2+n_3}\Gamma(-n + n_3)\Gamma(-m + n_2)\Gamma(b + n + 1)\Gamma(a + m + 1)}{\Gamma(n + 1)\Gamma(m + 1)\Gamma(b + n_3 + 1)\Gamma(a + n_2 + 1)\Gamma(-m)\Gamma(-n)}$$
$$\times \langle a + b + n_1 + n_2 + n_3 + 1 \rangle.$$

There are three choices of free/fixed variables in the solution of the linear system coming from the vanishing of the bracket:

Case 1: With n_1 and n_2 free, the resulting series is zero and makes no contribution.

$$\sum_{n_1,n_2} \frac{(-1)^{a+b+2n_1+n_2+1}\Gamma(-m+n_2)\Gamma(b+n+1)\Gamma(a+m+1)}{\Gamma(n_2+1)\Gamma(n_1+1)\Gamma(n+1)\Gamma(m+1)\Gamma(a+n_2+1}$$

$$\times \frac{\Gamma(a+b+n_1+n_2+1)\Gamma(-a-b-n-n_1-n_2-1)}{)\Gamma(-a-n_1-n_2)\Gamma(-m)\Gamma(-n)}$$

$$= \frac{(-1)^{a+b}\Gamma(1+a+b)\Gamma(-m)\Gamma(1+a+m)\Gamma(1+b+n)\Gamma(-a-m+n)}{\pi^2 \sin((a+b+n)\pi)\Gamma(1+n)\Gamma(1+b-m+n)}$$

$$\times \sin(a\pi)\sin(m\pi)\sin(n\pi) = 0.$$

Case 2: With n_1 and n_3 free, the result is the same as in Case 1.

Case 3: With n_2 and n_3 free, the resulting series can be evaluated as follows to match the value in the table:

$$\sum_{n_2}\sum_{n_3}$$

$$\times \frac{\Gamma(-n+n_3)\Gamma(-m+n_2)\Gamma(b+n+1)\Gamma(a+m+1)\Gamma(a+b+n_2+n_3+1)}{\Gamma(n_3+1)\Gamma(n_2+1)\Gamma(n+1)\Gamma(m+1)\Gamma(b+n_3+1)\Gamma(a+n_2+1)\Gamma(-m)\Gamma(-n)}$$

$$= \sum_{n_2} \frac{\Gamma(-m+n_2)\Gamma(b+n+1)\Gamma(a+m+1)}{\Gamma(n_2+1)\Gamma(n+1)\Gamma(m+1)\Gamma(a+n_2+1)\Gamma(-m)\Gamma(-n)}$$

$$\times \frac{\Gamma(-n)\Gamma(a+b+n_2+1)}{\Gamma(b+1)} \,_2F_1(-n,1+a+b+n_2;1+b;1)$$

$$= \sum_{n_2} \frac{\Gamma(-m+n_2)\Gamma(b+n+1)\Gamma(a+m+1)\Gamma(a+b+n_2+1)}{\Gamma(n_2+1)\Gamma(n+1)\Gamma(m+1)\Gamma(a+n_2+1)\Gamma(-m)\Gamma(b+1)}$$

$$\times \frac{\Gamma(1+b)\Gamma(-a+n-n_2)}{\Gamma(1+b+n)\Gamma(-a-n_2)}$$

$$= \frac{\Gamma(a+m+1)\Gamma(1+a+b)}{\Gamma(n+1)\Gamma(m+1)} \sum_{n_2} \frac{(-m)_{n_2}(1+a+b)_{n_2}(-\pi\csc\pi(a-n+n_2))}{n_2!(-\pi\csc\pi(a+n_2))\Gamma(1+a-n+n_2)}$$

$$= \frac{\Gamma(a+m+1)\Gamma(1+a+b)}{\Gamma(n+1)\Gamma(m+1)\Gamma(1+a-n)} \sum_{n_2} \frac{(-m)_{n_2}(1+a+b)_{n_2}(\sin a\pi)(-1)^{n_2}}{n_2!(\sin a\pi)(-1)^{-n+n_2}(1+a-n)_{n_2}}$$

$$= \frac{(-1)^n\Gamma(a+m+1)\Gamma(1+a+b)}{\Gamma(n+1)\Gamma(m+1)\Gamma(1+a-n)} \,_2F_1(-m,1+a+b;1+a-n;1)$$

$$= \frac{(-1)^n\Gamma(a+m+1)\Gamma(1+a+b)}{\Gamma(n+1)\Gamma(m+1)\Gamma(1+a-n)} \frac{\Gamma(1+a-n)\Gamma(m-n-b)}{\Gamma(1+a-n+m)\Gamma(-n-b)}$$

$$= (-1)^{m+n}\Gamma(a+b+1)\frac{\Gamma(a+m+1)}{\Gamma(n+1)\Gamma(a+m-n+1)}\frac{\Gamma(b+n+1)}{\Gamma(m+1)\Gamma(b+n-m+1)}.$$

14.6. The goal is to minimize the index

In this section the last part of Rule P_3 is illustrated. Given a specific definite integral, it has been conjectured by the authors that the optimal solution by the method of brackets is the one with minimal index.

Observe that the index of an integral may be affected by the representation of the integrand or the order of expansion into series.

Example 6.1. Entry **3.331.1** gives the evaluation of

$$(14.6.1) \qquad \int_0^\infty e^{-\beta e^{-x} - \mu x} \, dx = \beta^{-\mu} \gamma(\mu, \beta).$$

Here $\gamma(\mu, \beta)$ is the *incomplete gamma function* defined by

$$(14.6.2) \qquad \gamma(\alpha, x) = \int_0^x e^{-t} t^{\alpha-1} \, dt.$$

Method 1. The integrand is associated a bracket series via

$$
\begin{aligned}
e^{-\beta e^{-x}} e^{-\mu x} &= \sum_{n_1 \geq 0} \phi_{n_1} (\beta e^{-x})^{n_1} e^{-\mu x} \\
&= \sum_{n_1 \geq 0} \phi_{n_1} \beta^{n_1} e^{-(n_1+\mu)x} \\
&= \sum_{n_1 \geq 0} \beta^{n_1} \sum_{n_2 \geq 0} \phi_{n_2} (n_1 + \mu)^{n_2} x^{n_2}.
\end{aligned}
$$

The final step is to produce the bracket $\langle n_2 + 1 \rangle$ appearing from integration of x^{n_2}. Therefore, the bracket series associated with this representation of the integral is

$$(14.6.3) \qquad \int_0^\infty e^{-\beta e^{-x} - \mu x} \, dx \mapsto \sum_{n_1 \geq 0} \sum_{n_2 \geq 0} \phi_{n_1 n_2} \beta^{n_1} (n_1 + \mu)^{n_2} \langle n_2 + 1 \rangle.$$

This representation has index $+1$.

The vanishing of the bracket yields that n_2 is fixed as $n_2^* = -1$ and n_1 must be free. It follows that the integral is

$$
\begin{aligned}
\sum_{n_1 \geq 0} \phi_{n_1} \beta^{n_1} (n_1 + \mu)^{-1} \Gamma(1) &= \frac{\Gamma(\mu)}{\Gamma(\mu+1)} \, {}_1F_1 \left(\begin{matrix} \mu \\ \mu+1 \end{matrix} \Big| \beta \right) \\
&= \frac{1}{\beta^\mu} \gamma(\mu, \beta).
\end{aligned}
$$

Method 2. A second representation is produced as follows:

$$
\begin{aligned}
e^{-\beta e^{-x}} e^{-\mu x} &= \sum_{n_1 \geq 0} \phi_{n_1} (\beta e^{-x})^{n_1} e^{-\mu x} \\
&= \sum_{n_1 \geq 0} \phi_{n_1} \beta^{n_1} e^{-n_1 x} e^{-\mu x} \\
&= \sum_{n_1 \geq 0} \phi_{n_1} \beta^{n_1} \sum_{n_2 \geq 0} \phi_{n_2} n_1^{n_2} x^{n_2} \sum_{n_3 \geq 0} \phi_{n_3} \mu^{n_3} x^{n_3} \\
&= \sum_{n_1 \geq 0} \sum_{n_2 \geq 0} \sum_{n_3 \geq 0} \phi_{n_1,n_2,n_3} \beta^{n_1} n_1^{n_2} \mu^{n_3} x^{n_2+n_3}.
\end{aligned}
$$

The final step is now to replace the power $x^{n_2+n_3}$ by the bracket $\langle n_2 + n_3 + 1 \rangle$ to produce

$$
(14.6.4) \qquad \int_0^\infty e^{-\beta e^{-x} - \mu x} \, dx \mapsto \sum_{n_1 \geq 0} \sum_{n_2 \geq 0} \sum_{n_3 \geq 0} \phi_{n_1,n_2,n_3} \beta^{n_1} n_1^{n_2} \mu^{n_3} \langle n_2 + n_3 + 1 \rangle.
$$

This representation has index $+2$.

The vanishing of the brackets shows that n_1 is free and either n_2 or n_3 is fixed.

Case 1: n_1 free and $n_3^* = -n_2 - 1$:

$$
\begin{aligned}
S_1 &= \sum_{n_1 \geq 0} \sum_{n_2 \geq 0} \phi_{n_1,n_2} \beta^{n_1} n_1^{n_2} \mu^{-1-n_2} \Gamma(n_2 + 1) \\
&= \sum_{n_1 \geq 0} \frac{(-\beta)^{n_1}}{\mu \Gamma(n_1 + 1)} \sum_{n_2 \geq 0} \left(-\frac{n_1}{\mu} \right)^{n_2} \\
&= \sum_{n_1 \geq 0} \frac{(-\beta)^{n_1}}{\mu \Gamma(n_1 + 1)} \left(1 + \frac{n_1}{\mu} \right)^{-1} \\
&= \frac{\Gamma(\mu)}{\Gamma(\mu + 1)} {}_1 F_1 \left(\left. \begin{matrix} \mu \\ \mu + 1 \end{matrix} \right| -\beta \right) \\
&= \beta^{-\mu} \gamma(\mu, \beta).
\end{aligned}
$$

Case 2: n_1 free and $n_2^* = -n_3 - 1$:

$$
\begin{aligned}
S_2 &= \sum_{n_1 \geq 0} \sum_{n_3 \geq 0} \phi_{n_1, n_3} \beta^{n_1} n_1^{-n_3 - 1} \mu^{n_3} \Gamma(n_3 + 1) \\
&= \sum_{n_1 \geq 0} \frac{(-\beta)^{n_1}}{n_1 \Gamma(n_1 + 1)} \sum_{n_3 \geq 0} \left(-\frac{\mu}{n_1} \right)^{n_3} \\
&= \sum_{n_1 \geq 0} \frac{(-\beta)^{n_1}}{n_1 \Gamma(n_1 + 1)} \left(1 + \frac{\mu}{n_1} \right)^{-1} \\
&= \frac{\Gamma(\mu)}{\Gamma(\mu + 1)} {}_1F_1 \left(\begin{array}{c} \mu \\ \mu + 1 \end{array} \middle| -\beta \right) \\
&= \beta^{-\mu} \gamma(\mu, \beta).
\end{aligned}
$$

Rule E_3 would return the sum $S_1 + S_2$, but this is *twice* the correct value. Between these two methods for producing bracket series, the one giving the minimal index should be chosen. This doubling phenomena has appeared in other examples where the minimal index is not chosen. The reason behind this phenomenon remains to be elucidated.

Example 6.2. Entry **3.451.1** in [**35**] states that

$$
(14.6.5) \qquad \int_0^\infty x e^{-x} \sqrt{1 - e^{-x}} \, dx = \frac{4}{9} (4 - 3 \ln 2).
$$

To evaluate this entry by classical methods, observe that the integral is $-h'(1)$, where

$$
(14.6.6) \qquad h(a) = \int_0^\infty e^{-ax} \sqrt{1 - e^{-x}} \, dx.
$$

The change of variables $t = e^{-x}$ gives

$$
(14.6.7) \qquad h(a) = \int_0^1 t^{a-1} (1 - t)^{1/2} \, dt = B(a, \tfrac{3}{2}),
$$

where B is the beta function. Differentiation yields

$$
(14.6.8) \qquad h'(a) = h(a) \left[\psi(a) - \psi(a + \tfrac{3}{2}) \right],
$$

where $\psi = \Gamma'/\Gamma$ is the polygamma function. This gives

$$
(14.6.9) \qquad \int_0^\infty x e^{-x} \sqrt{1 - e^{-x}} \, dx = -\frac{\Gamma(1)\Gamma(\tfrac{3}{2})}{\Gamma(\tfrac{5}{2})} \left[\psi(1) - \psi(\tfrac{5}{2}) \right].
$$

The values $\psi(1) = -\gamma$ (the Euler constant) and $\psi(\tfrac{5}{2}) = -\gamma - 2\ln 2 + \tfrac{8}{3}$ give the result. To obtain this last special value, use

$$
(14.6.10) \qquad \psi(n) = -\gamma + \sum_{k=1}^{n-1} \frac{1}{k} \quad \text{and} \quad 2\psi(2x) = 2\ln 2 + \psi(x) + \psi(x + \tfrac{1}{2}).
$$

This last relation follows by differentiaton of the duplication formula for the gamma function $\Gamma(2x) = 2^{2x-1} \Gamma(x) \Gamma(x + \tfrac{1}{2})/\sqrt{\pi}$.

The evaluation of (14.6.5) is now obtained by the method of brackets.

Method 1. The exponential term is replaced by

$$(14.6.11) \qquad e^{-x} = \sum_{n=0}^{\infty} \frac{(-1)^n x^n}{n!} \mapsto \sum_{n_1 \geq 0} \phi_1 x^{n_1}$$

and Rule P_2 is employed to produce

$$(14.6.12) \qquad \sqrt{1 - e^{-x}} = \sum_{n_2 \geq 0} \sum_{n_3 \geq 0} \phi_{2,3} 1^{n_2} (-e^{-x})^{n_3} \frac{\langle -\frac{1}{2} + n_2 + n_3 \rangle}{\Gamma(-1/2)}.$$

Now expand the exponential terms $e^{-n_3 x}$ and replace the integral by the corresponding bracket to obtain the series

$$(14.6.13) \quad \sum_{n_1 \geq 0} \sum_{n_2 \geq 0} \sum_{n_3 \geq 0} \sum_{n_4 \geq 0} \phi_{1,2,3,4} \frac{(-1)^{n_3} n_3^{n_4}}{\Gamma(-1/2)} \langle n_1 + n_4 + 2 \rangle \langle n_2 + n_3 - \tfrac{1}{2} \rangle.$$

This gives a representation of index $+2$.

Case 1: n_1, n_2 free. Then $n_4^* = -n_1 - 2$ and $n_3^* = -n_2 + \frac{1}{2}$. The corresponding determinant is -1 and the series becomes

$$\sum_{n_1 \geq 0} \sum_{n_2 \geq 0} \frac{-(-n_2 + 1/2)^{-n_1 - 2} (-1)^{n_1 + 1/2} \Gamma(n_2 - 1/2) \Gamma(n_1 + 2)}{2\sqrt{\pi} \Gamma(n_2 + 1) \Gamma(n_1 + 1)}.$$

This result is *purely imaginary* and therefore discarded.

Case 2: n_1, n_3 free. Then $n_4^* = -n_1 - 2$ and $n_2^* = -n_3 + \frac{1}{2}$. The determinant is -1 and the series becomes

$$
\begin{aligned}
S_2 &= \sum_{n_1} \sum_{n_3} \frac{-(-1)^{n_1} n_3^{-n_1 - 2} \Gamma(n_3 - 1/2) \Gamma(n_1 + 2)}{2\sqrt{\pi} \Gamma(n_3 + 1) \Gamma(n_1 + 1)} \\
&= \sum_{n_3} \frac{-\Gamma(n_3 - 1/2)}{2\sqrt{\pi} \Gamma(n_3 + 1)} \sum_{n_1} \frac{(-n_3)^{n_1} \Gamma(n_1 + 2)}{\Gamma(n_1 + 1)} \\
&= \sum_{n_3} \frac{-\Gamma(n_3 - 1/2)}{2\sqrt{\pi} \Gamma(n_3 + 1)(1 + n_3)^2} \\
&= {}_3F_2 \left(\begin{matrix} 1, 1, -\frac{1}{2} \\ 2, 2 \end{matrix} \Big| 1 \right) \\
&= \frac{\psi\left(\frac{5}{2}\right) + \gamma}{\frac{3}{2}} \\
&= \tfrac{4}{9}(4 - 3\ln 2).
\end{aligned}
$$

Case 3: n_2, n_4 free. This case is similar to Case 1 so it is also discarded.

Case 4: n_3, n_4 free. This case is analogous to Case 2 with value $\frac{2}{3}\left(\frac{8}{3} - \ln 4\right)$.

Summing the results from Cases 2 and 4 would result in *twice* the correct value. This doubling is due to the fact that the series in each of these cases converge on the boundary.

Method 2. An alternative is to obtain the bracket series for $(1 - e^{-x})^{-1/2}$
to produce

$$\int_0^\infty xe^{-x}(1 - e^{-x})^{-1/2}\, dx$$

$$\mapsto \int_0^\infty xe^{-x} \sum_{n_1 \geq 0} \sum_{n_2 \geq 0} \phi_{1,2} 1^{n_1}(-e^{-x})^{n_2} \frac{\langle -\frac{1}{2} + n_1 + n_2 \rangle}{\Gamma(-1/2)}\, dx,$$

which can be written as

$$\int_0^\infty xe^{-x}(1 - e^{-x})^{-1/2}\, dx$$

$$\mapsto \int_0^\infty \sum_{n_1 \geq 0} \sum_{n_2 \geq 0} \phi_{1,2}(-1)^{n_2} xe^{-(1+n_2)x} \frac{\langle -\frac{1}{2} + n_1 + n_2 \rangle}{\Gamma(-1/2)}\, dx.$$

The exponential term is now expanded to produce the representation

$$\int_0^\infty xe^{-x}(1 - e^{-x})^{-1/2}\, dx$$

$$\mapsto \frac{1}{\Gamma(-1/2)} \sum_{n_1 \geq 0} \sum_{n_2 \geq 0} \sum_{n_2 \geq 0} \phi_{1,2,3}(-1)^{n_2}(1 + n_2)^{n_3} \langle n_3 + 2 \rangle \langle n_1 + n_2 - \tfrac{1}{2} \rangle.$$

This is a representation of index $+1$.

The value $n_3^* = -2$ is determined and the indices n_1 and n_2 are free.

Case 1. n_1 is free. Then $n_2^* = \frac{1}{2} - n_1$ leads to the contribution

(14.6.14) $$\sum_{n_1 \geq 0} \phi_1 \frac{(-1)^{\frac{1}{2} - n_1}(\frac{3}{2} - n_1)^{-2}}{\Gamma(-1/2)} \Gamma(2)\Gamma(n_1 - \tfrac{1}{2}).$$

This is discarded because it is purely imaginary.

Case 2. n_2 is free. Then $n_1^* = \frac{1}{2} - n_2$ produces to the contribution

(14.6.15) $$\sum_{n_2 \geq 0} \phi_2 \frac{(-1)^{n_2}(1 + n_2)^{-2}}{\Gamma(-1/2)} \Gamma(2)\Gamma(n_2 - \tfrac{1}{2}).$$

Therefore, the method of brackets shows that

$$\int_0^\infty xe^{-x}(1 - e^{-x})^{-1/2}\, dx = -\frac{1}{2\sqrt{\pi}} \sum_{n_2 \geq 0} \phi_2 \frac{(-1)^{n_2} \Gamma(n_2 - \frac{1}{2})}{(1 + n_2)^2}$$

$$= {}_3F_2\left(\begin{matrix} 1, 1, \frac{1}{2} \\ 2, 2 \end{matrix} \middle| 1\right)$$

$$= \frac{\psi\left(\frac{5}{2}\right) + \gamma}{\frac{3}{2}}$$

$$= \tfrac{4}{9}(4 - 3\ln 2).$$

This verifies (14.6.5). As in the first method for this integral, the series converged on the boundary, but it was counted only once in this evaluation using the bracket series of index $+1$.

14.7. The evaluation of a Mellin transform

Several entries in [35] are instances of the Mellin transform

$$(14.7.1) \qquad \mathcal{M}(f) := \int_0^\infty x^{s-1} f(x)\, dx.$$

For instance, Entry **3.764.2**

$$(14.7.2) \qquad \int_0^\infty x^p \cos(ax + b)\, dx = -\frac{1}{a^{p+1}} \Gamma(p+1) \sin\left(b + \frac{\pi p}{2}\right)$$

is of this form with $p = s - 1$. The reader will find in [7] an elementary proof of this evaluation.

The evaluation of (14.7.2) by the method of brackets uses the hypergeometric representation

$$\cos(ax + b) = {}_0F_1\left(\begin{matrix} - \\ \frac{1}{2} \end{matrix} \left| -\frac{(ax+b)^2}{4}\right.\right) = \sum_{n_1=0}^\infty \frac{\Gamma(\frac{1}{2})}{n_1!\,\Gamma(n + \frac{1}{2})} \left(-\frac{(ax+b)^2}{4}\right)^{n_1}.$$

Therefore,

$$(14.7.3) \qquad \int_0^\infty x^p \cos(ax+b)\, dx = \int_0^\infty x^p \sum_{n_1=0}^\infty \phi_{n_1} \frac{\Gamma(\frac{1}{2})}{4^{n_1}\,\Gamma(n_1 + \frac{1}{2})} (ax+b)^{2n_1}\, dx.$$

The bracket expansion

$$(14.7.4) \qquad (ax + b)^{2n_1} = \sum_{n_2=0}^\infty \sum_{n_3=0}^\infty \phi_{2,3}\, a^{n_2} b^{n_3} x^{n_2} \frac{\langle -2n_1 + n_2 + n_3\rangle}{\Gamma(-2n_1)}$$

gives

$$(14.7.5) \qquad \int_0^\infty x^p \cos(ax + b)\, dx =$$

$$\sum_{n_1,n_2,n_3 \geq 0} \phi_{1,2,3}\, \frac{\Gamma(\frac{1}{2})\, a^{n_2} b^{n_3}}{4^{n_1} \Gamma(n_1 + \frac{1}{2}) \Gamma(-2n_1)} \langle -2n_1 + n_2 + n_3\rangle \langle n_2 + p + 1\rangle.$$

The vanishing of the bracket $\langle n_2 + p + 1\rangle$ determines $n_2^* = -p - 1$. There is one sum and two possible choices for a free index.

Case 1: n_1 is free. Then $n_3^* = 2n_1 - p - 1$ and the corresponding determinant is -1. The contribution to the integral is given by

$$\begin{aligned}
S_1 &= \sum_{n_1 \geq 0} \phi_{n_1} \frac{1}{|-1|} \frac{\Gamma(\frac{1}{2}) a^{n_2^*} b^{n_3^*}}{4^{n_1} \Gamma(n_1 + \frac{1}{2}) \Gamma(-2n_1)} \Gamma(-n_2^*) \Gamma(-n_3^*) \\
&= \sum_{n_1 \geq 0} \phi_{n_1} \frac{\Gamma(\frac{1}{2}) a^{-p-1} b^{2n_1+p+1}}{4^{n_1} \Gamma(n_1 + \frac{1}{2}) \Gamma(-2n_1)} \Gamma(p+1) \Gamma(-2n_1 - p - 1).
\end{aligned}$$

Each term $1/\Gamma(-2n_1)$ vanishes; it follows that $S_1 = 0$.

Case 2: n_3 is free. Then $n_1^* = \frac{1}{2}(n_3 - p - 1)$ and the contribution to the integral is

$$
\begin{aligned}
S_2 &= \sum_{n_3 \geq 0} \phi_{n_3} \frac{1}{2} \frac{\Gamma(\frac{1}{2}) a^{n_2^*} b^{n_3}}{4^{n_1^*} \Gamma(n_1^* + \frac{1}{2}) \Gamma(-2n_1^*)} \Gamma(-n_1^*) \Gamma(-n_2^*) \\
&= \frac{\Gamma\left(\frac{1}{2}\right) \Gamma(p+1) 2^p}{a^{p+1}} \sum_{n_3=0}^{\infty} \phi_{n_3} \frac{b^{n_3} \Gamma(\frac{1}{2}(-n_3 + p + 1))}{2^{n_3} \Gamma(\frac{1}{2}(n_3 - p)) \Gamma(-n_3 + p + 1)}.
\end{aligned}
$$

The factors in the last summand can be simplified to produce

$$
\begin{aligned}
S_2 &= \frac{\Gamma(p+1)}{a^{p+1}} \left[-\sin\frac{\pi p}{2} \sum_{k=0}^{\infty} \frac{(-b^2)^k}{\Gamma(2k+1)} - b\cos\frac{\pi p}{2} \sum_{k=0}^{\infty} \frac{(-b^2)^k}{\Gamma(2k+2)} \right] \\
&= \frac{\Gamma(p+1)}{a^{p+1}} \left[-\sin\frac{\pi p}{2} \cos b - b\cos\frac{\pi p}{2} \sin b \right] \\
&= -\frac{\Gamma(p+1)}{a^{p+1}} \sin\left(\frac{\pi p}{2} + b\right).
\end{aligned}
$$

Adding all the finite contributions of free indices gives the evaluation

$$
(14.7.6) \qquad \int_0^{\infty} x^p \cos(ax + b)\, dx = -\frac{\Gamma(p+1)}{a^{p+1}} \sin\left(\frac{\pi p}{2} + b\right).
$$

14.8. The introduction of a parameter

This section illustrates the evaluation of Entry **3.249**

$$
(14.8.1) \qquad \int_0^{\infty} \left[e^{-x} - (1+x)^{-\mu} \right] \frac{dx}{x} = \psi(\mu), \qquad \text{for } \operatorname{Re}\mu > 0.
$$

A classical evaluation of this entry appears in [**47**].

To apply the method of brackets, consider first the integral

$$
(14.8.2) \qquad I(\varepsilon) = \int_0^{\infty} \frac{\exp(-x) - (1+x)^{-\mu}}{x^{1-\varepsilon}}\, dx.
$$

The result is obtained by letting $\varepsilon \to 0$. Now compute the bracket series associated to the integrand in (14.8.2) to obtain

$$
(14.8.3) \qquad I(\varepsilon) = \sum_{k \geq 0} \phi_k \left[1 - \frac{\Gamma(\mu + n)}{\Gamma(\mu)} \right] \langle k + \varepsilon \rangle.
$$

Therefore,

$$
(14.8.4) \qquad I(\varepsilon) = \Gamma(\varepsilon) \left[1 - \frac{\Gamma(\mu - \varepsilon)}{\Gamma(\mu)} \right].
$$

To obtain the value of (14.8.1), simply use the expansion

$$\Gamma(\varepsilon)\left[1 - \frac{\Gamma(\mu - \varepsilon)}{\Gamma(\mu)}\right] = \psi(\mu) - \left(\frac{\psi'(\mu)}{2} - \frac{1}{2}\psi^2(\mu) + \gamma\psi(\mu)\right)\varepsilon + O(\varepsilon^2)$$

as $\varepsilon \to 0$.

Conclusions. The examples given here, all taken from the classical table of integrals by I. S. Gradshteyn and I. M. Ryzhik, have been evaluated using the method of brackets. This illustrates the great flexibility of this method. The rules for evaluation have been partially justified via Ramanujan's Master Theorem.

CHAPTER 15

The list of integrals

15.1. The list

This is the list of formulas in Gradshteyn and Ryzhik [**35**] that are established in this volume. The sections are named as in the table.

Section 2.14. Forms containing the binomial $1 \pm x^n$

Subsection 2.148

2.148.3 $\quad \displaystyle\int \frac{dx}{(1+x^2)^n} = \frac{1}{2n-2} \frac{x}{(1+x^2)^{n-1}} + \frac{2n-3}{2n-2} \int \frac{dx}{(1+x^2)^{n-1}}$ $\qquad$ **28**

Section 2.42 Powers of $\sinh x$, $\cosh x$, $\tanh x$, and $\coth x$

Subsection 2.423

2.423.9 $\quad \displaystyle\int \frac{dx}{\cosh x} = 2\arctan(e^x)$ $\qquad\qquad\qquad\qquad$ **73**

Section 3.13–3.17. Expressions that can be reduced to square roots of third- and fourth-degree polynomials and their products with rational functions

Subsection 3.166

3.166.16 $\quad \displaystyle\int_0^1 \frac{dx}{\sqrt{1-x^4}} = \frac{1}{4\sqrt{2\pi}} \left\{ \Gamma\left(\frac{1}{4}\right) \right\}^2$ $\qquad\qquad$ **2**

3.166.18 $\quad \displaystyle\int_0^1 \frac{x^2\,dx}{\sqrt{1-x^4}} = \frac{1}{\sqrt{2\pi}} \left\{ \Gamma\left(\frac{3}{4}\right) \right\}^2$ $\qquad\qquad$ **2**

Section 3.19–3.23. Combinations of powers of x and powers of binomials of the form $(\alpha + \beta x)$

Subsection 3.194

$$\int_0^u \frac{x^{\mu-1}\,dx}{(1+\beta x)^\nu} = \frac{u^\mu}{\mu}\,{}_2F_1\left(\nu,\mu;1+\mu;-\beta u\right) \qquad\qquad \text{3.194.1 \quad 63}$$

$$\int_u^\infty \frac{x^{\mu-1}\,dx}{(1+\beta x)^\nu} = \frac{u^{\mu-\nu}}{\beta^u(\nu-\mu)}\,{}_2F_1$$
$$\times\left(\nu,\nu-\mu;\nu-\mu+1;-\frac{1}{\beta u}\right) \qquad \text{3.194.2 \quad 68}$$

$$\int_0^\infty \frac{x^{\mu-1}\,dx}{(1+\beta x)^\nu} = \beta^{-\nu}B(\mu,\nu-\mu) \qquad\qquad \text{3.194.3 \quad 172}$$

$$\int_0^u \frac{x^{\mu-1}\,dx}{1+\beta x} = \frac{u^\mu}{\mu}\,{}_2F_1(1,\mu;1+\mu;-\beta\mu) \qquad\qquad \text{3.194.5 \quad 63}$$

$$\int_0^1 \frac{x^{n-1}\,dx}{(1+x)^m} = 2^{-n}\sum_{k=0}^\infty \binom{m-n-1}{k}\frac{(-2)^{-k}}{n+k} \qquad \text{3.194.8 \quad 140}$$

Subsection 3.196

$$\int_0^u (x+\beta)^\nu(u-x)^{\mu-1}\,dx = \frac{\beta^\nu u^\mu}{\mu}\,{}_2F_1\left(1,-\nu;1+\mu;-\frac{u}{\beta}\right) \quad \text{3.196.1 \quad 63}$$

Subsection 3.197

$$\int_u^\infty x^{-\lambda}(x+\beta)^\nu(x-u)^{\mu-1}\,dx$$
$$= \frac{(\beta+u)^{\mu+\nu}}{u^\lambda}B(\lambda-\mu-\nu,\mu)\,{}_2F_1$$
$$\times\left(\lambda,\mu;\lambda-\mu;-\frac{\beta}{u}\right) \qquad\qquad \text{3.197.2 \quad 69}$$

$$\int_0^1 x^{\lambda-1}(1-x)^{\mu-1}(1-\beta x)^{-\nu}\,dx$$
$$= B(\lambda,\mu)\,{}_2F_1(\nu,\lambda;\lambda+\mu;\beta) \qquad\qquad \text{3.197.3 \quad 62}$$

$$\int_0^1 x^{\mu-1}(1-x)^{\nu-1}(1+ax)^{-\mu-\nu}\,dx$$
$$= (1+a)^{-\mu}B(\mu,\nu) \qquad\qquad \text{3.197.4 \quad 62}$$

$$\int_0^\infty x^{\lambda-1}(1+x)^\nu(1+\alpha x)^\mu\,dx$$
$$= B(\lambda,-\mu-\nu-\lambda)\,{}_2F_1\left(-\mu,\lambda;-\mu-\nu;1-\alpha\right) \qquad \text{3.197.5 \quad 68}$$

$$\int_1^\infty x^{\lambda-\nu}(x-1)^{\nu-\mu-1}(\alpha x-1)^{-\lambda}\,dx$$

$$= \frac{B(\mu,\nu-\mu)}{\alpha^\lambda}\,{}_2F_1(\nu,\mu;\lambda;\alpha^{-1}) \qquad\qquad \textbf{3.197.6} \quad \textbf{68}$$

$$\int_0^\infty x^{\mu-1/2}(x+a)^{-\mu}(x+b)^{-\mu}\,dx$$

$$= \sqrt{\pi}(\sqrt{a}+\sqrt{b})^{1-2\mu}\frac{\Gamma(\mu-1/2)}{\Gamma(\mu)} \qquad\qquad \textbf{3.197.7} \quad \textbf{69}$$

$$\int_0^u x^{\nu-1}(x+\alpha)^\lambda(u-x-)^{\mu-1}\,dx$$

$$= \alpha^\lambda u^{\mu+\nu-1}B(\mu,\nu){}_2F_1\left(-\lambda,\nu;\mu+\nu;-\frac{u}{\alpha}\right) \qquad\qquad \textbf{3.197.8} \quad \textbf{68}$$

$$\int_0^\infty x^{\lambda-1}(1+x)^{-\mu+\nu}(x+\beta)^{-\nu}\,dx$$

$$= B(\mu-\lambda,\lambda)\,{}_2F_1(\nu,\mu-\lambda;\mu;1-\beta) \qquad\qquad \textbf{3.197.9} \quad \textbf{68}$$

$$\int_0^1 \frac{x^{q-1}\,dx}{(1-x)^q(1+px)}$$

$$= \frac{\pi}{(1+p)^q}\operatorname{cosec}\pi q \qquad\qquad \textbf{3.197.10} \quad \textbf{62}$$

Subsection 3.198

$$\int_0^1 \frac{x^{\mu-1}(1-x)^{\nu-1}\,dx}{[ax+b(1-x)+x]^{\mu+\nu}} = \frac{B(\mu,\nu)}{(a+c)^\mu\,(b+c)^\nu} \qquad\qquad \textbf{3.198} \quad \textbf{63}$$

Subsection 3.199

$$\int_a^b \frac{(x-a)^{\mu-1}(b-x)^{\nu-1}}{(x-c)^{\mu+\nu}}\,dx = \frac{(b-a)^{\mu+\nu-1}}{(b-c)^\mu(a-c)^\nu}B(\mu,\nu) \qquad\qquad \textbf{3.199} \quad \textbf{64}$$

Subsection 3.227

$$\int_0^\infty \frac{x^{\nu-1}(\beta+x)^{1-\mu}}{\gamma+x}\,dx$$

$$= \frac{\gamma^{\nu-1}}{\beta^{\mu-1}}B(\nu,\mu-\nu){}_2F_1\left(\mu-1,\nu;\mu;1-\frac{\gamma}{\beta}\right) \qquad\qquad \textbf{3.227.1} \quad \textbf{68}$$

Subsection 3.231

$$\int_0^1 \frac{x^{p-1} - x^{-p}}{1-x}\, dx = \pi \cot \pi p \qquad\qquad \textbf{3.231.1}\quad 171$$

$$\int_0^1 \frac{x^{\mu-1} - x^{\nu-1}}{1-x}\, dx = \psi(\nu) - \psi(\mu) \qquad\qquad \textbf{3.231.5}\quad 78$$

$$\int_0^\infty \frac{x^{p-1} - x^{q-1}}{1-x}\, dx = \pi(\cot \pi p - \cot \pi q) \qquad\qquad \textbf{3.231.6}\quad 169$$

Section 3.24–3.27. Powers of x, of binomials of the form $\alpha + \beta x^p$ and of polynomials in x

Subsection 3.241

$$\mathrm{PV} \int_0^\infty \frac{x^{p-1}\, dx}{1 - x^q} = \frac{\pi}{q} \cot \frac{\pi p}{q} \qquad\qquad \textbf{3.241.3}\quad 172$$

Subsection 3.265

$$\int_0^1 \frac{1 - x^{\mu-1}}{1-x} = \psi(\mu) + \mathbf{C} \qquad\qquad \textbf{3.265}\quad 90$$

Subsection 3.271

$$\int_0^\infty \left(\frac{x^p - x^{-p}}{1-x} \right)^2 dx = 2(1 - 2p\pi \cot 2p\pi) \qquad\qquad \textbf{3.271.1}\quad 5$$

Section 3.3 − 3.4. Exponential functions

Subsection 3.31. Exponential functions

Subsection 3.310.

$$\int_0^\infty e^{-px}\, dx = \frac{1}{p} \qquad\qquad \textbf{3.310}\quad 195$$

Subsection 3.311.

$$\int_0^\infty \frac{dx}{1 + e^{px}} = \frac{\ln 2}{p} \qquad\qquad \textbf{3.311.1}\quad 197$$

$$\int_0^\infty \frac{e^{-qx}\, dx}{1 - ae^{-px}} = \sum_{k=0}^\infty \frac{a^k}{q + kp} \qquad\qquad \textbf{3.311.4}\quad 139$$

Subsection 3.312.

$$\int_0^\infty (1 - e^{-x})^{\nu-1}(1 - \beta e^{-x})^{-\rho} e^{-\mu x}\, dx$$
$$= B(\mu, \nu)\, {}_2F_1(\rho, \mu; \mu + \nu; \beta) \hspace{3cm} \textbf{3.312.3} \quad \textbf{69}$$

Subsection 3.32–3.34. Exponentials of more complicated arguments

Subsection 3.321.

$$\int_0^u e^{-q^2 x^2}\, dx = \frac{\sqrt{\pi}}{2q} \Phi(qu) \hspace{4cm} \textbf{3.321.2} \quad \textbf{44}$$

$$\int_0^\infty e^{-q^2 x^2}\, dx = \frac{\sqrt{\pi}}{2q} \hspace{4.5cm} \textbf{3.321.3} \quad \textbf{45}$$

$$\int_0^u x e^{-q^2 x^2}\, dx = \frac{1}{2q^2}\left[1 - e^{-q^2 u^2}\right] \hspace{2.8cm} \textbf{3.321.4} \quad \textbf{45}$$

$$\int_0^u x^2 e^{-q^2 x^2}\, dx = \frac{1}{2q^3}\left[\frac{\sqrt{\pi}}{2}\Phi(qu) - que^{-q^2 u^2}\right] \hspace{1cm} \textbf{3.321.5} \quad \textbf{45}$$

$$\int_0^u x^3 e^{-q^2 x^2}\, dx = \frac{1}{2q^4}\left[1 - (1 + q^2 u^2)e^{-q^2 u^2}\right] \hspace{1cm} \textbf{3.321.6} \quad \textbf{45}$$

$$\int_0^u x^4 e^{-q^2 x^2}\, dx = \frac{1}{2q^5}\left[\frac{3\sqrt{\pi}}{4}\Phi(qu) - \left(\tfrac{3}{2} + q^2 u^2\right) que^{-q^2 u^2}\right] \hspace{0.5cm} \textbf{3.321.7} \quad \textbf{45}$$

Subsection 3.322.

$$\int_u^\infty \exp\left(-\frac{x^2}{4\beta} - \gamma x\right) dx = \sqrt{\pi\beta}e^{\beta\gamma^2}\left[1 - \Phi\left(\gamma\sqrt{\beta} + \frac{u}{2\sqrt{\beta}}\right)\right] \hspace{0.3cm} \textbf{3.322.1} \quad \textbf{45}$$

$$\int_0^\infty \exp\left(-\frac{x^2}{4\beta} - \gamma x\right) dx = \sqrt{\pi\beta}e^{\beta\gamma^2}\left[1 - \Phi\left(\gamma\sqrt{\beta}\right)\right] \hspace{1.5cm} \textbf{3.322.2} \quad \textbf{45}$$

Subsection 3.323.

$$\int_1^\infty \exp(-qx - x^2)\, dx = \frac{\sqrt{\pi}}{2}e^{q^2/4}\left[1 - \Phi\left(1 + \frac{q}{2}\right)\right] \hspace{1cm} \textbf{3.323.1} \quad \textbf{46}$$

$$\int_{-\infty}^\infty \exp(-p^2 x^2 \pm qx)\, dx = \frac{\sqrt{\pi}}{p}\exp\left(\frac{q^2}{4p^2}\right) \hspace{2cm} \textbf{3.323.2} \quad \textbf{46}$$

$$\int_0^\infty \exp(-\beta^2 x^4 - 2\gamma^2 x^2)\, dx = 2^{-3/2}\frac{\gamma}{\beta}e^{\gamma^4/2\beta^2} K_{\frac{1}{4}}\left(\frac{\gamma^4}{2\beta^2}\right) \hspace{0.8cm} \textbf{3.323.3} \quad \textbf{107}$$

Subsection 3.324.

$$\int_0^\infty \exp\left(-\frac{\beta}{4x} - \gamma x\right) dx = \sqrt{\frac{\beta}{\gamma}} K_1(\sqrt{\beta\gamma})$$

3.324.1   94

Subsection 3.325.

$$\int_0^\infty \exp\left(-ax^2 - \frac{b}{x^2}\right) dx = \frac{1}{2}\sqrt{\frac{\pi}{a}}\exp(-2\sqrt{ab})$$

3.325   47

Subsection 3.327–3.334. Exponentials of exponentials.
Subsection 3.327.

$$\int_0^\infty \exp\left(-ae^{nx}\right) dx = -\frac{1}{n}\mathrm{Ei}(-a)$$

3.327   161

Subsection 3.331.

$$\int_0^\infty \exp\left(-\beta e^{-x} - \mu x\right) dx = \beta^{-\mu}\gamma(\mu, \beta)$$

3.331.1   204

Subsection 3.333.

$$\int_{-\infty}^\infty \frac{e^{-\mu x}\, dx}{\exp\left(e^{-x}\right) - 1} = \Gamma(\mu)\zeta(\mu)$$

3.333.1   24

$$\int_{-\infty}^\infty \frac{e^{-\mu x}\, dx}{\exp\left(e^{-x}\right) + 1} = (1 - 2^{1-\mu})\Gamma(\mu)\zeta(\mu)$$

3.333.2   24

Subsection 3.335–3.337. Exponentials of hyperbolic functions
Subsection 3.337.

$$\int_{-\infty}^\infty \exp(-\alpha x - \beta \cosh x)\, dx = 2K_\alpha(\beta)$$

3.337.1   97

Subsection 3.338–3.342.   Exponentials of trigonometric functions and logarithms
Subsection 3.342.

$$\int_0^1 \exp(-px \ln x)\, dx = \int_0^1 x^{-px}\, dx = \sum_{k=1}^\infty \frac{p^{k-1}}{k^k}$$

3.342   143

Subsection 3.35. Combinations of exponentials and rational functions

Subsection 3.351.

$$\int_u^\infty \frac{e^{-px}\,dx}{x^{n+1}} = (-1)^{n+1}\frac{p^n\operatorname{Ei}(-pu)}{n!}$$
$$+ \frac{e^{-pu}}{u^n}\sum_{k=0}^{n-1}\frac{(-1)^k p^k u^k}{n(n-1)\cdots(n-k)} \qquad\qquad \textbf{3.351.4} \quad \textbf{154}$$

$$\int_1^\infty \frac{e^{-\mu x}}{x}\,dx = -\operatorname{Ei}(-\mu) \qquad\qquad \textbf{3.351.5} \quad \textbf{152}$$

$$\int_{-\infty}^u \frac{e^x}{x}\,dx = \operatorname{li}(e^u) = \operatorname{Ei}(u) \qquad\qquad \textbf{3.351.6} \quad \textbf{151}$$

Subsection 3.352.

$$\int_u^\infty \frac{e^{-\mu x}\,dx}{x+\beta} = -e^{\beta\mu}\operatorname{Ei}(-\mu u - \mu\beta) \qquad\qquad \textbf{3.352.2} \quad \textbf{152}$$

$$\int_u^v \frac{e^{-\mu x}\,dx}{x+\alpha} = e^{\alpha\mu}\left\{\operatorname{Ei}[-(\alpha+v)\mu] - \operatorname{Ei}[-(\alpha+u)\mu]\right\} \qquad\qquad \textbf{3.352.3} \quad \textbf{152}$$

$$\int_0^\infty \frac{e^{-\mu x}\,dx}{x+\beta} = -e^{\beta\mu}\operatorname{Ei}(-\mu\beta) \qquad\qquad \textbf{3.352.4} \quad \textbf{152}$$

$$\int_u^\infty \frac{e^{-px}\,dx}{a-x} = e^{-pa}\operatorname{Ei}(pa - pu) \qquad\qquad \textbf{3.352.5} \quad \textbf{153}$$

$$\int_0^\infty \frac{e^{-\mu x}\,dx}{a-x} = e^{-\mu a}\operatorname{Ei}(a\mu) \qquad\qquad \textbf{3.352.6} \quad \textbf{153}$$

Subsection 3.353.

$$\int_0^\infty \frac{e^{-px}\,dx}{(a+x)^2} = pe^{ap}\operatorname{Ei}(-ap) + \frac{1}{a} \qquad\qquad \textbf{3.353.3} \quad \textbf{153}$$

$$\int_0^1 \frac{xe^x\,dx}{(1+x)^2} = \frac{e}{2} - 1 \qquad\qquad \textbf{3.353.4} \quad \textbf{154}$$

Subsection 3.354.

$$\int_0^\infty \frac{e^{-\mu x}\,dx}{\beta^2 + x^2} = \frac{1}{\beta}\left[\text{ci}(\beta\mu)\sin\beta\mu - \text{si}(\beta\mu)\cos\beta\mu\right] \qquad \textbf{3.354.1} \quad \textbf{155}$$

$$\int_0^\infty \frac{xe^{-\mu x}\,dx}{\beta^2 + x^2} = -\text{ci}(\beta\mu)\cos\beta\mu - \text{si}(\beta\mu)\sin\beta\mu \qquad \textbf{3.354.2} \quad \textbf{155}$$

$$\int_0^\infty \frac{e^{-\mu x}\,dx}{\beta^2 - x^2} = \frac{1}{2\beta}\left[e^{-\beta\mu}\,\text{Ei}(\beta\mu) - e^{\beta\mu}\,\text{Ei}(-\beta\mu)\right] \qquad \textbf{3.354.3} \quad \textbf{154}$$

$$\int_0^\infty \frac{xe^{-\mu x}\,dx}{\beta^2 - x^2} = \frac{1}{2}\left[e^{-\beta\mu}\,\text{Ei}(\beta\mu) + e^{\beta\mu}\,\text{Ei}(-\beta\mu)\right] \qquad \textbf{3.354.4} \quad \textbf{155}$$

Subsection 3.355.

$$\int_0^\infty \frac{xe^{-\mu x}\,dx}{(\beta^2 + x^2)^2}$$
$$= \frac{1}{2\beta^2}\left\{1 - \beta\mu\left[\text{ci}(\beta\mu)\sin\beta\mu - \text{si}(\beta\mu)\cos\beta\mu\right]\right\} \qquad \textbf{3.355.2} \quad \textbf{155}$$

$$\int_0^\infty \frac{e^{-px}\,dx}{(a^2 - x^2)^2}$$
$$= \frac{1}{4a^3}\left[(ap - 1)e^{ap}\text{Ei}(-ap) + (1 + ap)e^{-ap}\text{Ei}(ap)\right] \qquad \textbf{3.355.3} \quad \textbf{156}$$

$$\int_0^\infty \frac{xe^{-px}\,dx}{(a^2 - x^2)^2}$$
$$= \frac{1}{4a^2}\left\{-2 + ap\left[e^{-ap}\text{Ei}(ap) - e^{ap}\text{Ei}(-ap)\right]\right\} \qquad \textbf{3.555.4} \quad \textbf{156}$$

Subsection 3.36–3.37. Combinations of exponentials and algebraic functions

Subsection 3.361.

$$\int_0^u \frac{e^{-qx}}{\sqrt{x}}\,dx = \sqrt{\frac{\pi}{q}}\,\Phi(\sqrt{qu}) \qquad \textbf{3.361.1} \quad \textbf{49}$$

$$\int_0^\infty \frac{e^{-qx}}{\sqrt{x}}\,dx = \sqrt{\frac{\pi}{q}} \qquad \textbf{3.361.2} \quad \textbf{49}$$

$$\int_{-1}^\infty \frac{e^{-qx}}{\sqrt{1+x}}\,dx = e^q\sqrt{\frac{\pi}{q}} \qquad \textbf{3.361.3} \quad \textbf{49}$$

Subsection 3.362.

$$\int_1^\infty \frac{e^{-\mu x}}{\sqrt{x-1}} \, dx = e^{-\mu} \sqrt{\frac{\pi}{\mu}} \qquad\qquad\qquad \textbf{3.362.1} \quad \textbf{49}$$

$$\int_0^\infty \frac{e^{-\mu x}}{\sqrt{x+\beta}} \, dx = e^{\beta\mu} \sqrt{\frac{\pi}{\mu}} \left[1 - \Phi(\sqrt{\beta\mu})\right] \qquad \textbf{3.362.2} \quad \textbf{52}$$

Subsection 3.363.

$$\int_u^\infty \frac{\sqrt{x-u}}{x} e^{-\mu x} \, dx = e^{-u\mu} \sqrt{\frac{\pi}{\mu}} - \pi\sqrt{u}\left[1 - \Phi(\sqrt{u\mu})\right] \qquad \textbf{3.363.1} \quad \textbf{50}$$

$$\int_u^\infty \frac{e^{-\mu x} \, dx}{x\sqrt{x-u}} = \frac{\pi}{\sqrt{u}}\left[1 - \Phi(\sqrt{u\mu})\right] \qquad\qquad \textbf{3.363.2} \quad \textbf{52}$$

Subsection 3.364.

$$\int_0^\infty \frac{e^{-px} \, dx}{\sqrt{x(x+a)}} = e^{ap/2} K_0\left(\frac{ap}{2}\right) \qquad\qquad \textbf{3.364.3} \quad \textbf{100}$$

Subsection 3.365.

$$\int_u^\infty \frac{xe^{-\mu x} \, dx}{\sqrt{x^2 - u^2}} = uK_1(u\mu) \qquad\qquad\qquad \textbf{3.365.2} \quad \textbf{105}$$

Subsection 3.366.

$$\int_0^\infty \frac{(x+\beta)\,e^{-\mu x} \, dx}{\sqrt{x^2 + 2\beta x}} = \beta e^{\beta\mu} K_1(\beta\mu) \qquad\qquad \textbf{3.366.2} \quad \textbf{106}$$

Subsection 3.369.

$$\int_0^\infty \frac{e^{-\mu x} \, dx}{\sqrt{(x+a)^3}} = \frac{2}{\sqrt{a}} - 2\sqrt{\pi\mu}e^{a\mu}\left(1 - \Phi(\sqrt{a\mu})\right) \qquad \textbf{3.369} \quad \textbf{53}$$

Subsection 3.372.

$$\int_0^\infty x^{n-1/2}(2+x)^{n-1/2}e^{-px} \, dx = \frac{(2n-1)!!}{p^n} e^p K_n(p) \qquad \textbf{3.372} \quad \textbf{101}$$

Subsection 3.38–3.39. Combinations of exponentials and arbitrary powers

Subsection 3.382.

$$\int_0^\infty (x+\beta)^\nu e^{-\mu x}\, dx = \mu^{-\nu-1} e^{\beta\mu}\, \Gamma(\nu+1, \beta\mu) \qquad\qquad \textbf{3.382.4} \quad \textbf{53}$$

Subsection 3.383.

$$\int_u^\infty x^{\mu-1}(x-u)^{\mu-1} e^{-\beta x}\, dx$$

$$= \frac{1}{\sqrt{\pi}}\left(\frac{u}{\beta}\right)^{\mu-1/2} \Gamma(\mu)\exp\left(-\frac{\beta u}{2}\right) K_{\mu-\frac{1}{2}}\left(\frac{\beta u}{2}\right) \qquad \textbf{3.383.3} \quad \textbf{101}$$

$$\int_0^\infty x^{\nu-1}(x+\beta)^{\nu-1} e^{-\mu x}\, dx$$

$$= \frac{1}{\sqrt{\pi}}\left(\frac{\beta}{\mu}\right)^{\nu-\frac{1}{2}} e^{\beta\mu/2}\, \Gamma(\nu) K_{\frac{1}{2}-\nu}\left(\frac{\beta\mu}{2}\right) \qquad\qquad \textbf{3.383.8} \quad \textbf{101}$$

Subsection 3.387.

$$\int_{-1}^1 (1-x^2)^{\nu-1} e^{-\mu x}\, dx$$

$$= \sqrt{\pi}\left(\frac{2}{\nu}\right)^{\nu-1/2} \Gamma(\nu)\, I_{\nu-1/2|}(\mu) \qquad\qquad \textbf{3.387.1} \quad \textbf{94}$$

$$\int_1^\infty (x^2-1)^{\nu-1} e^{-\mu x}\, dx$$

$$= \frac{1}{\sqrt{\pi}}\left(\frac{2}{\nu}\right)^{\nu-1/2} \Gamma(\nu)\, K_{\nu-1/2|}(\mu) \qquad\qquad \textbf{3.387.3} \quad \textbf{98}$$

$$\int_u^\infty (x^2-u^2)^{\nu-1} e^{-\mu x}\, dx$$

$$= \frac{1}{\sqrt{\pi}}\left(\frac{2u}{\nu}\right)^{\nu-1/2} \Gamma(\nu)\, K_{\nu-1/2|}(u\mu) \qquad\qquad \textbf{3.387.6} \quad \textbf{100}$$

Subsection 3.388.

$$\int_0^\infty (2\beta x + x^2)^{\nu-1} e^{-\mu x}\, dx$$

$$= \frac{1}{\sqrt{\pi}} \left(\frac{2\beta}{\mu}\right)^{\nu-1/2} e^{\beta\mu}\, \Gamma(\nu)\, K_{\mu-1/2}(\beta\mu) \qquad\qquad \textbf{3.388.2}\quad 102$$

Subsection 3.389.

$$\int_u^\infty x(x^2 - u^2)^{\nu-1} e^{-\mu x}\, dx$$

$$= \frac{2^{\nu-1/2}}{\sqrt{\pi}} \mu^{1/2-\nu} u^{\nu+1/2}\, \Gamma(\nu)\, K_{\nu+1/2}(u\mu) \qquad\qquad \textbf{3.389.4}\quad 104$$

Subsection 3.395.

$$\int_1^\infty \frac{(\sqrt{x^2-1}+x)^\nu + (\sqrt{x^2-1}+x)^{-\nu}}{\sqrt{x^2-1}}\, dx = 2K_\nu(\mu) \qquad \textbf{3.395.1}\quad 96$$

Subsection 3.41–3.44. Combinations of rational functions of powers and exponentials

Subsection 3.411.

$$\int_0^\infty \frac{x^{\nu-1}\, dx}{e^{\mu x} - 1} = \frac{1}{\mu^\nu}\Gamma(\nu)\zeta(\nu) \qquad\qquad \textbf{3.411.1}\quad 16$$

$$\int_0^\infty \frac{x^{2n-1}\, dx}{e^{px} - 1} = (-1)^{n-1}\left(\frac{2\pi}{p}\right)^{2n}\frac{B_{2n}}{4n} \qquad\qquad \textbf{3.411.2}\quad 17$$

$$\int_0^\infty \frac{x^{\nu-1}\, dx}{e^{\mu x} + 1} = \frac{1}{\mu^\nu}(1 - 2^{1-\nu})\,\Gamma(\nu)\zeta(\nu) \qquad\qquad \textbf{3.411.3}\quad 17$$

$$\int_0^\infty \frac{x^{2n-1}\, dx}{e^{px} + 1} = (1 - 2^{1-2n})\left(\frac{2\pi}{p}\right)^{2n}\frac{|B_{2n}|}{4n} \qquad\qquad \textbf{3.411.4}\quad 17$$

$$\int_0^\infty \frac{x^{\nu-1}e^{-\mu x}}{1 - \beta e^{-x}}\, dx = \Gamma(\nu)\sum_{n=0}^\infty \frac{\beta^n}{(\mu+n)^\nu} \qquad\qquad \textbf{3.411.6}\quad 18$$

$$\int_0^\infty \frac{x^{\nu-1}e^{-\mu x}}{1 - e^{-\beta x}}\, dx = \frac{1}{\beta^\nu}\Gamma(\nu)\,\zeta\left(\nu, \frac{\mu}{\beta}\right) \qquad\qquad \textbf{3.411.7}\quad 18$$

$$\int_0^\infty \frac{x^{n-1} e^{-px}}{1 + e^x} = (n-1)! \sum_{k=1}^\infty \frac{(-1)^{k-1}}{(p+k)^n}$$

3.411.8　20

$$\int_0^\infty \frac{xe^{-x}\, dx}{e^x - 1} = \frac{\pi^2}{6} - 1$$

3.411.9　19

$$\int_0^\infty \frac{xe^{-2x}\, dx}{e^{-x} + 1} = 1 - \frac{\pi^2}{12}$$

3.411.10　21

$$\int_0^\infty \frac{xe^{-3x}\, dx}{e^{-x} + 1} = \frac{\pi^2}{12} - \frac{3}{4}$$

3.411.11　21

$$\int_0^\infty \frac{xe^{-(2n-1)x}}{1 + e^x} = -\frac{\pi^2}{12} + \sum_{k=1}^{2n-1} \frac{(-1)^{k-1}}{k^2}$$

3.411.12　21

$$\int_0^\infty \frac{xe^{-2nx}}{1 + e^x} = \frac{\pi^2}{12} + \sum_{k=1}^{2n} \frac{(-1)^k}{k^2}$$

3.411.13　21

$$\int_0^\infty \frac{x^2 e^{-nx}}{1 - e^x} = 2 \sum_{k=n}^\infty \frac{1}{k^3} = 2\left(\zeta(3) - \sum_{k=1}^n \frac{1}{k^3} \right)$$

3.411.14　19

$$\int_0^\infty \frac{x^2 e^{-nx}}{1 + e^{-x}} = 2 \sum_{k=n}^\infty \frac{(-1)^{n+k}}{k^3} = (-1)^{n+1}$$

$$\times \left(\frac{3}{2}\zeta(3) + 2 \sum_{k=1}^{n-1} \frac{(-1)^k}{k^3} \right)$$

3.411.15　21

$$\int_0^\infty \frac{x^3 e^{-nx}\, dx}{1 - e^{-x}} = \frac{\pi^4}{15} - 6 \sum_{k=1}^{n-1} \frac{1}{k^4}$$

3.411.17　19

$$\int_0^\infty \frac{x^3 e^{-nx}\, dx}{1 + e^{-x}} = 6 \sum_{k=n}^\infty \frac{(-1)^{n+k}}{k^4} = (-1)^{n+1}$$

$$\times \left(\frac{7}{120}\pi^4 + 6 \sum_{k=1}^{n-1} \frac{(-1)^k}{k^4} \right)$$

3.411.18　21

$$\int_0^\infty x^{n-1} \frac{1 - e^{-mx}}{1 - e^x}\, dx = -(n-1)! \sum_{k=1}^m \frac{1}{k^m}$$

3.411.21　19

$$\int_0^\infty \frac{x^{p-1}\, dx}{e^{rx} - q} = \frac{1}{qr^p} \Gamma(p) \sum_{k=1}^\infty \frac{q^k}{k^p}$$

$$= \Gamma(p) r^{-p} \Phi(q, p, 1)$$

3.411.22　18

$$\int_0^\infty x \frac{1 + e^{-x}}{e^x - 1}\, dx = \frac{\pi^2}{3} - 1$$

3.411.25　19

$$\int_0^\infty xe^{-x} \frac{1 - e^{-x}}{1 + e^{-3x}}\, dx = \frac{2\pi^2}{27}$$

3.411.26　26

Subsection 3.417.

$$\int_{-\infty}^{\infty} \frac{x\,dx}{a^2 e^x + b^2 e^{-x}} = \frac{\pi}{2ab} \ln \frac{b}{a}$$

$$\int_{-\infty}^{\infty} \frac{x\,dx}{a^2 e^x - b^2 e^{-x}} = \frac{\pi^2}{4ab}$$

Subsection 3.418.

$$\int_0^{\infty} \frac{x\,dx}{e^x + e^{-x} - 1} = \frac{1}{3} \left[\psi'\left(\frac{1}{3}\right) - \frac{2}{3}\pi^2 \right]$$

Subsection 3.423.

$$\int_0^{\infty} \frac{x^{\nu-1}\,dx}{(e^x - 1)^2} = \Gamma(\nu)\left[\zeta(\nu-1) - \zeta(\nu)\right]$$

Subsection 3.45. Combinations of powers and algebraic functions of exponentials

Subsection 3.451.

$$\int_0^{\infty} xe^{-x}\sqrt{1 - e^{-x}}\,dx = \frac{4}{3}\left(\frac{4}{3} - \ln 2\right)$$

$$\int_0^{\infty} xe^{-x}\sqrt{1 - e^{-2x}}\,dx = \frac{\pi}{4}\left(\frac{1}{2} + \ln 2\right)$$

Subsection 3.452.

$$\int_0^{\infty} \frac{x\,dx}{\sqrt{e^x - 1}} = 2\pi \ln 2$$

Subsection 3.46–3.48. Combinations of exponentials of more complicated arguments and powers

Subsection 3.461.

$$\int_u^\infty e^{-\mu x^2} \frac{dx}{x^2} = \frac{1}{u} e^{-\mu u^2} - \sqrt{\mu\pi}\,[1 - \Phi(u\sqrt{\mu})] \qquad\qquad \textbf{3.461.5} \quad \textbf{49}$$

$$\int_0^\infty \exp\left(-a\sqrt{x^2 + b^2}\right)\, dx = b\, K_1(ab) \qquad\qquad \textbf{3.461.6} \quad \textbf{106}$$

$$\int_0^\infty x^2 \exp\left(-a\sqrt{x^2 + b^2}\right)\, dx$$
$$= \frac{2b}{a^2} K_1(ab) + \frac{b^2}{a} K_0(ab) \qquad\qquad \textbf{3.461.7} \quad \textbf{106}$$

$$\int_0^\infty x^4 \exp\left(-a\sqrt{x^2 + b^2}\right)\, dx$$
$$= \frac{12b^2}{a^3} K_2(ab) + \frac{3b^3}{a^2} K_1(ab) \qquad\qquad \textbf{3.461.8} \quad \textbf{106}$$

$$\int_0^\infty x^6 \exp\left(-a\sqrt{x^2 + b^2}\right)\, dx$$
$$= \frac{90b^3}{a^4} K_3(ab) + \frac{15b^4}{a^3} K_2(ab) \qquad\qquad \textbf{3.461.9} \quad \textbf{106}$$

Subsection 3.462.

$$\int_0^\infty x e^{-\mu x^2 - 2\nu x}\, dx$$
$$= \frac{1}{2\mu} - \frac{\nu}{2\mu}\sqrt{\frac{\pi}{\mu}} e^{\nu^2/\mu}\left[1 - \mathrm{erf}\left(\frac{\nu}{\sqrt{\mu}}\right)\right] \qquad\qquad \textbf{3.462.5} \quad \textbf{50}$$

$$\int_{-\infty}^\infty x e^{-px^2 + 2qx}\, dx = \frac{q}{p}\sqrt{\frac{\pi}{p}}\exp\left(\frac{q^2}{p}\right) \qquad\qquad \textbf{3.462.6} \quad \textbf{50}$$

$$\int_{-\infty}^\infty x^2 e^{-\mu x^2 + 2\nu x}\, dx = \frac{1}{2\mu}\sqrt{\frac{\pi}{\mu}}\left(1 + \frac{2\nu^2}{\mu}\right)e^{\nu^2/\mu} \qquad\qquad \textbf{3.462.8} \quad \textbf{40}$$

$$\int_0^\infty \frac{\exp(-a\sqrt{x + b^2})}{\sqrt{x^2 + b^2}}\, dx = K_0(ab) \qquad\qquad \textbf{3.462.20} \quad \textbf{103}$$

$$\int_0^\infty \frac{x^2 \exp(-a\sqrt{x + b^2})}{\sqrt{x^2 + b^2}}\, dx$$
$$= \frac{b}{a} K_1(ab) \qquad\qquad \textbf{3.462.21} \quad \textbf{103}$$

$$\int_0^\infty \frac{x^4 \exp(-a\sqrt{x+b^2})}{\sqrt{x^2+b^2}}\, dx = \frac{3b^2}{a^2}K_2(ab) \qquad \textbf{3.462.22} \quad \textbf{103}$$

$$\int_0^\infty \frac{x^6 \exp(-a\sqrt{x+b^2})}{\sqrt{x^2+b^2}}\, dx = \frac{15b^3}{a^3}K_3(ab) \qquad \textbf{3.462.23} \quad \textbf{103}$$

$$\int_0^\infty \frac{x^{2n} \exp(-a\sqrt{x+b^2})}{\sqrt{x^2+b^2}}\, dx = (2n-1)!! \left(\frac{b}{a}\right)^n K_n(ab) \qquad \textbf{3.462.24} \quad \textbf{102}$$

$$\int_0^\infty \frac{\exp(-px^2)}{\sqrt{a^2+x^2}}\, dx = \frac{1}{2}\exp\left(\frac{a^2p}{2}\right) K_0\left(\frac{a^2p}{2}\right) \qquad \textbf{3.462.25} \quad \textbf{100}$$

Subsection 3.464.

$$\int_0^\infty \left(e^{-\mu x^2} - e^{-x}\right) \frac{dx}{x} = \frac{1}{2}\mathbf{C} \qquad \textbf{3.464} \quad \textbf{52}$$

Subsection 3.466.

$$\int_0^\infty \frac{e^{-\mu^2 x^2}\, dx}{x^2+\beta^2} = [1 - \Phi(\beta\mu)]\, \frac{\pi}{2\beta}e^{\beta^2\mu^2} \qquad \textbf{3.466.1} \quad \textbf{52}$$

$$\int_0^\infty \frac{x^2 e^{-\mu^2 x^2}\, dx}{x^2+\beta^2} = \frac{\sqrt{\pi}}{2\mu} - \frac{\pi\beta}{2}e^{\mu^2\beta^2}\, [1 - \Phi(\beta\mu)] \qquad \textbf{3.466.2} \quad \textbf{43}$$

$$\int_0^1 \frac{e^{x^2}-1}{x^2}\, dx = \sum_{k=1}^\infty \frac{1}{k!\,(2k-1)} \qquad \textbf{3.466.3} \quad \textbf{143}$$

Subsection 3.469.

$$\int_0^\infty e^{-\mu x^4 - 2\nu x^2}\, dx = \frac{1}{4}\sqrt{\frac{2\nu}{\mu}}\exp\left(\frac{\nu^2}{2\mu}\right) K_{\frac{1}{4}}\left(\frac{\nu^2}{2\mu}\right) \qquad \textbf{3.469.1} \quad \textbf{107}$$

Subsection 3.471.

$$\int_0^u x^{-2\mu}(u-x)^{\mu-1}e^{-\beta/x}\, dx$$

$$= \frac{1}{\sqrt{\pi u}}\beta^{1/2-\mu}e^{-\beta/2u}\Gamma(\mu)K_{\mu-\frac{1}{2}}\left(\frac{\beta}{2u}\right) \qquad \textbf{3.471.4} \quad \textbf{102}$$

$$\int_0^u x^{-2\mu}(u^2-x^2)^{\mu-1}e^{-\beta/x}\,dx$$

$$= \frac{1}{\sqrt{\pi}}\left(\frac{2}{\beta}\right)^{\mu-1/2}u^{\mu-3/2}\Gamma(\mu)K_{\mu-\frac{1}{2}}\left(\frac{\beta}{u}\right) \qquad \textbf{3.471.8} \quad \textbf{102}$$

$$\int_0^\infty x^{\nu-1}e^{-\beta/x-\gamma x}\,dx$$

$$= 2\left(\frac{\beta}{\gamma}\right)^{\nu/2}K_\nu(2\sqrt{\beta\nu}) \qquad \textbf{3.471.9} \quad \textbf{97}$$

$$\int_0^\infty x^{\nu-1}\exp\left(-x-\frac{\mu^2}{4x}\right)\,dx$$

$$= 2\left(\frac{\mu}{2}\right)^\nu K_{-\nu}(\mu) \qquad \textbf{3.471.12} \quad \textbf{96}$$

$$\int_0^\infty x^{-1/2}e^{-\gamma x-\beta/x}\,dx$$

$$= \sqrt{\frac{\pi}{\gamma}}e^{-2\sqrt{\beta\gamma}} \qquad \textbf{3.471.15} \quad \textbf{48}$$

$$\int_0^\infty x^{n-1/2}e^{-px-q/x}\,dx$$

$$= (-1)^n\sqrt{\pi}\frac{\partial^n}{\partial p^n}\left(p^{-1/2}e^{-2\sqrt{pq}}\right) \qquad \textbf{3.471.16} \quad \textbf{48}$$

Subsection 3.472.

$$\int_0^\infty \left(\exp\left(-\frac{a}{x^2}\right)-1\right)e^{-\mu x^2}\,dx$$

$$= \frac{1}{2}\sqrt{\frac{\pi}{\mu}}\,[\exp(-2\sqrt{a\mu})-1] \qquad \textbf{3.472.1} \quad \textbf{48}$$

Subsection 3.478.

$$\int_0^\infty x^{\nu-1}\exp\left(-\beta x^p-\gamma x^{-p}\right)\,dx$$

$$= \frac{2}{p}\left(\frac{\gamma}{\beta}\right)^{\frac{\nu}{2p}}K_{\frac{\nu}{p}}(2\sqrt{\beta\gamma}) \qquad \textbf{3.478.4} \quad \textbf{98}$$

Subsection 3.479.

$$\int_0^\infty \frac{x^{\nu-1}\exp(-\beta\sqrt{1+x})}{\sqrt{1+x}}\, dx = \frac{2}{\sqrt{\pi}}\left(\frac{\beta}{2}\right)^{\frac{1}{2}-\nu}\Gamma(\nu)\, K_{\frac{1}{2}-\nu}(\beta) \qquad \textbf{3.479.1} \quad \textbf{100}$$

Section 3.5. Hyperbolic functions

Subsection 3.51. Hyperbolic functions

Subsection 3.511.

$$\int_0^\infty \frac{dx}{\cosh ax} = \frac{\pi}{2a} \qquad \textbf{3.511.1} \quad \textbf{73}$$

$$\int_0^\infty \frac{\sinh ax}{\sinh bx}\, dx = \frac{\pi}{2b}\tan\frac{\pi a}{2b} \qquad \textbf{3.511.2} \quad \textbf{78}$$

$$\int_0^\infty \frac{\cosh ax}{\cosh bx}\, dx = \frac{\pi}{2b}\sec\frac{\pi a}{2b} \qquad \textbf{3.511.4} \quad \textbf{79}$$

$$\int_0^\infty \frac{\sinh ax\,\cosh bx}{\sinh cx}\, dx = \frac{\pi}{2c}\frac{\sin\frac{\pi a}{c}}{\cos\frac{\pi a}{c}+\cos\frac{\pi b}{c}} \qquad \textbf{3.511.5} \quad \textbf{89}$$

$$\int_0^\infty \frac{dx}{\cosh^2 x} = 1 \qquad \textbf{3.511.8} \quad \textbf{75}$$

Subsection 3.512.

$$\int_0^\infty \frac{\cosh 2\beta x}{\cosh^{2\nu} ax}\, dx = \frac{4^{\nu-1}}{a}B\left(\nu+\frac{\beta}{a},\nu-\frac{\beta}{a}\right) \qquad \textbf{3.512.1} \quad \textbf{87}$$

$$\int_0^\infty \frac{\sinh^\mu x}{\cosh^\nu x}\, dx = \frac{1}{2}B\left(\frac{\mu+1}{2},\frac{\nu-\mu}{2}\right) \qquad \textbf{3.512.2} \quad \textbf{88}$$

Subsection 3.514.

$$\int_0^\infty \frac{dx}{\cosh ax+\cos t} = \frac{t}{a\sin t} \qquad \textbf{3.514.1} \quad \textbf{136}$$

Subsection 3.52–3.53. Combinations of hyperbolic functions and algebraic functions

Subsection 3.521.

$$\int_0^\infty \frac{x\, dx}{\sinh ax} = \frac{\pi^2}{4a^2} \qquad \textbf{3.521.1} \quad \textbf{76}$$

$$\int_0^\infty \frac{x\, dx}{\cosh x} = 2\mathbf{G} = \pi\ln 2 - 4L\left(\frac{\pi}{4}\right) \qquad \textbf{3.521.2} \quad \textbf{76}$$

Subsection 3.522.

$$\int_0^\infty \frac{dx}{(b^2 + x^2)\cosh ax} = \frac{2\pi}{b}\sum_{k=1}^\infty \frac{(-1)^{k-1}}{2ab + (2k-1)\pi} \qquad\qquad \textbf{3.522.3}\quad \textbf{82}$$

$$\int_0^\infty \frac{dx}{(1 + x^2)\cosh \pi x} = 2 - \frac{\pi}{2} \qquad\qquad \textbf{3.522.6}\quad \textbf{82}$$

$$\int_0^\infty \frac{dx}{(1 + x^2)\cosh \frac{\pi x}{2}} = \ln 2 \qquad\qquad \textbf{3.522.8}\quad \textbf{83}$$

$$\int_0^\infty \frac{dx}{(1 + x^2)\cosh \frac{\pi x}{4}} = \frac{1}{\sqrt{2}}\left[\pi - 2\ln(\sqrt{2} + 1)\right] \qquad\qquad \textbf{3.522.10}\quad \textbf{83}$$

Subsection 3.523.

$$\int_0^\infty \frac{x^{\beta-1}\, dx}{\sinh ax} = \frac{2^\beta - 1}{2^{\beta-1} a^\beta}\,\Gamma(\beta)\,\zeta(\beta) \qquad\qquad \textbf{3.523.1}\quad \textbf{76}$$

$$\int_0^\infty \frac{x^{2n-1}\, dx}{\sinh ax} = \frac{2^{2n} - 1}{2n}\left(\frac{\pi}{a}\right)^{2n}|B_{2n}| \qquad\qquad \textbf{3.523.2}\quad \textbf{76}$$

$$\int_0^\infty \frac{x^{\beta-1}\, dx}{\cosh ax} = \frac{2}{a^\beta}\Gamma(\beta)\sum_{k=0}^\infty \frac{(-1)^k}{(2k+1)^\beta} \qquad\qquad \textbf{3.523.3}\quad \textbf{77}$$

$$\int_0^\infty \frac{x^{2n}\, dx}{\cosh ax} = \left(\frac{\pi}{2a}\right)^{2n+1}|E_{2n}| \qquad\qquad \textbf{3.523.4}\quad \textbf{77}$$

$$\int_0^\infty \frac{x^2\, dx}{\cosh x} = \frac{\pi^3}{8} \qquad\qquad \textbf{3.523.5}\quad \textbf{77}$$

$$\int_0^\infty \frac{x^3\, dx}{\sinh x} = \frac{\pi^4}{8} \qquad\qquad \textbf{3.523.6}\quad \textbf{76}$$

$$\int_0^\infty \frac{x^4\, dx}{\cosh x} = \frac{5\pi^5}{32} \qquad\qquad \textbf{3.523.7}\quad \textbf{77}$$

$$\int_0^\infty \frac{x^5\, dx}{\sinh x} = \frac{\pi^6}{4} \qquad\qquad \textbf{3.523.8}\quad \textbf{76}$$

$$\int_0^\infty \frac{x^6\, dx}{\cosh x} = \frac{61\pi^7}{128} \qquad\qquad \textbf{3.523.9}\quad \textbf{78}$$

$$\int_0^\infty \frac{x^7\, dx}{\sinh x} = \frac{17\pi^8}{16} \qquad\qquad \textbf{3.523.10}\quad \textbf{76}$$

$$\int_0^\infty \frac{x^{1/2}\, dx}{\cosh x} = \sqrt{\pi}\sum_{k=0}^\infty \frac{(-1)^k}{(2k+1)^{3/2}} \qquad\qquad \textbf{3.523.11}\quad \textbf{76}$$

$$\int_0^\infty \frac{dx}{x^{1/2}\cosh x} = 2\sqrt{\pi}\sum_{k=0}^\infty \frac{(-1)^k}{(2k+1)^{1/2}} \qquad\qquad \textbf{3.523.12}\quad \textbf{75}$$

Subsection 3.524.

$$\int_0^\infty x^{2m} \frac{\sinh ax}{\sinh bx} \, dx = \frac{\pi}{2b} \frac{d^{2m}}{da^{2m}} \left(\tan \frac{\pi a}{2b} \right)$$

        3.524.2 79

$$\int_0^\infty x^{2m+1} \frac{\sinh ax}{\cosh bx} \, dx = \frac{\pi}{2b} \frac{d^{2m+1}}{da^{2m+1}} \left(\sec \frac{\pi a}{2b} \right)$$

        3.524.4 80

$$\int_0^\infty x^{2m} \frac{\cosh ax}{\cosh bx} \, dx = \frac{\pi}{2b} \frac{d^{2m}}{da^{2m}} \left(\sec \frac{\pi a}{2b} \right)$$

        3.524.6 80

$$\int_0^\infty x^2 \frac{\sinh ax}{\sinh bx} \, dx = \frac{\pi^3}{4b^3} \sin \frac{\pi a}{2b} \sec^3 \frac{\pi a}{2b}$$

        3.524.9 79

$$\int_0^\infty x^4 \frac{\sinh ax}{\sinh bx} \, dx = 8 \left(\frac{\pi}{2b} \sec \frac{\pi a}{2b} \right)^5 \left(2 + \sin^2 \frac{\pi a}{2b} \right) \sin \frac{\pi a}{2b}$$

        3.524.10 79

$$\int_0^\infty x \frac{\sinh ax}{\cosh bx} \, dx = \frac{\pi^2}{4b^2} \sec^2 \frac{\pi a}{2b} \sin \frac{\pi a}{2b}$$

        3.524.12 80

$$\int_0^\infty x^3 \frac{\sinh ax}{\cosh bx} \, dx = \left(\frac{\pi}{2b} \sec \frac{\pi a}{2b} \right)^4 \left(6 - \cos^2 \frac{\pi a}{2b} \right) \sin \frac{\pi a}{2b}$$

        3.524.13 80

$$\int_0^\infty x \frac{\cosh ax}{\sinh bx} \, dx = \left(\frac{\pi}{2b} \sec \frac{\pi a}{2b} \right)^2$$

        3.524.16 79

$$\int_0^\infty x^3 \frac{\cosh ax}{\sinh bx} \, dx = 2 \left(\frac{\pi}{2b} \sec \frac{\pi a}{2b} \right)^4 \left(1 + 2 \sin^2 \frac{\pi a}{2b} \right)$$

        3.524.17 79

$$\int_0^\infty x^5 \frac{\cosh ax}{\sinh bx} \, dx$$
$$= 8 \left(\frac{\pi}{2b} \sec \frac{\pi a}{2b} \right)^6 \left(15 - 15 \cos^2 \frac{\pi a}{2b} + 2 \cos^4 \frac{\pi a}{2b} \right)$$

        3.524.18 79

$$\int_0^\infty x^2 \frac{\cosh ax}{\cosh bx} \, dx = \frac{\pi^3}{8b^3} \left(2 \sec^3 \frac{\pi a}{2b} - \sec \frac{\pi a}{2b} \right)$$

        3.524.20 80

$$\int_0^\infty x^4 \frac{\cosh ax}{\cosh bx} \, dx$$
$$= \left(\frac{\pi}{2b} \sec \frac{\pi a}{2b} \right)^5 \left(24 - 20 \cos^2 \frac{\pi a}{2b} + \cos^4 \frac{\pi a}{2b} \right)$$

        3.524.21 80

$$\int_0^\infty x^6 \frac{\cosh ax}{\cosh bx} \, dx = \left(\frac{\pi}{2b} \sec \frac{\pi a}{2b} \right)^7$$
$$\times \left(720 - 840 \cos^2 \frac{\pi a}{2b} + 182 \cos^4 \frac{\pi a}{2b} - \cos^6 \frac{\pi a}{2b} \right)$$

        3.524.22 80

$$\int_0^\infty \frac{\sinh ax}{\cosh bx} \frac{dx}{x} = \ln \left(\frac{\pi a}{4b} + \frac{\pi}{4} \right)$$

        3.524.23 81

Subsection 3.525.

$$\int_0^\infty \frac{\sinh ax}{\sinh \pi x} \frac{dx}{1+x^2}$$
$$= -\frac{a}{2}\cos a + \frac{1}{2}\sin a \ln[2(1+\cos a)] \qquad\qquad \textbf{3.525.1} \quad 84$$

$$\int_0^\infty \frac{\sinh ax}{\sinh \frac{\pi}{2}x} \frac{dx}{1+x^2}$$
$$= \frac{\pi}{2}\sin a + \frac{1}{2}\cos a \ln \frac{1-\sin a}{1+\sin a} \qquad\qquad \textbf{3.525.2} \quad 83$$

$$\int_0^\infty \frac{\cosh ax}{\sinh \pi x} \frac{x\,dx}{1+x^2}$$
$$= \frac{1}{2}(a\sin a - 1) + \frac{1}{2}\cos a \ln[2(1+\cos a)] \qquad\qquad \textbf{3.525.3} \quad 84$$

$$\int_0^\infty \frac{\cosh ax}{\sinh \frac{\pi}{2}x} \frac{x\,dx}{1+x^2}$$
$$= \frac{\pi}{2}\cos a - 1 + \frac{1}{2}\sin a \ln \frac{1+\sin a}{1-\sin a} \qquad\qquad \textbf{3.525.4} \quad 84$$

$$\int_0^\infty \frac{\sinh ax}{\cosh \pi x} \frac{x\,dx}{1+x^2}$$
$$= -2\sin \frac{a}{2} + \frac{\pi}{2}\sin a - \cos a \ln \tan \frac{a+\pi}{4} \qquad\qquad \textbf{3.525.5} \quad 84$$

$$\int_0^\infty \frac{\cosh ax}{\cosh \pi x} \frac{dx}{1+x^2}$$
$$= 2\cos \frac{a}{2} - \frac{\pi}{2}\cos a - \sin a \ln \tan \frac{a+\pi}{4} \qquad\qquad \textbf{3.525.6} \quad 84$$

$$\int_0^\infty \frac{\sinh ax}{\sinh bx} \frac{dx}{c^2+x^2} = \frac{\pi}{c}\sum_{k=1}^\infty \frac{\sin \frac{k(b-a)}{b}\pi}{bc+k\pi} \qquad\qquad \textbf{3.525.7} \quad 89$$

$$\int_0^\infty \frac{\cosh ax}{\sinh bx} \frac{x\,dx}{c^2+x^2} = \frac{\pi}{2bc} + \pi\sum_{k=1}^\infty \frac{\cos \frac{k(b-a)}{b}\pi}{bc+k\pi} \qquad\qquad \textbf{3.525.8} \quad 89$$

Subsection 3.527.

$$\int_0^\infty \frac{x^{\mu-1}\,dx}{\sinh^2 ax} = \frac{4}{(2a)^\mu}\Gamma(\mu)\zeta(\mu-1) \qquad\qquad \textbf{3.527.1} \quad 85$$

$$\int_0^\infty \frac{x^{2m}\,dx}{\sinh^2 ax} = \frac{\pi^{2m}}{a^{2m+1}}|B_{2m}| \qquad\qquad \textbf{3.527.2} \quad 85$$

$$\int_0^\infty \frac{x^{\mu-1}\,dx}{\cosh^2 ax} = \frac{4}{(2a)^\mu}\left(1-2^{2-\mu}\right)\Gamma(\mu)\zeta(\mu-1) \qquad\qquad \textbf{3.527.3} \quad 86$$

$$\int_0^\infty \frac{x\,dx}{\cosh^2 ax} = \frac{\ln 2}{a^2} \qquad\qquad\qquad \textbf{3.527.4} \quad 86$$

$$\int_0^\infty \frac{x^{2m}\,dx}{\cosh^2 ax} = \frac{(2^{2m}-2)\pi^{2m}}{a(2a)^{2m}}|B_{2m}| \qquad\qquad \textbf{3.527.5} \quad 86$$

$$\int_0^\infty x^{\mu-1}\frac{\sinh ax}{\cosh^2 ax}\,dx = \frac{2\Gamma(\mu)}{a^\mu}\sum_{k=0}^\infty \frac{(-1)^k}{(2k+1)^{\mu-1}} \qquad \textbf{3.527.6} \quad 81$$

$$\int_0^\infty \frac{x\sinh ax}{\cosh^2 ax}\,dx = \frac{\pi}{2a^2} \qquad\qquad\qquad \textbf{3.527.7} \quad 81$$

$$\int_0^\infty x^{2m+1}\frac{\sinh ax}{\cosh^2 ax}\,dx = \frac{2m+1}{a}\left(\frac{\pi}{2a}\right)^{2m+1}|E_{2m}| \qquad \textbf{3.527.8} \quad 81$$

$$\int_0^\infty x^{2m+1}\frac{\cosh ax}{\sinh^2 ax}\,dx = \frac{2^{2m+1}-1}{a^2(2a)^{2m}}(2m+1)!\,\zeta(2m+1) \qquad \textbf{3.527.9} \quad 86$$

$$\int_0^\infty x^{2m}\frac{\cosh ax}{\sinh^2 ax}\,dx = \frac{2^{2m}-1}{a}\left(\frac{\pi}{a}\right)^{2m}|B_{2m}| \qquad \textbf{3.527.10} \quad 86$$

$$\int_0^\infty \frac{x\sinh ax}{\cosh^{2\mu+1} ax}\,dx = \frac{\sqrt\pi}{4\mu a^2}\frac{\Gamma(\mu)}{\Gamma\left(\mu+\frac12\right)} \qquad \textbf{3.527.11} \quad 88$$

$$\int_{-\infty}^\infty \frac{x^2\,dx}{\sinh^2 x} = \frac{\pi^2}{3} \qquad\qquad\qquad \textbf{3.527.12} \quad 85$$

$$\int_0^\infty x^2\frac{\cosh ax}{\sinh^2 ax}\,dx = \frac{\pi^2}{2a^3} \qquad\qquad\qquad \textbf{3.527.13} \quad 86$$

$$\int_0^\infty x^2\frac{\sinh x}{\cosh^2 x}\,dx = 4\mathbf{G} \qquad\qquad\qquad \textbf{3.527.14} \quad 78$$

$$\int_0^\infty \frac{\tanh\frac{x}{2}\,dx}{\cosh x} = \ln 2 \qquad\qquad\qquad \textbf{3.527.15} \quad 74$$

$$\int_0^\infty x^{\mu-1}\frac{\cosh ax}{\sinh^2 ax}\,dx = \frac{2\Gamma(\mu)\zeta(\mu-1)}{a^\mu}(1-2^{1-\mu}) \qquad \textbf{3.527.16} \quad 86$$

Subsection 3.531.

$$\int_0^\infty \frac{x\,dx}{2\cosh x - 1} = \frac{4}{\sqrt 3}\left[\frac{\pi}{3}\ln 2 - L\left(\frac{\pi}{3}\right)\right] \qquad \textbf{3.531.1} \quad 136$$

$$\int_0^\infty \frac{x^2\,dx}{\cosh 2x + \cos 2t} = \frac{t\ln 2 - L(t)}{\sin 2t} \qquad \textbf{3.531.2} \quad 134$$

$$\int_0^\infty \frac{x^2\,dx}{\cosh x + \cos t} = \frac{t(\pi^2 - t^2)}{3\sin t} \qquad \textbf{3.531.3} \quad 137$$

$$\int_0^\infty \frac{x^4\,dx}{\cosh x + \cos t} = \frac{t(\pi^2 - t^2)(7\pi^2 - 3t^2)}{15\sin t}$$
3.531.4   137

$$\int_0^\infty \frac{x^{2m}\,dx}{\cosh x - \cos 2a\pi} = \frac{2(2m)!}{\sin 2\pi a}\sum_{k=1}^\infty \frac{\sin 2ka\pi}{k^{2m+1}}$$
3.531.5   138

$$\int_0^\infty \frac{x^{\mu-1}\,dx}{\cosh x - \cos t}$$
$$= \frac{i\Gamma(\mu)}{\sin t}\left[e^{-it}\Phi\left(e^{-it},\mu,1\right) - e^{it}\Phi\left(e^{it},\mu,1\right)\right]$$
3.531.6   137

$$\int_0^\infty \frac{x^\mu\,dx}{\cosh x + \cos t}$$
$$= \frac{2\Gamma(\mu+1)}{\sin t}\sum_{k=1}^\infty(-1)^{k-1}\frac{\sin kt}{k^{\mu+1}}$$
3.531.7   134

Subsection 3.54. Combinations of hyperbolic functions and exponentials

Subsection 3.543.

$$\int_0^\infty \frac{e^{-\mu x}\,dx}{\cosh x - \cos t} = \frac{2}{\sin t}\sum_{k=1}^\infty \frac{\sin kt}{\mu + k}$$
3.543.2   89

Subsection 3.547.

$$\int_0^\infty \exp(-\beta\cosh x)\sinh\gamma x\,\sinh x\,dx \doteq \frac{\gamma}{\beta}K_\gamma(\beta)$$
3.547.2   98

$$\int_0^\infty \exp(-\beta\cosh x)\cosh\gamma x\,dx = K_\gamma(\beta)$$
3.547.4   96

$$\int_0^\infty \exp(-\beta\cosh x)\sinh^{2\nu} x\,dx$$
$$= \frac{1}{\sqrt{\pi}}\left(\frac{2}{\beta}\right)^\nu \Gamma\left(\nu + \tfrac{1}{2}\right)K_\nu(\beta)$$
3.547.9   100

Section 3.6–4.1. Trigonometric functions

Subsection 3.62. Powers of trigonometric functions

Subsection 3.621.

$$\int_0^{\pi/2} \sin^{2m} x \, dx = \frac{(2m-1)!!}{(2m)!!} \frac{\pi}{2}$$

3.621.3 177

$$\int_0^{\pi/2} \sin^{\mu-1} x \, \cos^{\nu-1} x \, dx = \frac{1}{2} B\left(\frac{\mu}{2}, \frac{\nu}{2}\right)$$

3.621.5 174

Subsection 3.72–3.74. Combinations of trigonometric and rational functions

Subsection 3.747.

$$\int_0^{\pi/2} x \cot x \, dx = \frac{\pi}{2} \ln 2$$

3.747.7 147

Subsection 3.76–3.77. Combinations of trigonometric and powers

Subsection 3.764.

$$\int_0^\infty x^p \cos(ax+b) \, dx = -\frac{1}{a^{p+1}} \Gamma(1+p) \sin\left(b + \frac{\pi p}{2}\right)$$

3.764.2 209

Subsection 3.82–3.83. Powers of trigonometric functions combined with other powers

Subsection 3.821.

$$\int_0^\infty \frac{\sin^{2n+1} x}{x} \, dx = \frac{(2n-1)!!}{(2n)!!} \frac{\pi}{2}$$

3.821.7 7

Subsection 3.84. Integrals containing $\sqrt{1 - k^2 \sin^2 x}$, $\sqrt{1 - k^2 \cos^2 x}$, and similar expressions

Subsection 3.841.

$$\int_0^\infty \sin x \sqrt{1 - k^2 \sin^2 x} \, \frac{dx}{x} = \mathbf{E}(k) \qquad\qquad \textbf{3.841.1 \quad 5}$$

$$\int_0^\infty \sin x \sqrt{1 - k^2 \cos^2 x} \, \frac{dx}{x} = \mathbf{E}(k) \qquad\qquad \textbf{3.841.2 \quad 5}$$

$$\int_0^\infty \tan x \sqrt{1 - k^2 \sin^2 x} \, \frac{dx}{x} = \mathbf{E}(k) \qquad\qquad \textbf{3.841.3 \quad 6}$$

$$\int_0^\infty \tan x \sqrt{1 - k^2 \cos^2 x} \, \frac{dx}{x} = \mathbf{E}(k) \qquad\qquad \textbf{3.841.4 \quad 6}$$

Subsection 3.842.

$$\int_0^\infty \frac{\sin x}{\sqrt{1 - k^2 \sin^2 x}} \, \frac{dx}{x} = \mathbf{K}(k) \qquad\qquad \textbf{3.842.3a \quad 5}$$

$$\int_0^\infty \frac{\tan x}{\sqrt{1 - k^2 \sin^2 x}} \, \frac{dx}{x} = \mathbf{K}(k) \qquad\qquad \textbf{3.842.3b \quad 5}$$

$$\int_0^\infty \frac{\sin x}{\sqrt{1 - k^2 \cos^2 x}} \, \frac{dx}{x} = \mathbf{K}(k) \qquad\qquad \textbf{3.842.3c \quad 5}$$

$$\int_0^\infty \frac{\tan x}{\sqrt{1 - k^2 \cos^2 x}} \, \frac{dx}{x} = \mathbf{K}(k) \qquad\qquad \textbf{3.842.3d \quad 5}$$

$$\int_0^\infty \frac{x \sin x \cos x}{\sqrt{1 - k^2 \sin^2 x}} \, dx = \frac{1}{2k^2} \left[-\pi k' + 2\mathbf{E}(k) \right] \qquad\qquad \textbf{3.842.4 \quad 9}$$

Subsection 4.11–4.12. Combinations involving trigonometric and hyperbolic functions and powers

Subsection 4.113.

$$\int_0^\infty \frac{\sin ax}{\sinh \pi x} \, \frac{dx}{1 + x^2} = -\frac{a}{2} \cosh a + \sinh a \, \ln \left(2 \cosh \frac{a}{2} \right) \qquad \textbf{4.113.3 \quad 91}$$

Section 4.2–4.4. Logarithmic functions

Section 4.21. Logarithmic functions

Subsection 4.211.

$$\int_e^\infty \frac{dx}{\ln x} = \infty \qquad\qquad \textbf{4.211.1 \quad 151}$$

$$\int_0^u \frac{dx}{\ln x} = \operatorname{li} u \qquad\qquad \textbf{4.211.2 \quad 151}$$

Subsection 4.212.

$$\int_0^1 \frac{dx}{a + \ln x} = e^{-a}\text{Ei}(a) \qquad\qquad\qquad \textbf{4.212.1} \quad \textbf{159}$$

$$\int_0^1 \frac{dx}{a - \ln x} = -e^a\text{Ei}(-a) \qquad\qquad\qquad \textbf{4.212.2} \quad \textbf{159}$$

$$\int_0^1 \frac{dx}{(a + \ln x)^2} = -\frac{1}{a} + e^{-a}\text{Ei}(a) \qquad\qquad\qquad \textbf{4.212.3} \quad \textbf{160}$$

$$\int_0^1 \frac{dx}{(a - \ln x)^2} = \frac{1}{a} + e^a\text{Ei}(-a) \qquad\qquad\qquad \textbf{4.212.4} \quad \textbf{160}$$

$$\int_0^1 \frac{\ln x \, dx}{(a + \ln x)^2} = 1 + (1 - a)e^{-a}\text{Ei}(a) \qquad\qquad\qquad \textbf{4.212.5} \quad \textbf{160}$$

$$\int_0^1 \frac{\ln x \, dx}{(a - \ln x)^2} = 1 + (1 + a)e^a\text{Ei}(-a) \qquad\qquad\qquad \textbf{4.212.6} \quad \textbf{160}$$

$$\int_1^e \frac{\ln x \, dx}{(1 + \ln x)^2} = \frac{e}{2} - 1 \qquad\qquad\qquad \textbf{4.212.7} \quad \textbf{115}$$

$$\int_0^1 \frac{dx}{(a + \ln x)^n}$$

$$= \frac{1}{(n-1)!} e^{-a}\text{Ei}(a) - \frac{1}{(n-1)!} \sum_{k=1}^{n-1} (n - k - 1)! \, a^{k-n} \qquad \textbf{4.212.8} \quad \textbf{160}$$

$$\int_0^1 \frac{dx}{(a - \ln x)^n}$$

$$= \frac{(-1)^n}{(n-1)!} e^a\text{Ei}(-a) + \frac{1}{(n-1)!} \sum_{k=1}^{n-1} (n - k - 1)! \, (-a)^{k-n} \qquad \textbf{4.212.9} \quad \textbf{160}$$

Section 4.22. Logarithms of more complicated functions

Subsection 4.221.

$$\int_0^1 \ln x \ln(1 - x) \, dx = 2 - \frac{\pi^2}{6} \qquad\qquad\qquad \textbf{4.221.1} \quad \textbf{141}$$

$$\int_0^1 \ln x \ln(1 + x) \, dx = 2 - \frac{\pi^2}{12} - 2\ln 2 \qquad\qquad\qquad \textbf{4.221.2} \quad \textbf{141}$$

$$\int_0^1 \ln \frac{1 - ax}{1 - a} \frac{dx}{\ln x} = -\sum_{k=1}^{\infty} a^k \frac{\ln(1 + k)}{k} \qquad\qquad\qquad \textbf{4.221.3} \quad \textbf{142}$$

Subsection 4.224.

$$\int_0^{\pi/4} \ln \sin x \, dx = -\frac{\pi}{4} \ln 2 - \frac{1}{2} \mathbf{G}$$ 4.224.2   177

$$\int_0^{\pi/4} \ln \cos x \, dx = -\frac{\pi}{4} \ln 2 + \frac{1}{2} \mathbf{G}$$ 4.224.5   120

$$\int_0^{\pi/2} \ln \cos x \, dx = -\frac{\pi}{2} \ln 2$$ 4.224.6   120

Subsection 4.225.

$$\int_0^{\pi/4} \ln(\cos x - \sin x) \, dx = -\frac{\pi}{8} \ln 2 - \frac{1}{2} \mathbf{G}$$ 4.225.1   120

$$\int_0^{\pi/4} \ln(\cos x + \sin x) \, dx = -\frac{\pi}{8} \ln 2 + \frac{1}{2} \mathbf{G}$$ 4.225.2   120

Section 4.23. Combinations of logarithms and rational functions

Subsection 4.231.

$$\int_0^1 \frac{\ln x}{1 + x} \, dx = -\frac{\pi^2}{12}$$ 4.231.1   23

$$\int_0^1 \frac{\ln x}{1 - x} \, dx = -\frac{\pi^2}{6}$$ 4.231.2   131

$$\int_0^\infty \frac{\ln x \, dx}{a^2 + b^2 x^2} = \frac{\pi}{2ab} \ln \frac{a}{b}$$ 4.231.8   132

$$\int_0^\infty \frac{\ln px \, dx}{q^2 + x^2} = \frac{\pi}{2q} \ln pq$$ 4.231.9   121

$$\int_0^\infty \frac{\ln x \, dx}{a^2 - b^2 x^2} = -\frac{\pi^2}{4ab}$$ 4.231.10   131

$$\int_0^a \frac{\ln x \, dx}{x^2 + a^2} = \frac{\pi \ln a}{4a} - \frac{\mathbf{G}}{a}$$ 4.231.11   115

$$\int_0^1 \frac{\ln x \, dx}{1 + x^2} = -\mathbf{G}$$ 4.231.12a   75

$$\int_1^\infty \frac{\ln x \, dx}{1 + x^2} = \mathbf{G}$$ 4.231.12b   75

$$\int_0^1 \ln x \, \frac{1 - x^{2n+2}}{(1 - x^2)^2} \, dx = -\frac{(n+1)\pi^2}{8} + \sum_{k=1}^n \frac{n - k + 1}{(2k - 1)^2}$$ 4.231.16   130

$$\int_0^1 \ln x \, \frac{1 + (-1)^n x^{n+1}}{(1+x)^2} \, dx$$

$$= -\frac{(n+1)\pi^2}{12} - \sum_{k=1}^n (-1)^k \frac{n-k+1}{k^2} \qquad\qquad \textbf{4.231.17} \quad \textbf{130}$$

$$\int_0^1 \ln x \, \frac{1 - x^{n+1}}{(1-x)^2} \, dx$$

$$= -\frac{(n+1)\pi^2}{6} + \sum_{k=1}^n \frac{n-k+1}{k^2} \qquad\qquad \textbf{4.231.18} \quad \textbf{129}$$

Subsection 4.233.

$$\int_0^1 \frac{\ln x \, dx}{1 + x + x^2} = \frac{2}{9}\left[\frac{2\pi^2}{3} - \psi'\left(\frac{1}{3}\right)\right] \qquad\qquad \textbf{4.233.1} \quad \textbf{115}$$

$$\int_0^\infty \frac{\ln x \, dx}{x^2 + 2ax\cos t + a^2} = \frac{\ln a}{a} \frac{t}{\sin t} \qquad\qquad \textbf{4.233.5} \quad \textbf{165}$$

Subsection 4.234.

$$\int_0^\infty \frac{1 + x^2}{(1 - x^2)^2} \ln x \, dx = 0 \qquad\qquad \textbf{4.234.3} \quad \textbf{116}$$

$$\int_0^\infty \frac{1 - x^2}{(1 + x^2)^2} \ln x \, dx = -\frac{\pi}{2} \qquad\qquad \textbf{4.234.4} \quad \textbf{163}$$

$$\int_0^1 \frac{x^2 \ln x \, dx}{(1 - x^2)(1 + x^4)} = -\frac{\pi^2}{16(2 + \sqrt{2})} \qquad\qquad \textbf{4.234.5} \quad \textbf{164}$$

$$\int_0^\infty \frac{\ln x \, dx}{(a^2 + b^2 x^2)(1 + x^2)} = \frac{\pi b}{2a(b^2 - a^2)} \ln \frac{a}{b} \qquad\qquad \textbf{4.234.6} \quad \textbf{121}$$

$$\int_0^\infty \frac{\ln x}{x^2 + a^2} \frac{dx}{1 + b^2 x^2} = \frac{\pi}{2(1 - a^2 b^2)} \left(\frac{\ln a}{a} + b \ln b\right) \qquad\qquad \textbf{4.234.7} \quad \textbf{121}$$

$$\int_0^\infty \frac{x^2 \ln x \, dx}{(a^2 + b^2 x^2)(1 + x^2)} = \frac{\pi a}{2b(b^2 - a^2)} \ln \frac{b}{a} \qquad\qquad \textbf{4.234.8} \quad \textbf{121}$$

Subsection 4.235.

$$\int_0^\infty \frac{(1-x)x^{n-2}}{1-x^{2n}} \ln x\, dx = -\frac{\pi^2}{4n^2} \tan^2 \frac{\pi}{2n}$$

4.235.1   170

$$\int_0^\infty \frac{(1-x^2)x^{m-1}}{1-x^{2n}} \ln x\, dx$$

$$= -\frac{\pi^2 \sin\left(\frac{(m+1)\pi}{n}\right) \sin\left(\frac{\pi}{n}\right)}{4n^2 \sin^2\left(\frac{\pi m}{2n}\right) \sin^2\left(\frac{(m+2)\pi}{2n}\right)}$$

qquad4.235.2   170

$$\int_0^\infty \frac{(1-x^2)x^{n-1}}{1-x^{2n}} \ln x\, dx = -\frac{\pi^2}{4n^2} \tan^2 \left(\frac{\pi}{n}\right)$$

4.235.3   170

$$\int_0^1 \frac{x^{m-1} + x^{n-m-1}}{1-x^n} \ln x\, dx = -\frac{\pi^2}{n^2 \sin^2\left(\frac{\pi m}{n}\right)}$$

4.235.4   170

Section 4.24. Combinations of logarithms and algebraic functions

Subsection 4.241.

$$\int_0^1 \ln x \sqrt{(1-x^2)^{2n-1}}\, dx$$

$$= -\frac{(2n-1)!!}{4(2n)!!} \pi \left[\psi(n+1) + \mathbf{C} + \ln 4\right]$$

4.241.5   174

$$\int_0^1 \frac{\ln x\, dx}{\sqrt{1-x^2}} = -\frac{\pi}{2} \ln 2$$

4.241.7   174

$$\int_0^1 \frac{\ln x\, dx}{\sqrt{x(1-x^2)}} = -\frac{\sqrt{2\pi}}{8} \left[\Gamma\left(\frac{1}{4}\right)\right]^2$$

4.241.11   174

Subsection 4.242.

$$\int_0^\infty \frac{\ln x\, dx}{\sqrt{(x^2+a^2)(x^2+b^2)}} = \frac{\ln ab}{2a} \mathbf{K}\left(\frac{\sqrt{a^2-b^2}}{a}\right)$$

4.242.1   13

Section 4.25. Combinations of logarithms and powers
Subsection 4.251.

$$\int_0^\infty \frac{x^{\mu-1} \ln x}{x + \beta} \, dx = \frac{\pi \beta^{\mu-1}}{\sin \pi \mu} \left(\ln \beta - \pi \cot \pi \mu \right)$$
4.251.1   172

$$\int_0^\infty \frac{x^{\mu-1} \ln x}{a - x} \, dx = \pi a^{\mu-1} \left(\ln a \cot \pi \mu - \frac{\pi}{\sin^2 \pi \mu} \right)$$
4.251.2   171

$$\int_0^1 \frac{x^{2n}}{1 + x} \ln x \, dx = -\frac{\pi^2}{12} + \sum_{k=1}^{2n} \frac{(-1)^{k-1}}{k^2}$$
4.251.5   149

$$\int_0^1 \frac{x^{2n-1}}{1 + x} \ln x \, dx = \frac{\pi^2}{12} + \sum_{k=1}^{2n-1} \frac{(-1)^k}{k^2}$$
4.251.6   149

Subsection 4.252.

$$\int_0^\infty \frac{x^{\mu-1} \ln x \, dx}{(x + \beta)(x - 1)}$$
$$= \frac{\pi}{(\beta + 1) \sin^2 \pi \mu} \left[\pi - \beta^{\mu-1} (\ln \beta \sin \pi \mu - \pi \cos \pi \mu) \right]$$
4.252.2   171

$$\int_0^\infty \frac{x^{p-1} \ln x}{1 - x^2} \, dx = -\frac{\pi^2}{4} \csc^2 \frac{\pi p}{2}$$
4.252.3   171

$$\int_0^\infty \frac{x^{\mu-1} \ln x}{(x + a)^2} \, dx$$
$$= \frac{(1 - \mu) a^{\mu-2} \pi}{\sin \pi \mu} \left(\ln a - \pi \cot \pi \mu + \frac{1}{\mu - 1} \right)$$
4.252.4   172

Subsection 4.254.

$$\int_0^\infty \frac{x^{p-1} \ln x}{1 - x^q} \, dx = -\frac{\pi^2}{q^2 \sin^2 \frac{\pi p}{q}}$$
4.254.2   169

$$\int_0^\infty \frac{\ln x}{x^q - 1} \frac{dx}{x^p} = \frac{\pi^2}{q^2 \sin^2 \left(\frac{(p-1)\pi}{q} \right)}$$
4.254.3   172

Subsection 4.255.

$$\int_0^1 \frac{(1 + x^2) x^{p-2}}{1 - x^{2p}} \ln x \, dx = -\left(\frac{\pi}{2p} \right)^2 \sec^2 \left(\frac{\pi}{2p} \right)$$
4.255.2   173

$$\int_0^\infty \frac{(1 - x^p)}{1 - x^2} \ln x \, dx = \frac{\pi^2}{4} \tan^2 \left(\frac{\pi p}{2} \right)$$
4.255.3   171

Subsection 4.257.

$$\int_0^\infty \frac{x^\nu \ln \frac{x}{\beta} \, dx}{(x+\beta)(x+\gamma)}$$

$$= \frac{\pi}{(\gamma - \beta) \sin \pi\nu} \left[\gamma^\nu \ln \frac{\gamma}{\beta} + \pi \left(\beta^\nu - \gamma^\nu\right) \cot \pi\nu \right] \qquad \textbf{4.257.1 \quad 172}$$

Section 4.26–4.27. Combinations involving powers of the logarithm and other powers

Subsection 4.261.

$$\int_0^1 (\ln x)^2 \frac{dx}{\sqrt{1-x^2}} = \frac{\pi}{2} \left[\ln^2 2 + \frac{\pi^2}{12} \right] \qquad \textbf{4.261.9 \quad 174}$$

$$\int_0^1 \frac{x^n}{1-x} \ln^2 x \, dx = 2 \left(\zeta(3) - \sum_{k=1}^n \frac{1}{k^3} \right) \qquad \textbf{4.261.12 \quad 22}$$

$$\int_0^1 \frac{x^{2n}}{1-x^2} \ln^2 x \, dx = \frac{7}{4}\zeta(3) - 2 \sum_{k=1}^n \frac{1}{(2k-1)^3} \qquad \textbf{4.261.13 \quad 23}$$

Subsection 4.262.

$$\int_0^1 \frac{\ln^3 x}{1+x} \, dx = -\frac{7}{120}\pi^4 \qquad \textbf{4.262.1 \quad 23}$$

$$\int_0^1 \frac{\ln^3 x}{1-x} \, dx = -\frac{\pi^4}{15} \qquad \textbf{4.262.2 \quad 22}$$

$$\int_0^1 \frac{x^n}{1-x} \ln^3 x \, dx = -\frac{\pi^4}{15} + 6 \sum_{k=0}^{n-1} \frac{1}{(k+1)^4} \qquad \textbf{4.262.5 \quad 22}$$

$$\int_0^1 \frac{x^{2n}}{1-x^2} \ln^3 x \, dx = -\frac{\pi^4}{16} + 6 \sum_{k=0}^{n-1} \frac{1}{(2k+1)^4} \qquad \textbf{4.262.6 \quad 22}$$

$$\int_0^1 \frac{1-x^{n+1}}{(1-x)^2} \ln^3 x \, dx = -\frac{(n+1)\pi^4}{15} + 6 \sum_{k=1}^n \frac{n-k+1}{k^4} \qquad \textbf{4.262.7 \quad 129}$$

$$\int_0^1 \frac{1 + (-1)^n x^{n+1}}{(1+x)^2} \ln^3 x \, dx$$

$$= -\frac{7(n+1)\pi^4}{120} + 6 \sum_{k=1}^{n} (-1)^{k-1} \frac{n-k+1}{k^4}$$

4.262.8   131

$$\int_0^1 \frac{1 - x^{2n+2}}{(1-x^2)^2} \ln^3 x \, dx$$

$$= -\frac{(n+1)\pi^4}{16} + 6 \sum_{k=1}^{n} \frac{n-k+1}{(2k-1)^4}$$

4.262.9   131

Subsection 4.264.

$$\int_0^1 \frac{\ln^5 x}{1+x} \, dx = -\frac{31\pi^6}{252}$$

4.264.1   23

$$\int_0^1 \frac{\ln^5 x}{1-x} \, dx = -\frac{8\pi^6}{63}$$

4.264.2   22

Subsection 4.266.

$$\int_0^1 \frac{\ln^7 x}{1+x} \, dx = -\frac{127\pi^8}{240}$$

4.266.1   23

$$\int_0^1 \frac{\ln^7 x}{1-x} \, dx = -\frac{8\pi^8}{15}$$

4.266.2   22

Subsection 4.269.

$$\int_0^1 \sqrt{\ln \frac{1}{x}} \, \frac{dx}{1+x^2} = \frac{\sqrt{\pi}}{2} \sum_{k=0}^{\infty} \frac{(-1)^k}{(2k+1)^{3/2}}$$

4.269.1   150

$$\int_0^1 \frac{dx}{(1+x^2)\sqrt{\ln \frac{1}{x}}} = \sqrt{\pi} \sum_{k=0}^{\infty} \frac{(-1)^k}{\sqrt{2k+1}}$$

4.269.2   150

Subsection 4.271.

$$\int_0^1 (\ln x)^{2n} \frac{dx}{1+x} = \frac{2^{2n}-1}{2^{2n}} (2n)! \zeta(2n+1)$$

4.271.1   24

$$\int_0^1 (\ln x)^{2n-1} \frac{dx}{1+x} = \frac{1 - 2^{2n-1}}{2n} \pi^{2n} |B_{2n}|$$

4.271.2   24

Section 4.29–4.32. Combinations of logarithmic functions of more complicated arguments and powers

Subsection 4.291.

$$\int_0^1 \frac{\ln(1+x)}{x}\, dx = \frac{\pi^2}{12} \qquad\qquad\qquad 4.291.1 \quad 115$$

$$\int_0^1 \frac{\ln(1-x)}{x}\, dx = -\frac{\pi^2}{6} \qquad\qquad\qquad 4.291.2 \quad 115$$

$$\int_0^{1/2} \frac{\ln(1-x)}{x}\, dx = \frac{1}{2}\ln^2 2 - \frac{\pi^2}{12} \qquad\qquad 4.291.3 \quad 116$$

$$\int_0^1 \ln\left(1-\frac{x}{2}\right)\frac{dx}{x} = \frac{1}{2}\ln^2 2 - \frac{\pi^2}{12} \qquad\qquad 4.291.4 \quad 116$$

$$\int_0^1 \ln\left(\frac{1+x}{2}\right)\frac{dx}{1-x} = \frac{1}{2}\ln^2 2 - \frac{\pi^2}{12} \qquad\qquad 4.291.5 \quad 116$$

$$\int_0^1 \frac{\ln(1+x)}{1+x}\, dx = \frac{1}{2}\ln^2 2 \qquad\qquad\qquad 4.291.6 \quad 117$$

$$\int_0^\infty \frac{\ln(1+ax)}{1+x^2}\, dx = \frac{\pi}{4}\ln(1+a^2) - \int_0^a \frac{\ln u\, du}{1+u^2} \qquad 4.291.7 \quad 126$$

$$\int_0^1 \frac{\ln(1+x)}{1+x^2}\, dx = \frac{\pi}{8}\ln 2 \qquad\qquad\qquad 4.291.8 \quad 119$$

$$\int_0^\infty \frac{\ln(1+x)}{1+x^2}\, dx = \frac{\pi}{4}\ln 2 + \mathbf{G} \qquad\qquad 4.291.9 \quad 120$$

$$\int_0^1 \frac{\ln(1-x)}{1+x^2}\, dx = \frac{\pi}{8}\ln 2 - \mathbf{G} \qquad\qquad 4.291.10 \quad 120$$

$$\int_1^\infty \frac{\ln(x-1)}{1+x^2}\, dx = \frac{\pi}{8}\ln 2 \qquad\qquad\qquad 4.291.11 \quad 120$$

$$\int_0^1 \frac{\ln(1+x)}{x(1+x)}\, dx = \frac{\pi^2}{12} - \frac{1}{2}\ln^2 2 \qquad\qquad 4.291.12 \quad 117$$

$$\int_0^\infty \frac{\ln(1+x)}{x(1+x)}\, dx = \frac{\pi^2}{6} \qquad\qquad\qquad 4.291.13 \quad 117$$

$$\int_0^1 \frac{\ln(1+x)}{(ax+b)^2}\, dx = \frac{1}{a(a-b)}\ln\frac{a+b}{b} + \frac{2\ln 2}{b^2-a^2} \qquad 4.291.14a \quad 118$$

$$\int_0^1 \frac{\ln(1+x)}{(x+1)^2}\, dx = \frac{1}{2}(1-\ln 2) \qquad\qquad 4.291.14b \quad 118$$

$$\int_0^\infty \frac{\ln(1+x)}{(ax+b)^2} = \frac{\ln a - \ln b}{a(a-b)} \qquad\qquad 4.291.15 \quad 119$$

$$\int_0^1 \ln(a+x) \frac{dx}{a+x^2} = \frac{1}{2\sqrt{a}} \operatorname{arccot}\sqrt{a} \ln[a(1+a)]$$

4.291.16   126

$$\int_0^\infty \frac{\ln(a+x)}{(b+x)^2} dx = \frac{a\ln a - b\ln b}{b(a-b)}$$

4.291.17   119

$$\int_0^a \frac{\ln(1+ax)}{1+x^2} dx = \frac{1}{2}\arctan a \ln(1+a^2)$$

4.291.18   125

$$\int_0^1 \frac{\ln(1+ax)}{1+ax^2} dx = \frac{1}{2\sqrt{a}}\arctan\sqrt{a} \ln(1+a)$$

4.291.19   126

$$\int_0^1 \frac{\ln(ax+b)}{(1+x)^2} dx$$
$$= \frac{1}{a-b}\left[\frac{1}{2}(a+b)\ln(a+b) - b\ln b - a\ln 2\right]$$

4.291.20   118

$$\int_0^\infty \frac{\ln(ax+b)}{(1+x)^2} dx = \frac{a\ln a - b\ln b}{a-b}$$

4.291.21   119

$$\int_0^\infty \frac{x}{(b^2+x^2)^2} \ln(a+x)\, dx$$
$$= \frac{1}{2(a^2+b^2)}\left(\ln b + \frac{\pi a}{2b} + \frac{a^2}{b^2}\ln a\right)$$

4.291.22   128

$$\int_0^1 \frac{1+x^2}{(1+x)^4} \ln(1+x)\, dx = -\frac{1}{3}\ln 2 + \frac{23}{72}$$

4.291.23   122

$$\int_0^\infty \ln(1+x)\frac{1-x^2}{(ax+b)^2(bx+a)^2} dx$$
$$= \frac{1}{ab(a^2-b^2)}\ln\frac{b}{a}$$

4.291.26   119

$$\int_0^\infty \frac{b^2-x^2}{(b^2+x^2)^2} \ln(a+x)\, dx$$
$$= \frac{1}{a^2+b^2}\left(a\ln\frac{b}{a} - \frac{\pi b}{2}\right)$$

4.291.28   123

$$\int_0^\infty \ln^2(a-x)\frac{b^2-x^2}{(b^2+x^2)^2} dx$$
$$= \frac{2}{a^2+b^2}\left(a\ln\frac{a}{b} - \frac{\pi b}{2}\right)$$

4.291.29   124

$$\int_0^\infty \ln^2(a-x)\frac{x\, dx}{(b^2+x^2)^2} dx$$
$$= \frac{1}{a^2+b^2}\left(\ln b - \frac{\pi a}{2b} + \frac{a^2}{b^2}\ln a\right)$$

4.291.30   125

Subsection 4.297.

$$\int_0^1 \ln \frac{1+ax}{1-ax} \, \frac{dx}{x\sqrt{1-x^2}} = \pi \arcsin a \qquad\qquad \textbf{4.297.8} \quad 177$$

Section 4.33–4.34. Combinations of logarithms and exponentials

Subsection 4.331.

$$\int_1^\infty e^{-\mu x} \ln x \, dx = -\frac{1}{\mu}\mathrm{Ei}(-\mu) \qquad\qquad \textbf{4.331.2} \quad 161$$

Subsection 4.337.

$$\int_0^\infty e^{-\mu x} \ln(\beta+x) \, dx = \frac{1}{\mu}\left[\ln\beta - e^{\mu\beta}\mathrm{Ei}(-\beta\mu)\right] \qquad\qquad \textbf{4.337.1} \quad 161$$

$$\int_0^\infty e^{-\mu x} \ln(1+\beta x) \, dx = -\frac{1}{\mu}e^{\mu/\beta}\,\mathrm{Ei}\left(-\frac{\mu}{\beta}\right) \qquad\qquad \textbf{4.337.2} \quad 161$$

$$\int_0^\infty e^{-\mu x} \ln|a-x| \, dx = \frac{1}{\mu}\left[\ln a - e^{-a\mu}\,\mathrm{Ei}\,(a\mu)\right] \qquad\qquad \textbf{4.337.3} \quad 158$$

$$\int_0^\infty e^{-\mu x} \ln\left|\frac{\beta}{\beta-x}\right| \, dx = \frac{1}{\mu}\left[e^{-\beta\mu}\,\mathrm{Ei}(\beta\mu)\right] \qquad\qquad \textbf{4.337.4} \quad 158$$

Section 4.38–4.41. Logarithms and trigonometric functions

Subsection 4.395.

$$\int_0^{\pi/2} \frac{\ln\tan x \, dx}{\sqrt{1-k^2\sin^2 x}} = -\frac{1}{2}\ln k' \, \mathbf{K}(k) \qquad\qquad \textbf{4.395.1} \quad 10$$

Subsection 4.414.

$$\int_0^{\pi/2} \ln(1-k^2\sin^2 x) \, \frac{dx}{\sqrt{1-k^2\sin^2 x}} = \ln k' \, \mathbf{K}(k) \qquad\qquad \textbf{4.414.1} \quad 5$$

Section 4.42–4.43. Combinations of logarithms, trigonometric functions, and powers

Subsection 4.432.

$$\int_0^\infty \ln(1-k^2\sin^2 x) \, \frac{\sin x}{\sqrt{1-k^2\sin^2 x}} \, \frac{dx}{x} = \ln k' \, \mathbf{K}(k) \qquad\qquad \textbf{4.432.1a} \quad 5$$

$$\int_0^\infty \ln(1-k^2\cos^2 x) \, \frac{\sin x}{\sqrt{1-k^2\cos^2 x}} \, \frac{dx}{x} = \ln k' \, \mathbf{K}(k) \qquad\qquad \textbf{4.432.1b} \quad 5$$

Section 4.5. Inverse trigonometric functions

Section 4.52. Combinations of arcsines, arccosines, and powers

Subsection 4.522.

$$\int_0^1 \frac{x \arcsin x}{\sqrt{1-k^2 x^2}}\, dx = \frac{1}{k^2}\left[-\frac{\pi k'}{2} + \mathbf{E}(k)\right] \qquad \text{4.522.4} \quad 3$$

$$\int_0^1 \frac{x \arccos x}{\sqrt{1-k^2 x^2}}\, dx = \frac{1}{k^2}\left[\frac{\pi}{2} - \mathbf{E}(k)\right] \qquad \text{4.522.5} \quad 3$$

$$\int_0^1 \frac{x \arcsin x}{\sqrt{k'^2 + k^2 x^2}}\, dx = \frac{1}{k^2}\left[\frac{\pi}{2} - \mathbf{E}(k)\right] \qquad \text{4.522.6} \quad 3$$

$$\int_0^1 \frac{x \arccos x}{\sqrt{k'^2 + k^2 x^2}}\, dx = \frac{1}{k^2}\left[-\frac{\pi k'}{2} + \mathbf{E}(k)\right] \qquad \text{4.522.7} \quad 3$$

Section 4.53–4.54. Combinations of arctangents, arccotangents, and powers

Subsection 4.535.

$$\int_0^1 \frac{\arctan px}{1 + p^2 x^2}\, dx = \frac{1}{2p^2}\arctan p \ln(1 + p^2) \qquad \text{4.535.1} \quad 41$$

This is a list of entries that are too wide to fit in the previous scheme.

$$\int_0^\infty x^{\nu-1}(\beta + x)^{-\mu}(x + \gamma)^{-\rho}\, dx =$$
$$\frac{\gamma^{\nu-\rho}}{\beta^\nu} B(\nu, \mu-\nu+\rho)\,_2F_1\left(\mu, \nu; \mu + \rho; 1 - \frac{\gamma}{\beta}\right) \qquad \text{3.197.1} \quad 68$$

$$\int_0^1 \frac{x^{p-1/2}\, dx}{(1-x)^p(1+qx)^p} = \frac{2\Gamma(p+1/2)\Gamma(1-p)}{\sqrt{\pi}} \times$$
$$\cos^{2p}(\arctan \sqrt{q})\,\frac{\sin[(2p-1)\arctan(\sqrt{q})]}{(2p-1)\sin[\arctan(\sqrt{q})]} \qquad \text{3.197.11} \quad 64$$

$$\int_0^1 \frac{x^{p-1/2}\, dx}{(1-x)^p(1-qx)^p} = \frac{\Gamma(p+1/2)\Gamma(1-p)}{\sqrt{\pi}} \times$$
$$\frac{(1-\sqrt{q})^{1-2p} - (1+\sqrt{q})^{1-2p}}{(2p-1)\sqrt{q}} \qquad \text{3.197.12} \quad 64$$

$$\int_0^u x^{\lambda-1}(u-x)^{\mu-1}(x^2+\beta^2)^\nu \, dx = \beta^{2\nu} u^{\lambda+\mu-1} B(\lambda,\mu)$$

$$\times \, {}_3F_2\left(-\nu, \frac{\lambda}{2}, \frac{\lambda+1}{2}; \frac{\lambda+\mu}{2}, \frac{\lambda+\mu+1}{2}; -\frac{u^2}{\beta^2}\right) \qquad \textbf{3.254.1} \quad \textbf{65}$$

$$\int_u^\infty x^{-\lambda}(x-u)^{\mu-1}(x^2+\beta^2)^\nu \, dx = u^{\mu-\lambda+2\nu} \frac{\Gamma(\mu)\Gamma(\lambda-\mu-2\nu)}{\Gamma(\lambda-2\nu)}$$

$$\times \, {}_3F_2\left(-\nu, \frac{\lambda-\mu}{2}-\nu, \frac{1+\lambda-\mu}{2}-\nu; \frac{\lambda}{2}-\nu, \frac{1+\lambda}{2}-\nu; -\frac{\beta^2}{u^2}\right)$$

$$\textbf{3.254.2} \quad \textbf{66}$$

$$\int_0^u x^{\nu-1}(u-x)^{\mu-1}(x^m+\beta^m)^\lambda \, dx = \beta^{m\lambda} u^{\mu+\nu-1} B(\mu,\nu)_{m+1}F_m$$

$$\times \left(-\lambda, \frac{\nu}{m}, \frac{\nu+1}{m}, \cdots, \frac{\nu+m+1}{m};\right.$$

$$\left. \frac{\mu+\nu}{m}, \frac{\mu+\nu+1}{m}, \cdots, \frac{\mu+\nu+m-1}{m}; -\frac{u^m}{\beta^m}\right)$$

$$\textbf{3.259.2} \quad \textbf{66}$$

$$\frac{\sqrt{\pi}}{2}\Phi(u) = \frac{\sqrt{\pi}}{2}\operatorname{erf}(u) = \int_0^u e^{-x^2} \, dx$$

$$= \sum_{k=0}^\infty \frac{(-1)^k u^{2k+1}}{k!\,(2k+1)} = e^{-u^2} \sum_{k=0}^\infty \frac{2^k u^{2k+1}}{(2k+1)!!} \qquad \textbf{3.321.1} \quad \textbf{46}$$

$$\int_u^\infty \frac{e^{\mu x} \, dx}{(x+\beta)^n} = e^{-u\mu} \sum_{k=1}^{n-1} \frac{(k-1)!(-\mu)^{n-k-1}}{(n-1)!\,(u+\beta)^k}$$

$$- \frac{(-\mu)^{n-1}}{(n-1)!} e^{\beta\mu} \operatorname{Ei}\left[-(u+\beta)\mu\right] \qquad \textbf{3.353.1} \quad \textbf{153}$$

$$\int_0^\infty \frac{e^{-\mu x} \, dx}{(x+\beta)^n} = \frac{1}{(n-1)!} \sum_{k=1}^{n-1} (k-1)! \frac{(-\mu)^{n-k-1}}{\beta^k}$$

$$- \frac{(-\mu)^{n-1}}{(n-1)!} e^{\beta\mu} \operatorname{Ei}(-\beta\mu) \qquad \textbf{3.353.2} \quad \textbf{153}$$

$$\int_0^\infty \frac{x^n e^{-\mu x}}{x+\beta}\, dx = (-1)^{n-1}\beta^n e^{\beta\mu}\,\mathrm{Ei}(-\beta\mu)$$
$$+ \sum_{k=1}^n (k-1)!(-\beta)^{n-k}\mu^{-k} \qquad\qquad \textbf{3.353.5}\quad 154$$

$$\int_0^\infty \frac{e^{-\mu x}\, dx}{(\beta^2+x^2)^2} = \frac{1}{2\beta^3} \times \{\mathrm{ci}(\beta\mu)\sin\beta\mu - \mathrm{si}(\beta\mu)\cos\beta\mu$$
$$- \beta\mu\,[\mathrm{ci}(\beta\mu)\cos\beta\mu + \mathrm{si}(\beta\mu)\sin\beta\mu]\} \qquad \textbf{3.355.1}\quad 155$$

$$\int_0^\infty \frac{x^{2n+1}e^{-px}\, dx}{a^2+x^2} = (-1)^{n-1}a^{2n}\,[\mathrm{ci}(ap)\cos ap + \mathrm{si}(ap)\sin ap]$$
$$+ \frac{1}{p^{2n}} \sum_{k=1}^n (2n-2k+1)!(-a^2p^2)^{k-1} \qquad\qquad \textbf{3.356.1}\quad 156$$

$$\int_0^\infty \frac{x^{2n}e^{-px}}{a^2+x^2} = (-1)^n a^{2n-1}\,[\mathrm{ci}(ap)\sin ap - \mathrm{si}(ap)\cos ap]$$
$$+ \frac{1}{p^{2n-1}} \sum_{k=1}^n (2n-2k)!(-a^2p^2)^{k-1} \qquad\qquad \textbf{3.356.2}\quad 156$$

$$\int_0^\infty \frac{x^{2n+1}e^{-px}\, dx}{a^2-x^2} = \frac{1}{2}a^{2n}\,[e^{ap}\mathrm{Ei}(-ap) + e^{-ap}\mathrm{Ei}(ap)]$$
$$- \frac{1}{p^{2n}} \sum_{k=1}^n (2n-2k+1)!(a^2p^2)^{k-1} \qquad\qquad \textbf{3.356.3}\quad 156$$

$$\int_0^\infty \frac{x^{2n}e^{-px}\, dx}{a^2-x^2} = \frac{1}{2}a^{2n-1}\,[e^{-ap}\mathrm{Ei}(ap) - e^{ap}\mathrm{Ei}(-ap)]$$
$$- \frac{1}{p^{2n-1}} \sum_{k=1}^n (2n-2k)!(a^2p^2)^{k-1} \qquad\qquad \textbf{3.356.4}\quad 156$$

$$\int_0^\infty \frac{e^{-\mu x}\, dx}{a^3+a^2x+ax^2+x^3} = \frac{1}{2a^2}\{\mathrm{ci}(a\mu)(\sin a\mu + \cos a\mu)+$$
$$+ \mathrm{si}(a\mu)(\sin a\mu - \cos a\mu) - e^{a\mu}\mathrm{Ei}(-a\mu)\} \qquad \textbf{3.357.1}\quad 157$$

$$\int_0^\infty \frac{xe^{-\mu x}\, dx}{a^3+a^2x+ax^2+x^3} = \frac{1}{2a}\{\mathrm{ci}(a\mu)(\sin a\mu - \cos a\mu)$$
$$- \mathrm{si}(a\mu)(\sin a\mu + \cos a\mu) - e^{a\mu}\mathrm{Ei}(-a\mu)\} \qquad \textbf{3.357.2}\quad 158$$

$$\int_0^\infty \frac{x^2 e^{-\mu x}\, dx}{a^3 + a^2 x + ax^2 + x^3} = \frac{1}{2}\{-\mathrm{ci}(a\mu)(\sin a\mu + \cos a\mu)$$
$$- \mathrm{si}(a\mu)(\sin a\mu - \cos a\mu) - e^{a\mu}\mathrm{Ei}(-a\mu)\}$$

3.357.3   158

$$\int_0^\infty \frac{e^{-\mu x}\, dx}{a^3 - a^2 x + ax^2 - x^3} = \frac{1}{2a^2}\{\mathrm{ci}(a\mu)(\sin a\mu - \cos a\mu)$$
$$- \mathrm{si}(a\mu)(\sin a\mu + \cos a\mu) + e^{-a\mu}\mathrm{Ei}(a\mu)\}$$

3.357.4   158

$$\int_0^\infty \frac{x e^{-\mu x}\, dx}{a^3 - a^2 x + ax^2 - x^3} = \frac{1}{2a}\{-\mathrm{ci}(a\mu)(\sin a\mu + \cos a\mu)$$
$$- \mathrm{si}(a\mu)(\sin a\mu - \cos a\mu) + e^{-a\mu}\mathrm{Ei}(a\mu)\}$$

3.357.5   158

$$\int_0^\infty \frac{x^2 e^{-\mu x}\, dx}{a^3 - a^2 x + ax^2 - x^3} = \frac{1}{2}\{\mathrm{ci}(a\mu)(\cos a\mu - \sin a\mu)$$
$$+ \mathrm{si}(a\mu)(\cos a\mu + \sin a\mu) + e^{-a\mu}\mathrm{Ei}(a\mu)\}$$

3.357.6   158

$$\int_0^\infty \frac{e^{-px}\, dx}{a^4 - x^4} = \frac{1}{4a^3}\{e^{-ap}\mathrm{Ei}(ap) - e^{ap}\mathrm{Ei}(-ap)$$
$$+ 2\mathrm{ci}(ap)\sin ap - 2\mathrm{si}(ap)\cos ap\}$$

3.358.1   157

$$\int_0^\infty \frac{x e^{-px}\, dx}{a^4 - x^4} = \frac{1}{4a^2}\{e^{ap}\mathrm{Ei}(-ap) + e^{-ap}\mathrm{Ei}(ap)$$
$$- 2\mathrm{ci}(ap)\cos ap - 2\mathrm{si}(ap)\sin ap\}$$

3.358.2   157

$$\int_0^\infty \frac{x^2 e^{-px}\, dx}{a^4 - x^4} = \frac{1}{4a}\{e^{-ap}\mathrm{Ei}(ap) - e^{ap}\mathrm{Ei}(-ap)$$
$$- 2\mathrm{ci}(ap)\sin ap + 2\mathrm{si}(ap)\cos ap\}$$

3.358.3   157

$$\int_0^\infty \frac{x^3 e^{-px}\, dx}{a^4 - x^4} = \frac{1}{4}\{e^{ap}\mathrm{Ei}(-ap) + e^{-ap}\mathrm{Ei}(ap) +$$
$$2\mathrm{ci}(ap)\cos ap + 2\mathrm{si}(ap)\sin ap\}$$

3.358.4   157

$$\int_0^u x^{2\nu-1}(u^2-x^2)^{\rho-1}e^{\mu x}\,dx = \frac{B(\nu,\rho)}{2}u^{2\nu+2\rho-2}{}_1F_2\left(\nu;\tfrac{1}{2},\nu+\rho;\tfrac{1}{4}\mu^2u^2\right)$$
$$+\frac{\mu}{2}B\left(\nu+\tfrac{1}{2},\rho\right)u^{2\nu+2\rho-1}{}_1F_2\left(\nu+\tfrac{1}{2};\tfrac{3}{2},\nu+\rho+\tfrac{1}{2};\tfrac{1}{4}\mu^2u^2\right) \qquad \textbf{3.389.1} \quad \textbf{70}$$

$$\int_0^\infty \left[(\sqrt{x+2\beta}+\sqrt{x})^{2\nu}-(\sqrt{x+2\beta}-\sqrt{x})^{2\nu}\right]e^{-\mu x}\,dx$$
$$= 2^{\nu+1}\frac{\nu}{\mu}\beta^\nu e^{\beta\mu}K_\nu(\beta\mu) \qquad\qquad\qquad \textbf{3.391} \quad \textbf{97}$$

$$\int_0^\infty x^2 e^{-\mu x^2-2\nu x}\,dx$$
$$= -\frac{\nu}{2\mu^2}+\sqrt{\frac{\pi}{\mu^5}}\frac{2\nu^2+\mu}{4}e^{\nu^2/\mu}\left[1-\operatorname{erf}\left(\frac{\nu}{\sqrt{\mu}}\right)\right] \qquad \textbf{3.462.7} \quad \textbf{50}$$

$$\int_0^\infty x^6\frac{\sinh ax}{\sinh bx}\,dx = 16\left(\frac{\pi}{2b}\sec\frac{\pi a}{2b}\right)^7$$
$$\times \sin\frac{\pi a}{2b}\left(45-30\cos^2\frac{\pi a}{2b}+2\cos^4\frac{\pi a}{2b}\right) \qquad \textbf{3.524.11} \quad \textbf{79}$$

$$\int_0^\infty x^5\frac{\sinh ax}{\cosh bx}\,dx = \left(\frac{\pi}{2b}\sec\frac{\pi a}{2b}\right)^6$$
$$\times \left(120-60\cos^2\frac{\pi a}{2b}+\cos^4\frac{\pi a}{2b}\right)\sin\frac{\pi a}{2b} \qquad \textbf{3.524.14} \quad \textbf{80}$$

$$\int_0^\infty x^7\frac{\sinh ax}{\cosh bx}\,dx = \left(\frac{\pi}{2b}\sec\frac{\pi a}{2b}\right)^8$$
$$\times \left(5040-4200\cos^2\frac{\pi a}{2b}+546\cos^4\frac{\pi a}{2b}-\cos^6\frac{\pi a}{2b}\right)\sin\frac{\pi a}{2b} \qquad \textbf{3.524.15} \quad \textbf{80}$$

$$\int_0^\infty x^7\frac{\cosh ax}{\sinh bx}\,dx = 16\left(\frac{\pi}{2b}\sec\frac{\pi a}{2b}\right)^8$$
$$\times \left(315-420\cos^2\frac{\pi a}{2b}+126\cos^4\frac{\pi a}{2b}-4\cos^6\frac{\pi a}{2b}\right)\sin\frac{\pi a}{2b} \qquad \textbf{3.524.19} \quad \textbf{79}$$

$$\int_0^\infty x^{\nu-1}e^{-\gamma x-\beta x^2}\sin ax\,dx = -\frac{i}{2(2\beta)^{\mu/2}}\exp\left(\frac{\gamma^2-a^2}{8\beta}\right)\Gamma(\mu)$$
$$\times \left\{\exp\left(-\frac{ia\gamma}{4\beta}\right)D_{-\mu}\left(\frac{\gamma-ia}{\sqrt{2\beta}}\right)-\exp\left(\frac{ia\gamma}{4\beta}\right)D_{-\mu}\left(\frac{\gamma+ia}{\sqrt{2\beta}}\right)\right\} \quad \textbf{3.953.1} \quad \textbf{39}$$

$$\int_0^\infty \frac{x^{\mu-1} \ln x}{(x+\beta)(x+\gamma)} \, dx = \frac{\pi}{(\gamma-\beta)\sin\pi\mu}$$
$$\times \left[\beta^{\mu-1} \ln\beta - \gamma^{\mu-1} \ln\gamma - \pi\cot\pi\mu(\beta^{\mu-1} - \gamma^{\mu-1}] \right. \qquad\qquad \textbf{4.252.1} \quad \textbf{172}$$

$$\int_0^1 \frac{x^{2n}}{\sqrt{1-x^2}} \ln^2 x \, dx = \pi \frac{(2n-1)!!}{2(2n)!!}$$
$$\times \left\{ \frac{\pi^2}{12} + \sum_{k=1}^{2n} \frac{(-1)^k}{k^2} + \left[\sum_{k=1}^{2n} \frac{(-1)^k}{k} + \ln 2 \right]^2 \right\} \qquad \textbf{4.261.15} \quad \textbf{176}$$

$$\int_0^1 \frac{x^{2n+1}}{\sqrt{1-x^2}} \ln^2 x \, dx = \frac{(2n)!!}{(2n+1)!!}$$
$$\times \left\{ -\frac{\pi^2}{12} - \sum_{k=1}^{2n+1} \frac{(-1)^k}{k^2} + \left[\sum_{k=1}^{2n+1} \frac{(-1)^k}{k} + \ln 2 \right]^2 \right\} \qquad \textbf{4.261.16} \quad \textbf{177}$$

$$\int_0^1 \ln(1+x) \frac{1+x^2}{a^2+x^2} \frac{dx}{1+a^2x^2}$$
$$= \frac{1}{2a(1+a^2)} \left[\frac{\pi}{2} \ln(1+a^2) - 2\ln a \arctan a \right] \qquad \textbf{4.291.24} \quad \textbf{127}$$

$$\int_0^1 \ln(1+x) \frac{1-x^2}{(ax+b)^2} \frac{dx}{(bx+a)^2} = \frac{1}{a^2-b^2}$$
$$\times \left\{ \frac{1}{a-b} \left[\frac{a+b}{ab} \ln(a+b) - \frac{\ln b}{a} - \frac{\ln a}{b} \right] + \frac{4\ln 2}{b^2-a^2} \right\} \quad \textbf{4.291.25} \quad \textbf{118}$$

$$\int_0^1 \ln(1+ax) \frac{1-x^2}{(1+x^2)^2} \, dx$$
$$= \frac{1}{2} \frac{(1+a)^2}{1+a^2} \ln(1+a) - \frac{a\ln 2}{2(1+a^2)} - \frac{\pi a^2}{4(1+a^2)} \qquad \textbf{4.291.27} \quad \textbf{123}$$

References

[1] M. Abramowitz and I. Stegun. *Handbook of Mathematical Functions with Formulas, Graphs and Mathematical Tables*. Dover, New York, 1972.

[2] G. Almkvist and D. Zeilberger. The method of differentiating under the integral sign. *Jour. Symb. Comp.*, 10:571–591, 1990.

[3] T. Amdeberhan. Theorems, problems and conjectures. Available at http://www.math.tulane.edu/~tamdeberhan/conjectures.html.

[4] T. Amdeberhan, K. Boyadzhiev, and V. Moll. The integrals in Gradshteyn and Ryzhik. Part 17: The Riemann zeta function. *Scientia*, 20:61–71, 2011.

[5] T. Amdeberhan, O. Espinosa, I. Gonzalez, M. Harrison, V. Moll, and A. Straub. Ramanujan Master Theorem. *The Ramanujan Journal*, 29:103–120, 2012.

[6] T. Amdeberhan, O. Espinosa, V. Moll, and A. Straub. Wallis-Ramanujan-Schur-Feynman. *Amer. Math. Monthly*, 117:618–632, 2010.

[7] T. Amdeberhan, L. A. Medina, and V. Moll. The integrals in Gradshteyn and Ryzhik. Part 5: Some trigonometric integrals. *Scientia*, 15:47–60, 2007.

[8] T. Amdeberhan and V. Moll. The integrals in Gradshteyn and Ryzhik. Part 7: Elementary examples. *Scientia*, 16:25–40, 2008.

[9] T. Amdeberhan and V. Moll. The integrals in Gradshteyn and Ryzhik. Part 14: An elementary evaluation of entry 3.411.5. *Scientia*, 19:97–103, 2010.

[10] T. Amdeberhan, V. Moll, J. Rosenberg, A. Straub, and P. Whitworth. The integrals in Gradshteyn and Ryzhik. Part 9: Combinations of logarithmic, rational and trigonometric functions. *Scientia*, 17:27–44, 2009.

[11] C. Anastasiou, E. W. N. Glover, and C. Oleari. Application of the negative-dimension approach to massless scalar box integrals. *Nucl. Phys. B*, 565:445–467, 2000.

[12] C. Anastasiou, E. W. N. Glover, and C. Oleari. Scalar one-loop integrals using the negative-dimension approach. *Nucl. Phys. B*, 572:307–360, 2000.

[13] G. E. Andrews, R. Askey, and R. Roy. *Special Functions*, volume 71 of *Encyclopedia of Mathematics and Its Applications*. Cambridge University Press, New York, 1999.

[14] A. Apelblat. *Tables of Integrals and Series*. Verlag Harry Deutsch, Thun; Frankfurt am Main, 1996.

[15] W. N. Bailey. *Generalized Hypergeometric Series*. Cambridge University Press, 1935.

[16] R. Beals and R. Wong. *Special Functions. A Graduate Text*, volume 126 of *Cambridge Studies in Advanced Mathematics*. Cambridge University Press, New York, 2010.

[17] G. Boros and V. Moll. *Irresistible Integrals*. Cambridge University Press, New York, 1st edition, 2004.

[18] J. M. Borwein and P. B. Borwein. *Pi and the AGM—A Study in Analytic Number Theory and Computational Complexity*. Wiley, New York, 1st edition, 1987.

[19] P. Borwein, S. Choi, B. Rooney, and A. Weirathmueller. *The Riemann Hypothesis. A Resource for the Afficionado and Virtuoso Alike*. Canadian Mathematical Society, 1st edition, 2008.

[20] K. Boyadzhiev, L. Medina, and V. Moll. The integrals in Gradshteyn and Ryzhik. Part 11: The incomplete beta function. *Scientia*, 18:61–75, 2009.

[21] M. Bronstein. *Integration of Elementary Functions.* PhD thesis, University of California, Berkeley, California, 1987.

[22] Y. A. Brychkov. *Handbook of Special Functions. Derivatives, Integrals, Series and Other Formulas.* Taylor and Francis, Boca Raton, FL, 2008.

[23] G. W. Cherry. Integration in finite terms with special functions: the error function. *J. Symb. Comput.*, 1:283–302, 1985.

[24] F. Chyzak. *Fonctions holonomes en calcul formel.* PhD thesis, École polytechnique, 1998.

[25] F. Chyzak. An extension of Zeilberger's fast algorithm to general holonomic functions. *Discrete Mathematics*, 217(1-3):115–134, 2000.

[26] A. Devoto and D. W. Duke. Tables of integrals and formulae for Feynman diagram calculations. *Riv. Nuovo Cimento*, 7:1–39, 1984.

[27] H. M. Edwards. *Riemann's Zeta Function.* Academic Press, New York, 1974.

[28] A. Erdélyi. *Tables of Integral Transforms*, volume I. McGraw-Hill, New York, 1st edition, 1954.

[29] P. Flajolet, S. Gerhold, and B. Salvy. On the non-holonomic character of logarithms, powers, and the nth prime function. *Elec. Jour. Comb.*, 11:#A2, 2005.

[30] S. Gerhold. On some non-holonomic sequences. *Elec. Jour. Comb.*, 11:#R87, 2004.

[31] I. Gonzalez and V. Moll. Definite integrals by the method of brackets. Part 1. *Adv. Appl. Math.*, 45:50–73, 2010.

[32] I. Gonzalez, V. Moll, and A. Straub. The method of brackets. Part 2: Examples and applications. In T. Amdeberhan, L. Medina, and Victor H. Moll, editors, *Gems in Experimental Mathematics*, volume 517 of *Contemporary Mathematics*, pages 157–172. American Mathematical Society, 2010.

[33] I. Gonzalez and I. Schmidt. Optimized negative dimensional integration method (NDIM) and multiloop Feynman diagram calculation. *Nuclear Physics B*, 769:124–173, 2007.

[34] I. Gonzalez and I. Schmidt. Modular application of an integration by fractional expansion (IBFE) method to multiloop Feynman diagrams. *Phys. Rev. D*, 78:086003, 2008.

[35] I. S. Gradshteyn and I. M. Ryzhik. *Table of Integrals, Series, and Products.* Edited by D. Zwillinger and V. Moll. Academic Press, New York, 8th edition, 2015.

[36] I. G. Halliday and R. M. Ricotta. Negative dimensional integrals. I. Feynman graphs. *Phys. Lett. B*, 193:241, 1987.

[37] G. H. Hardy. The integral $\int_0^\infty \frac{\sin x}{x}\, dx$. *Math. Gazette*, 5:98–103, 1909.

[38] G. H. Hardy. Further remarks on the integral $\int_0^\infty \frac{\sin x}{x}\, dx$. *Math. Gazette*, 8:301–303, 1916.

[39] G. H. Hardy. *Ramanujan. Twelve Lectures on Subjects Suggested by His Life and Work.* Chelsea Publishing Company, New York, 3rd edition, 1978.

[40] K. Kohl. *Algorithmic Methods for Definite Integration.* PhD thesis, Tulane University, 2011.

[41] K. Kohl and V. Moll. The integrals in Gradshteyn and Ryzhik. Part 20: Hypergeometric functions. *Scientia*, 21:43–54, 2011.

[42] C. Koutschan. *Advanced Applications of the Holonomic Systems Approach.* PhD thesis, RISC, Johannes Kepler University, Linz, Austria, 2009.

[43] C. Koutschan. HolonomicFunctions (User's Guide). Technical Report 10-01, RISC Report Series, University of Linz, Austria, January 2010.

[44] C. Koutschan and V. Moll. The integrals in Gradshteyn and Ryzhik. Part 18: Some automatic proofs. *Scientia*, 20:93–111, 2011.

[45] H. P. McKean and V. Moll. *Elliptic Curves: Function Theory, Geometry, Arithmetic.* Cambridge University Press, New York, 1997.

[46] L. Medina and V. Moll. A class of logarithmic integrals. *Ramanujan Journal*, 20:91–126, 2009.

[47] L. Medina and V. Moll. The integrals in Gradshteyn and Ryzhik. Part 10: The digamma function. *Scientia*, 17:45–66, 2009.

[48] L. Medina and V. Moll. The integrals in Gradshteyn and Ryzhik. Part 23: Combinations of logarithms and rational functions. *Scientia*, 23:1–18, 2012.

[49] V. Moll. The integrals in Gradshteyn and Ryzhik. Part 1: A family of logarithmic integrals. *Scientia*, 14:1–6, 2007.

[50] V. Moll. The integrals in Gradshteyn and Ryzhik. Part 2: Elementary logarithmic integrals. *Scientia*, 14:7–15, 2007.

[51] V. Moll. The integrals in Gradshteyn and Ryzhik. Part 6: The beta function. *Scientia*, 16:9–24, 2008.

[52] V. Moll. The integrals in Gradshteyn and Ryzhik. Part 13: Trigonometric forms of the beta function. *Scientia*, 19:91–96, 2010.

[53] V. Moll. *Special Integrals of Gradshteyn and Ryzhik. The Proofs*, volume 1. CRC Press, Taylor and Francis Group, Chapman and Hall, 2015.

[54] V. Moll and R. Posey. The integrals in Gradshteyn and Ryzhik. Part 12: Some logarithmic integrals. *Scientia*, 18:77–84, 2009.

[55] M. Petkovsek, H. Wilf, and D. Zeilberger. *A=B*. A. K. Peters, 1st. edition, 1996.

[56] A. P. Prudnikov, Yu. A. Brychkov, and O. I. Marichev. *Integrals and Series. Five volumes*. Gordon and Breach Science Publishers, 1992.

[57] E. D. Rainville. *Special Functions*. The Macmillan Company, New York, 1960.

[58] R. H. Risch. The problem of integration in finite terms. *Trans. Amer. Math. Soc.*, 139:167–189, 1969.

[59] R. H. Risch. The solution of the problem of integration in finite terms. *Bull. Amer. Math. Soc.*, 76:605–608, 1970.

[60] J. F. Ritt. *Integration in Finite Terms. Liouville's Theory of Elementary Functions.* New York, 1948.

[61] A. T. Suzuki and A. G. M. Schmidt. An easy way to solve two-loop vertex integrals. *Phys. Rev. D*, 58:047701, 1998.

[62] A. T. Suzuki and A. G. M. Schmidt. Feynman integrals with tensorial structure in the negative dimensional integration scheme. *Eur. Phys. J.*, C-10:357–362, 1999.

[63] A. T. Suzuki and A. G. M. Schmidt. Negative dimensional approach for scalar two-loop three-point and three-loop two-point integrals. *Canad. Jour. Physics*, 78:769–777, 2000.

[64] N. Takayama. An algorithm of constructing the integral of a module—an infinite dimensional analog of Gröbner basis. *ISSAC'90: Proceedings of the International Symposium on Symbolic and Algebraic Computation*, pages 206–211, 1990.

[65] N. M. Temme. *Special Functions. An Introduction to the Classical Functions of Mathematical Physics*. John Wiley and Sons, New York, 1996.

[66] E. C. Titchmarsh. *The Theory of the Riemann Zeta Function*. Oxford University Press, 2nd edition, 1986.

[67] G. N. Watson. *A Treatise on the Theory of Bessel Functions*. Cambridge University Press, 1966.

[68] E. T. Whittaker and G. N. Watson. *Modern Analysis*. Cambridge University Press, 1962.

[69] D. Zeilberger. A holonomic systems approach to special function identities. *Journal of Computational and Applied Mathematics*, 32:321–368, 1990.

[70] D. Zwillinger, editor. *Handbook of Integration*. Jones and Barlett Publishers, Boston and London, 1st edition, 1992.

Index

Milton Keynes UK
Ingram Content Group UK Ltd.
UKHW040110071024
449327UK00019B/945

9 78036 577274